Modal Testing:
Theory, Practice and Application

SECOND EDITION

MECHANICAL ENGINEERING RESEARCH STUDIES

ENGINEERING DYNAMICS SERIES

Series Editor: **Professor J.B. Roberts**
University of Sussex, England

9.   Theoretical and Experimental Modal Analysis
     Edited by Nuno M. M. Maia *and* Júlio M. M. Silva

10. Modal Testing: Theory, Practice and Application
     SECOND EDITION
     D.J. Ewins

11. Structural Dynamics @ 2000:
     current status and future directions *
     Edited by D.J. Ewins and D.J. Inman

     * *forthcoming*

Photograph provided courtesy of The Boeing Company, Seattle, Washington, USA

Modal Testing:
Theory, Practice and Application

SECOND EDITION

D. J. Ewins
Professor of Vibration Engineering
Imperial College of Science, Technology and Medicine
London, England

RESEARCH STUDIES PRESS LTD.
Baldock, Hertfordshire, England

RESEARCH STUDIES PRESS LTD.

16 Coach House Cloisters, 10 Hitchin Street, Baldock, Hertfordshire, England, SG7 6AE

and

325 Chestnut Street, Philadelphia, PA 19106, USA

Copyright © 2000, by Research Studies Press Ltd.

Marketing:

UK, EUROPE & REST OF THE WORLD
Research Studies Press Ltd.
16 Coach House Cloisters, 10 Hitchin Street, Baldock, Hertfordshire, England, SG7 6AE

Distribution:

NORTH AMERICA
Taylor & Francis Inc.
International Thompson Publishing, Discovery Distribution Center, Receiving Dept., 2360 Progress Drive Hebron, Ky. 41048

ASIA-PACIFIC
Hemisphere Publication Services
Golden Wheel Building # 04-03, 41 Kallang Pudding Road, Singapore 349316

UK, EUROPE & REST OF THE WORLD
John Wiley & Sons Ltd.
Shripney Road, Bognor Regis, West Sussex, England, PO22 9SA

Library of Congress Cataloging-in-Publication Data

Available

British Library Cataloguing in Publication Data
A catalogue record for this book is available from the British Library.

ISBN 0 86380 218 4

Preface

Some 15 years and more have passed since the foreword to the first edition *of Modal Testing: Theory and Practice* was written. In some ways, little has changed since then concerning the origins of the subject, and as a result, much of the original book has survived the intervening years almost intact. At the same time, there have been great developments in the subject, with the result that some of today's technology would be largely unrecognisable to the practitioner of the early 1980s. It is to embrace these developments that the present book has been written, and it is because of the tremendous growth in the subject – not only in the relevant theory, but also in the practice of the subject and in the ever-widening range of applications – that this second edition is much longer than its predecessor.

It has to be admitted that the subject has become so large that one cannot hope to cover everything in a single book. For example, the topic of model updating – effectively introduced as part of the subject to the community in the first edition - has been the subject of perhaps 500-800 published papers in the past 15 years and a book in the last 5 years. Modal analysis methods have become a very advanced topic with sophisticated numerical analysis procedures which are beyond the scope of most modal testing practitioners. A huge number of papers have been published since the first edition *of Modal Testing*: there have been 15 IMACs (with about 400 papers in each), plus 10 ISMAs (with perhaps 200-250 papers presented at each) plus other more specialised conferences so that there must be well over 10000 papers published on modal testing and analysis since the first edition of this book appeared. It would be foolish to pretend that this new edition can have absorbed any more than a fraction of this volume of literature but, nevertheless, it does seek to bring the work somewhat more up to date. This book is still aimed at the serious practitioner, as well as the student to some of the "new" structural dynamics analysis methods. It may also still serve as a useful basis to the researcher in one or other of the various disciplines, of which the modal analysis and the model updating areas are two good examples where for each topic a whole new subject has grown out of part of the main one. For those seeking such advanced expositions, other texts now take on the story in greater detail.

In talking of references, and noting the vast number of relevant papers that have appeared in recent years, it should also be noted that in this text we have sought to provide the reader with a judicious selection of reference material which is not overwhelming by its very quantity. It is accepted that such a selection process is relatively subjective but the concerned reader can find literature reviews in many of the different areas covered by this book in some of the references that are cited here.

The refinements and additions to the text which result in the present edition have grown out of 15 years of using the text as the basis of a series of over 100 short courses, typically of between 20 and 40 hours concentrated instruction, that have been presented by the author in many countries around the world. In the time between these two editions of *Modal Testing*, the modal analysis community has grown to use the singular value decomposition as a routine tool, to use MIMO testing techniques on a standard basis, to expect much more quantitative performance from the applications of the results of its modal tests, and to be more ambitious in these applications it seeks to address. Reliability and speed of testing methods is demanded by industrial and commercial pressures which lead us towards greater use of simulation, test planning and 'virtual testing' in order that the tests we do conduct provide the quality and selection of data that are required to solve the problems being tackled.

However, behind all this progress and evolution is a re-assertion of the need for testing in structural dynamics in general and for modal testing in particular. One might have thought, 15 years ago, that by the end of the century the tremendous growth in computing technology would have rendered experimental testing more or less obsolete. However, even though our projections in 1984 of what would be possible by 2000 have almost certainly proved to be conservative, the need for testing is as deep rooted as ever. There are probably two main reasons for this - and it may be appropriate to bear these in mind throughout any testing exercise that one is involved in – and these are (1) that there are some parameters, quantities or effects that are effectively unpredictable, and likely to remain so for the foreseeable future, such as damping, friction and fatigue properties, as well as excitation forces

— and (2) no matter how much we improve our structural dynamic modelling and prediction capabilities (which is certainly a direct consequence of the dramatic advances in our computing technologies), we are always driven to seek better results. When we succeed in predicting a structure's natural frequencies to within 5%, or a response level to within 50%, then those targets will move to ones demanding an accuracy of 1%, or 10% in response, and so it will continue, and the only reference against which these predictions can be assessed is one that comes from experimental observation of what really happens in practice. And lest we imagine that we are 'close' to meeting the expectations of our designers in our ability to predict the dynamic behaviour of 'real' structures with sufficient reliability to permit paper designs which are "right first time", we have only to recall the current situation regarding the performance of our analysis tools at predicting the vibration response of a typical engineering structure which is composed of an assembly of separate components and subjected to various excitation forces generated by or in the operating environment for that structure to realise that we are still far from attaining the aspirations of our subject.

Hence, it is believed that the experimental branches of structural dynamics should be seen to have a very secure and long-term future, clearly justifying the investments that have been made in certain areas, and especially the one treated here of modal testing.

Acknowledgements

Although this book appears as the work of but one author, in fact it contains the contributions of a great many colleagues and fellow travellers who have accompanied that author along various parts of the road that its contents describe.

The book first appeared in the early 1980s as the result of a set of lecture notes prepared for some early short courses in modal testing. Over the ensuing 15 years, that course was developed and refined and presented by the author over 100 times in more than 20 countries around the world. The present second edition of the book is, to a large extent, the result of that development of teaching the subject, together with the parallel activity of researching, developing and practicing the technology which it describes. As a result, there are two groups of people who have contributed to this new edition. The first group comprises those who have worked with me in the development of the subject – mostly, research students and research assistants whose collected theses are listed in the bibliography section of the book. Clearly, their contributions are very significant for they constitute most of the currently-used techniques. The second group include all the colleagues who have participated in various ways in the courses which have played such an important role in the developments of the subject as will as its practice in a wide range of industries.

This important group of contributors includes, first and foremost, my colleagues at Imperial College – Peter Grootenhuis, David Robb and Mehmet Imregun – the first of whom encouraged me to undertake the voyage that 30 years of research in the field has become, while the others have played a major role in the running of the courses themselves and in the development of the MODENT software whose algorithms are a direct result of the same research. Their contributions to this book are to be found everywhere. There are also a band of colleagues who have promoted and facilitated the actual courses

x

themselves, and significant amongst these have been: Ron Eshelman, Dominique Bonnecase, Ole Doessing, Harry Zaveri, Menad Sidamed, Dominique Carreau as well as Dick DeMichele, the initiator of the Modal Analysis courses run at IMAC every year for the past 15 years.

I must record my thanks for the support, encouragement and incredible patience of those who have helped to realise this publication. First amongst these is Veronica Wallace who, in her role as publisher of my first book, encouraged and indulged me to the point where this second edition has actually been completed. Not – I have to admit – before she retired from RSP, as I had promised, but completed nevertheless. And that completion is due in some not inconsiderable measure to the persistence of her successor, Guy Robinson. I hope that they will find that their patience will be rewarded by the result. The actual mechanics of the production has also been made manageable for me by the constant help of my secretary, Liz Savage, and in the latter stages by Liz Hall and several of my current students, not forgetting the major editorial review carried out by my daughter, Caroline, at the critical time when the last phase had to be kick started back into action.

There are in addition to these named helpers a small number of other people whose support in all manner of ways has been the deciding factor in the long-running debate as to whether or not a second edition would ever appear: I know that they know who they are, and I acknowledge the support of each of them individually although they will remain properly anonymous to the book's readers. They have each played a critical role in providing me with the space and the support I needed to get to this point.

Contents

CHAPTER 1

Overview

1.1   INTRODUCTION TO MODAL TESTING

The experimental study of structural dynamics has always provided a major contribution to our efforts to understand and to control the many vibration phenomena encountered in practice. Today, at the turn of the millennium, even with the formidable growth in the use and capacity of computing power that we have witnessed in recent decades, the need for experimental measurement is as compelling as ever. Since the very early days of awareness of structural vibration, experimental observations have been necessary for the major objectives of (a) determining the nature and extent of vibration response levels in operation and (b) verifying theoretical models and predictions of the various dynamic phenomena which are collectively referred to as 'vibration'. There is also a third requirement, (c), which is for the measurement of the essential material properties under dynamic loading, such as damping capacity, friction and fatigue endurance.

Structural vibration problems continue to present a major hazard and design limitation for a very wide range of engineering products today. First, there are a number of structures, from turbine blades to suspension bridges, for which structural integrity is of paramount concern, and for which a thorough and precise knowledge of the dynamic characteristics is essential. Then, there is an even wider set of structural components or assemblies for which vibration is directly related to performance, either by virtue of causing temporary malfunction during excessive motion or by creating disturbance or discomfort, including that of noise. For all these examples, it is important that the vibration levels encountered in service or operation be anticipated and brought under satisfactory control.

The two major vibration measurement objectives indicated above represent two corresponding types of test. The first is one where vibration forces or, more usually, responses are measured during 'operation' of the machine or structure under study, while the second is a test where the structure or component is vibrated with a known excitation, often out of its normal service environment. This second type of test is generally made under much more closely-controlled conditions than the former and consequently yields more accurate and detailed information. This type of testing — including both the data acquisition and its subsequent analysis — is nowadays called 'Modal Testing' and is the subject of the following text. While we shall be defining the specific quantities and parameters used as we proceed, it is perhaps appropriate to state clearly at this point just what we mean by the term 'Modal Testing'. It is used here to encompass 'the processes involved in testing components or structures with the objective of obtaining a mathematical description of their dynamic or vibration behaviour'. The form of the 'mathematical description' or model varies considerably from one application to the next: it can be an estimate of natural frequency and damping factor in one case and a full mass-spring-dashpot model for the next.

Although the name is relatively new, the principles of modal testing were laid down many years ago. These have evolved through various phases when descriptions such as 'Resonance Testing' and 'Mechanical Impedance Methods' were used to describe the general area of activity. One of the more important milestones in the development of the subject was provided by the paper in 1947 by Kennedy and Pancu [1]*. The methods described there found application in the accurate determination of natural frequencies and damping levels in aircraft structures and were not out-dated for many years, until the rapid advance of electronic measurement and analysis techniques in the 1960s. This activity paved the way for more precise measurements and thus more powerful applications. A paper by Bishop and Gladwell in 1963 [2] described the state of the theory of resonance testing which, at that time, was considerably in advance of its practical implementation. Another work of the same period but from a totally different viewpoint was the book by Salter [3] in which a relatively non-analytical approach to the interpretation of measured data was proposed. Whilst more demanding of the user than today's computer-assisted automation of the same tasks, Salter's approach rewarded its practitioners with a considerable physical insight into the vibration of the structure thus studied. However, by 1970 there had been major advances in transducers, electronics and digital analysers and the current techniques of modal testing were established. There are a great many papers which relate to this period, as workers made further advances and applications, and a bibliography of several hundred such references now exists [4,5]. The following pages set out to bring together the major features of all aspects of the subject to provide a comprehensive guide to

---

* References are listed in pages 545-550.

both the theory and the practice of modal testing, including the more powerful applications.

In the almost two decades since the first edition of this text was written, there has been a veritable explosion of activity and publications in the subject. While much of this activity relates to advanced methods and applications (and is thus largely outside the scope of this book), nevertheless there have been a number of other texts aimed at explaining the basics of Modal Testing to a wider and wider audience. Notable amongst these are recent books by Maia et al [6], from the KUL in Leuven [7] and, in their respective languages, from Germany [8], China [9] and Romania [10]. Another important set of documents have been published in the UK by the Dynamic Testing Agency (DTA) in the form of (a) a Primer (an introductory overview for beginners and managers)[11], and (b) a Handbook of Best Practice, a detailed step-by-step guide to each of the individual procedures involved in performing a modal test [12]. More general texts which refer to modal testing, but which are not exclusively concerned with it, include the book by McConnell [13].

In addition to these specific works, there have been several thousand papers presented at the annual IMAC conferences since 1982 [14], and the bi-annual ISMA conferences in Belgium (since 1975)[15], as well as those published in several journals [16-18]. The present text is intended to be largely self-contained and so it is not planned to provide an extensive list of references. Readers who wish to be apprised of the full literature are advised to consult the proceedings of the aforementioned conferences, as well as [6].

1.2   APPLICATIONS OF MODAL TESTING

Before embarking on both summarised and detailed descriptions of the many aspects of the subject, it is important that we raise the question of why modal tests are undertaken. There are many applications to which the results from a modal test may be put and several of these are, in fact, quite powerful. However, we must remember that no single test or analysis procedure is 'best' for all cases and so it is very important that a clear objective is defined before any industrial test is undertaken so that the optimum methods or techniques may be used. This process is best dealt with by considering in some detail the following questions: what is the desired outcome from the study of which the modal test is a part? and, in what form are the results required in order to be of maximum use?

First, then, it is appropriate to review the major application areas in current use. In all cases, it is true to say that a modal test is undertaken in order to obtain a mathematical model of the structure but it is in the subsequent **use** of that model that the differences arise.

(a)   Perhaps the single most commonly used application is the measurement of a structure's vibration properties in order to compare these with corresponding data produced by a finite element or other theoretical model. This application is often borne out of a need or desire to validate the theoretical model prior to its use for predicting response levels to complex excitations, such as shock, or other further stages of analysis. It is generally felt that corroboration of the major modes of vibration by tests can provide reassurance of the basic validity of the theoretical model which may then be put to further use with confidence. For this specific application, all that we require from the test are: (i) accurate estimates of natural frequencies and (ii) descriptions of the mode shapes using just sufficient detail and accuracy to permit their identification and correlation with those from the theoretical model. At this stage, accurate mode shape data are not essential. It is generally not possible to 'predict' the damping in each mode of vibration from a theoretical model and so there is nothing with which to compare measurements of modal damping obtained from the tests. However, such information is useful as it can be incorporated into the theoretical model, albeit as an approximation, prior to that being called upon to predict specific response levels (which are often significantly influenced by the damping).

(b)   Many cases of experiment-theory comparison stop at the stage of obtaining a set of results by each route and simply comparing them. Sometimes, an attempt will be made to adjust or correct the theoretical model in order to bring its modal properties closer into line with the measured results. In the past, this was usually done using a trial-and-error approach but nowadays more formal reconciliation procedures known as model updating, or model refinement, are available.

A logical and necessary evolution of the procedure outlined above is the correlation, rather than the comparison, of experimental and theoretical results. By this is meant a process whereby the two sets of data are combined, quantitatively, in order to identify specifically the causes of the discrepancies between predicted and measured properties. Such an application is clearly more powerful than its less ambitious forerunner but, equally, will be more demanding in terms of the accuracy required in the data taken from the modal test. Specifically, a much more precise description of the mode shape data (the 'eigenvectors') is required than is generally necessary to depict or describe the general shape in pictorial form.

(c)   The next application area to be reviewed is that of using a modal test in order to produce a mathematical model of a component which may then be used to incorporate that component into a structural assembly. This is often referred to as a 'substructuring process' and is widely used in theoretical analysis of complex structures. Here again, it is a fully quantitative model that is sought — with accurate data required for natural frequencies, modal damping factors and mode shapes — and now has the added constraint that **all** modes must be included simultaneously. It is not sufficient to confine the model to certain individual modes — as may be done for the previous comparisons or correlations — since out-of-range modes will influence the structure's dynamic behaviour in a given frequency range of interest for the complete assembly. Also, it is not possible to ignore certain modes which exist in the range of interest but which may present some difficulties for measurement or analysis. This application is altogether more demanding than the previous ones and is often underestimated, and thus inappropriately tackled, with the result that the results do not always match up to expectations or, indeed, to those which are attainable with greater care.

(d)   There is a variant of the previous application which is becoming of considerable interest and potential and that is to the generation of a model which may be used for predicting the effects of modifications to the original structure, as tested. Theoretically, this falls into the same class of process as substructuring and has the same requirements of data accuracy and quantity. However, sometimes the modification procedure involves relatively minor changes to the original design, in order to fine tune a structure's dynamics, and this situation can relax the data requirements somewhat.

One particular consideration which applies to both this and the previous case concerns the need for information about rotational degrees of freedom (RDOFs), i.e. moments (as well as forces) and rotational (i.e. angular) displacements (as well as translational ones). These are automatically included in theoretical analyses but are generally ignored in experimentally-based studies for the simple reason that they are so much more difficult to measure. Nevertheless, they are generally an essential feature in coupling or modification applications.

(e)   A different application for the model produced by a modal test is that of force determination. There are a number of situations where knowledge of the dynamic forces causing vibration is required but where direct measurement of these forces is not practical. For these cases, one solution is offered by a process whereby measurements of the response caused by the forces are combined with a mathematical description of the transfer functions of the structure in order to deduce the forces. This process can be very sensitive to the accuracy and appropriateness of the model used for the structure and so it is often essential that the model itself be derived from measurements; in other words, via a modal test.

(f)   Lastly, in our review of applications, it is appropriate to note that whereas the normal procedure is (a) to measure, (b) to analyse the measured data and then (c) to derive a mathematical model of the structure, there are some cases where this is not the optimum procedure. The last step, (c), is usually taken in order to reduce a vast quantity of actual measurements to a small and efficient data set usually referred to as the 'modal model'. This reduction process has an additional benefit of eliminating small inconsistencies which will inevitably occur in measured data. However, it is sometimes found that the smoothing and averaging procedures involved in this step reduce the validity of the model and in applications where the subsequent analysis is very sensitive to the accuracy of the input data, this can present a problem. Examples of this problem may be found particularly in (e), force determination, and in (c), subsystem coupling. The solution adopted is to generate a model of the test structure using 'raw' measured data — unsmoothed and relatively unprocessed — which, in turn, may well demand additional measurements being made, and additional care to ensure the necessary accuracy.

1.3   PHILOSOPHY OF MODAL TESTING
One of the major requirements of the subject of modal testing is a thorough integration of three components:

(i)   the theoretical basis of vibration;
(ii)   accurate measurement of vibration; and
(iii)   realistic and detailed data analysis.

There has in the past been a tendency to regard these as three different specialist areas, with individual experts in each. However, the subject we are exploring now **demands** a high level of understanding and competence in all three areas and cannot achieve its full potential without the proper and judicious mixture of the necessary skills.

For example: when taking measurements of excitation and response levels, a full knowledge of how the measured data are to be processed can be essential if the correct decisions are to be made as to the quality and suitability of that data. Again, a thorough understanding of the various forms and trends adopted by plots of frequency response functions for complex structures can prevent the wasted effort of analysing incorrect measurements: there are many features of these plots that can be assessed very rapidly by the eyes of someone who understands their theoretical basis.

Throughout this work, we shall repeat and re-emphasise the need for this integration of theoretical and experimental skills. Indeed, the route chosen to develop and to explain the details of the subject takes us first through an extensive review of the necessary theoretical foundation of structural vibration. This theory is regarded as a necessary prerequisite to the subsequent studies of measurement techniques, signal processing and data analysis.

With an appreciation of both the theoretical basis (and if not with all the detail straight away, then at least with an awareness that many such details do exist), we can then turn our attention to the practical side; to the excitation of the test structure and to the measurement of both the input and response levels during the controlled testing conditions of FRF measurements. Already, here, there is a bewildering choice of test methods — harmonic, random, transient excitations, for example — vying for choice as the 'best' method in each application. If the experimenter is not to be left at the mercy of the sophisticated digital analysis equipment now widely available, he or she must fully acquaint him- or herself with the methods, limitations and implications of the various techniques now widely used in this measurement phase.

Next, we consider the 'analysis' stage where the measured data (almost invariably, frequency response functions) are subjected to a range of curve-fitting procedures in an attempt to find the mathematical model which provides the closest description of the actually-observed behaviour. There are many approaches, or algorithms, for this phase and, as is usually the case, no single one is ideal for all problems. Thus, an appreciation of the alternatives is a necessary requirement for the experimenter wishing to make optimum use of his time and resources.

Often, though not always, an analysis may be conducted on each measured curve individually. If this is the case, then there is a further step in the process which we refer to as 'Modelling'. This is the final stage where all the measured and processed data are combined to yield the most compact and efficient description of the end result — a mathematical model of the test structure — that is practicable.

This last phase, like the earlier ones, involves some type of averaging as the means by which a large quantity of measured data are reduced to a (relatively) small mathematical model. This is an operation which must be used with some care. The averaging process is a valid and valuable technique provided that the data thus treated contain **random** variations: data with systematic trends, such as are caused by poor testing practices or non-linearities in the test structure, should not be averaged in the same way. We shall discuss this problem later on.

Thus we have attempted to describe the underlying philosophy of our approach to modal testing and are now in a position to review the highlights of the three main phases: theory, measurement and analysis in order to provide a brief overview of the entire subject.

1.4   SUMMARY OF THEORY

In this and the following two sections an overview of the various aspects of the subject will be presented. This will highlight the key features of each area of activity and is included for a number of reasons. First, it provides the serious student with a non-detailed review of the different subjects to be dealt with, enabling him to see the context of each without being distracted by the minutiae, and thus acts as a useful introduction to the full study. Secondly, it provides him or her with a breakdown of the subject into identifiable topics which are then useful as milestones for the process of acquiring a comprehensive ability and understanding of the techniques. Lastly, it also serves to provide the non-specialist or manager with an explanation of the subject, trying to remove some of the mystery and folklore which may have developed.

We begin with the theoretical basis of the subject since, as has already been emphasised, a good grasp of this aspect is an essential prerequisite for successful modal testing.

It is very important that a clear distinction is made between free vibration and forced vibration analyses, these usually being two successive stages in a full vibration analysis. As usual with vibration studies, we start with the single degree-of-freedom (SDOF) system and use this familiar model to introduce the general notation and analysis procedures which are later extended to the more general multi degree-of-freedom (MDOF) systems. For the SDOF system, a free vibration analysis yields its natural frequency and damping factor, whereas a particular type of forced response analysis, assuming a harmonic excitation, leads to the definition of the frequency response function — such as mobility, the ratio of velocity response to force input. These two types of result are referred to as 'modal properties' and 'frequency response characteristics' respectively and they constitute the basis of all our studies. Before leaving the SDOF model, it is appropriate to consider the form which a plot of the mobility (of other type of frequency response function) takes. Three alternative ways of plotting this information are shown in Fig. 1.1 and, as will be discussed later, it is always helpful to seek the format which is best suited to the particular application to hand.

Next we consider the more general class of systems which have more than one degree-of-freedom. For these, it is customary that the spatial properties — the values of the mass, stiffness and damper elements which make up the model — be expressed as matrices. Those used throughout this work are $[M]$, the mass matrix, $[K]$, the stiffness

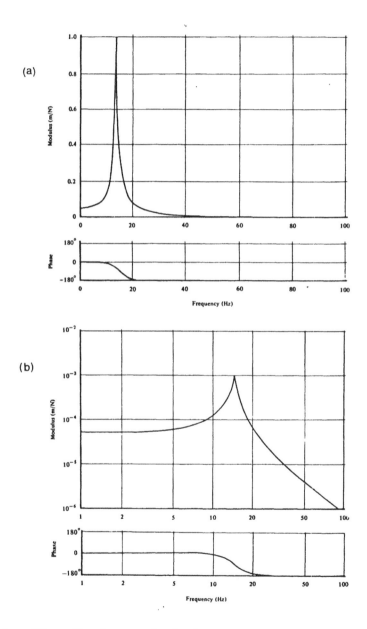

Fig. 1.1 Alternative formats for display of frequency response function
(FRF) of a single-degree-of-freedom (SDOF) system.
(a) Linear scales (Bode); (b) Logarithmic scales (Bode)

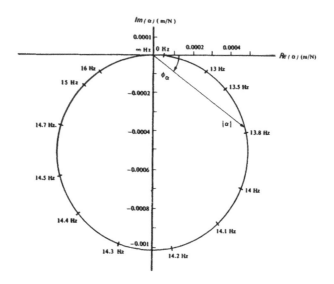

Fig. 1.1   Alternative formats for display of frequency response function
(FRF) of a single-degree-of-freedom (SDOF) system.
(c) Complex plane (Nyquist plot)

matrix, $[C]$, the viscous damping matrix, and $[D]$, the structural or
hysteretic damping matrix. The first phase (of three) in the vibration
analysis of such systems is that of setting up the governing equations of
motion which, in effect, means determining the elements of the above
matrices. (This is a process which does exist for the SDOF system but is
often trivial.) The second phase is that of performing a free vibration
analysis using the equations of motion. This analysis produces first a
set of N natural frequencies and damping factors (N being the number
of degrees of freedom, or equations of motion) and secondly a matching
set of N 'mode shape' vectors, each one of these being associated with a
specific natural frequency and damping factor. The complete free
vibration solution is conveniently contained in two matrices, $[\lambda^2]$ and
$[\Phi]$, which are again referred to as the 'modal properties' or,
sometimes, as the eigenvalue (natural frequency and damping) and
eigenvector (mode shape) matrices. One element from the diagonal
eigenvalue matrix (λ_r^2) contains both the natural frequency **and** the
damping factor for the rth normal mode of vibration of the system while
the corresponding column, $\{\phi\}_r$, from the full eigenvector matrix, $[\Phi]$,
describes the shape (the relative displacements of all parts of the
system) of that same mode of vibration. There are several detailed
variations to this general picture, depending upon the type and
distribution of damping, but all cases can be described in the same
general way.

The third and final phase of theoretical analysis is the forced response analysis, and in particular that for harmonic (or sinusoidal) excitation. By solving the equations of motion when harmonic forcing is applied, we are able to describe the complete solution by a single matrix, known as the 'frequency response matrix' $[H(\omega)]$, although unlike the previous two matrix descriptions, the elements of this matrix are not constants but are frequency-dependent, each element being itself a frequency response (or mobility) function. Thus, element $H_{jk}(\omega)$ represents the harmonic response, X_j, in one of the degrees of freedom, j, caused by a single harmonic force, F_k, applied at a different degree of freedom, k. Both these harmonic quantities are described using complex algebra to accommodate the magnitude **and** phase information, as also is $H_{jk}(\omega)$. Each such quantity is referred to as a frequency response function, or FRF for short.

The particular relevance of these specific response characteristics is the fact that they are the quantities which we are most likely to be able to measure in practice. It is, of course, possible to describe each individual frequency response function in terms of the various mass, stiffness and damping elements of the system but the relevant expressions tend to be extremely complex for practical structures. However, it transpires that the same expressions can be drastically simplified if we use the modal properties instead of the spatial properties and it is possible to write an expression for any FRF, $H_{jk}(\omega)$, which has the general form:

$$H_{jk}(\omega) = \frac{X_j}{F_k} = \sum_{r=1}^{N} \frac{{}_r A_{jk}}{\lambda_r^2 - \omega^2} \tag{1.1}$$

where λ_r^2 is the eigenvalue of the rth mode (its natural frequency and damping factor combined); ${}_r A_{jk}$, (the modal constant) is constructed from $\{\phi\}_r$; ϕ_{jr} is the jth element of the rth eigenvector $\{\phi\}_r$ (i.e. the relative displacement at that DOF during vibration in the rth mode); and N is the number of degrees of freedom (or modes).

This expression forms the foundation of modal analysis: it shows a direct connection between the modal properties of a system and its response characteristics. From a purely theoretical viewpoint it provides an efficient means of predicting response (by performing a free vibration analysis first) while from a more practical standpoint, it suggests that there may be means of determining modal properties from FRFs which are amenable to direct measurement. Here again, it is appropriate to consider the form which a plot of such an expression as (1.1) will take, and some examples are shown in Fig. 1.2 which can be deduced entirely from the equation itself, using different values for the

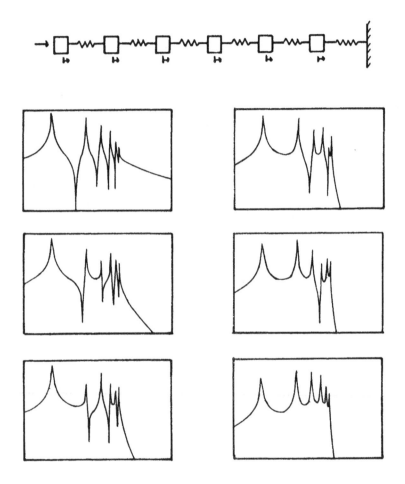

Fig. 1.2   Typical FRF plots for multi-degree-of-freedom (MDOF) system.
(a) Bode plots

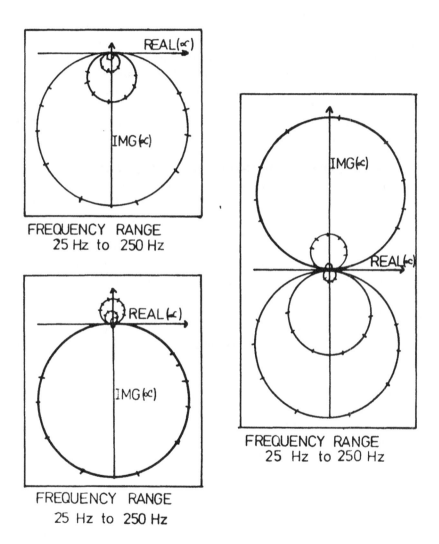

Fig. 1.2 Typical FRF plots for multi-degree-of-freedom (MDOF) system.
(b) Nyquist

modal properties or coefficients in the expression.

Thus we find that by making a thorough study of the theory of structural vibration, we are able to 'predict' what we might expect to find if we make FRF measurements on actual hardware. Indeed, we shall see later how these predictions can be quite detailed, to the point where it is possible to comment on the likely quality of measured data.

We have outlined above the major aspects of the 'theoretical route' of vibration analysis. There are also a number of topics which need to be covered dealing with aspects of signal processing, non-harmonic response characteristics and non linear behaviour, but these may be regarded as additional details which may be required in particular cases while the above-mentioned items are fundamental and central to any application of modal testing.

1.5   SUMMARY OF MEASUREMENT METHODS

In the previous section, we reviewed the major features of the appropriate theory and these all led up to the frequency response characteristics. Thus the main measurement techniques which must be devised and developed are those which will permit us to make direct measurements of the various FRF properties of the test structure.

In this review, we shall concentrate on the basic measurement system used for single-point excitation, the type of test best suited to FRF measurement, the main items of which are shown in Fig. 1.3. Essentially, there are three aspects of the measurement process which demand particular attention in order to ensure the acquisition of the high-quality data which are required for the next stage — data analysis. These are:

(i)   the mechanical aspects of supporting and (correctly) exciting the structure;

(ii)   the correct transduction of the quantities to be measured — force input and motion response; and

(iii)   the signal processing which is appropriate to the type of test used.

In the first category, we encounter questions as to how the testpiece should be suspended, or supported, and how it should be excited. Usually, one of three options is chosen for the support: **free**, or unrestrained, (which usually means suspended on very soft springs); **grounded**, which requires its rigid clamping at certain points; or **in situ**, where the testpiece is connected to some other structure or component which presents a non-rigid attachment. The choice itself will often be decided by various, sometimes conflicting, factors. Amongst these may be a desire to correlate the test results with theory

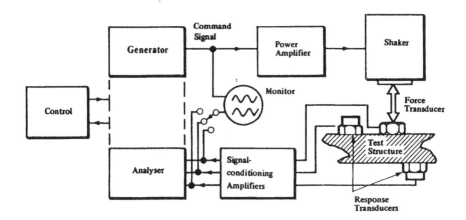

Fig. 1.3   Basic components of FRF measurement system

and in this case it should be remembered that free boundaries are much easier to simulate in the test condition than are clamped, or grounded ones. Also, if tests are being made on one component which forms part of an assembly, these may well be required for the free-free condition.

The mechanics of the excitation are achieved either by connecting a vibration generator, or shaker, or by using some form of transient input, such as a hammer blow or sudden release from a deformed position. Both approaches have advantages and disadvantages and it can be very important to choose the best one in each case.

Transducers are very important elements in the system as it is essential that accurate measurements be made of both the input to the structure and of its response. Nowadays, piezoelectric transducers are widely used to detect both force and acceleration and the major problems associated with them are to ensure that they interfere as little as possible with the test structure and that their performance is adequate for the ranges of frequency and amplitude of the test. Incorrect transducer selection can give rise to very large errors in the measured data upon which all the subsequent analysis is based.

The FRF parameters to be measured can be obtained directly by applying a harmonic excitation and then measuring the resulting harmonic response. This type of test is often referred to as 'sinewave testing' and it requires the attachment of a shaker to the structure. The frequency range is covered either by stepping from one frequency to the next, or by slowly sweeping the frequency continuously, in both cases allowing quasi-steady conditions to be attained. Alternative excitation procedures are now widely used. Transient (including burst signals) periodic, pseudo-random or random excitation signals often replace the sine-wave approach and are made practical by the existence of complex signal processing analysers which are capable of resolving the frequency content of both input and response signals, using Fourier analysis, and thereby deducing the mobility parameters required. A further extension of this development is possible using impulsive or transient excitations which may be applied without connecting a shaker to the structure. All of these latter options offer the possibility of shorter testing times but great care must be exercised in their use as there are many steps at which errors may be incurred by incorrect application. Once again, a sound understanding of the theoretical basis — this time of signal processing — is necessary to ensure successful use of these relatively advanced techniques.

As was the case with the theoretical review, the measurement process also contains many detailed features which will be described below. Here, we have just outlined the central and most important topics to be considered. One final observation which must be made is that in modal testing applications of vibration measurements, perhaps more than many others, accuracy of the measured data is of paramount importance. This is so because these data are generally to be submitted to a range of analysis procedures, outlined in the next section, in order to extract the results eventually sought. Some of these analysis processes are themselves quite complex and can seldom be regarded as insensitive to the accuracy of the input data. By way of a note of caution, Fig. 1.4 shows the extent of variations which may be obtained by using different measurement techniques on a particular test structure [19]. Since that survey was carried out, there have been further similar studies; two of particular note are (i) a survey which proposed a 'standard' or 'benchmark' rectangular plate structure that individual laboratories could acquire for themselves [20], and (ii) a European survey of modal testing capabilities under GARTEUR sponsorship, resulting in a contemporary assessment of the state-of-the-art of the technology in the mid-to-late 1990s [21].

1.6   SUMMARY OF MODAL ANALYSIS PROCESSES

The third skill required for modal testing is concerned with the analysis of the measured FRF data. This is quite separate from the signal processing which may be necessary to convert raw measurements into frequency responses. It is a procedure whereby the measured mobilities are analysed in such a way as to find a theoretical model which most closely resembles the behaviour of the actual testpiece. This process itself falls into two stages: first, to identify the appropriate type of model and second, to determine the appropriate parameters of the chosen model. Most of the effort goes into this second stage, which is widely referred to as 'modal parameter extraction' or, simply, as 'modal

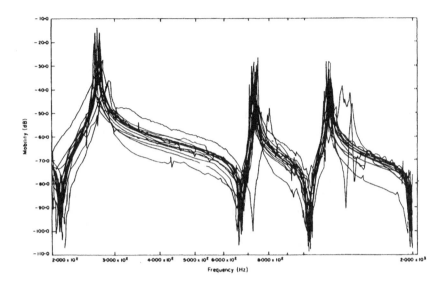

Fig. 1.4   Various measurements on standard benchmark testpiece

analysis'.

We have seen from our review of the theoretical aspects that we can 'predict' or, better, 'anticipate' the form of the FRF plots for a multi-degree-of-freedom system and we have also seen that these may be directly related to the modal properties of that system. The great majority of the modal analysis effort involves the matching or curve-fitting an expression such as equation (1.1) above to the measured FRFs and thereby finding the appropriate modal parameters.

A completely general curve-fitting approach is possible but generally inefficient. Mathematically, we can take an equation of the form

$$H(\omega) = \sum_{r=1}^{N} \frac{A_r}{\lambda_r^2 - \omega^2} \tag{1.2}$$

and curve-fit a set of measured values $H_m(\omega_1)$, $H_m(\omega_2)$, ..., etc. to this expression so that we obtain estimates for the coefficients A_1, A_2, ..., λ_1^2, λ_2^2, ..., etc. These coefficients are, of course, closely related to the modal properties of the system. However, although such approaches are made, they are inefficient and neither exploit the particular properties of resonant systems nor take due account of the unequal quality of the various measured points in the data set, $H_m(\omega_1)$, $H_m(\omega_2)$, ..., etc., both of which can have a significant influence on the overall analysis process. Thus there is no single modal analysis method but rather a selection, each being the most appropriate in differing conditions.

One of the most widespread and useful approaches is that known as the 'Single-Degree-of-Freedom Curve-Fit' or, often, the 'Circle Fit' procedure. This method uses the fact that at frequencies close to a natural frequency, the FRF can often be approximated to that of a single degree-of-freedom system plus a constant offset term (which approximately accounts for the existence of other modes). This assumption allows us to use the circular nature of a modulus/phase polar plot (the Nyquist plot) of the frequency response function of a SDOF system (see Fig. 1.1(c)) by curve-fitting a circle to just a few measured data points, as illustrated in Fig. 1.5. This process can be repeated for each resonance individually until the whole curve has been analysed. At this stage, a theoretical regeneration of the FRF is possible using the set of coefficients extracted, as illustrated in Fig. 1.6. More recent versions of the same basic approach — the so-called 'line-fit' methods — exploit the straight-line (rather than circular) nature of the FRF plot when presented as a reciprocal function.

These simple methods can be used for many of the cases encountered in practice but they become inadequate and inaccurate when the structure has modes which are 'close', a condition which is identified by the lack of an obviously-circular section on the Nyquist plot. Under these conditions it becomes necessary to use a more complex process which accepts the simultaneous influence of more than one mode. These latter methods are referred to as 'MDOF curve-fits' and are naturally more complicated and require more computation effort but, provided the data are accurate, they have the capability of producing more accurate estimates for the modal properties (or at least for the coefficients in equation (1.2)).

In this subject, again, there are many detailed refinements but the analysis process is always essentially the same: that of finding — by curve-fitting — a set of modal properties which best match the response characteristics of the tested structure. Some of the more detailed considerations include: compensating for slightly non-linear behaviour; simultaneously analysing more then one FRF and curve-fitting to actual time histories (rather than the processed frequency response functions).

As mentioned in the previous section on measurement techniques, there have been attempts to assess the reliability and consistency of the different modal analysis methods which are available. The most comprehensive is probably the one run by the Swedish Vibration Society (SVIB) whose results have been presented in reference [22].

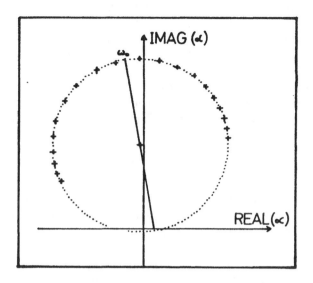

Fig. 1.5 Curve-fit to resonant FRF data

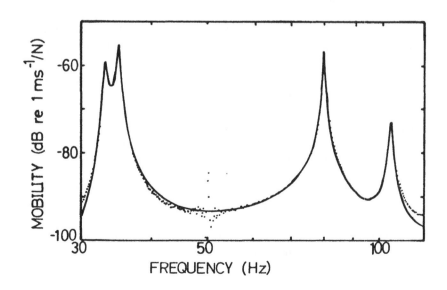

Fig. 1.6 Regeneration of mobility FRF curve from curve-fit data

1.7 REVIEW OF TEST PROCEDURES AND LEVELS

Having now outlined the major features of the three essential ingredients for modal testing, it is appropriate to end this introduction with a brief review of just how these capabilities are drawn together to conduct a modal test.

The overall objective of the test is to determine a set of modal properties for a structure. These consist of natural frequencies, damping factors and mode shapes. The procedure consists of three steps:

(i)   measure an appropriate set of mobilities, or FRFs;
(ii)   analyse these using appropriate curve-fitting procedures; and
(iii)   combine the results of the curve-fits to construct the required model. (In some modern implementations, (ii) and (iii) are performed as a single process.)

Using our knowledge of the theoretical relationship between FRF functions and modal properties, it is possible to show that an 'appropriate' set of FRFs to measure consists in most cases of just one row or one column in the FRF matrix, $[H(\omega)]$. In practice, this either means exciting the structure at one point and measuring responses at all points **or** measuring the response at one point while the excitation is applied separately at each point in turn. (This last option is most conveniently achieved using a hammer or other non-contacting excitation device.)

In practice, this relatively simple procedure will be embellished by various detailed additions, but the general method is always as described here.

As we consider each new modal test in some detail, we shall realise that, while we shall always follow the same overall procedure, there will be a different level of detail, or accuracy, required for each different application. This realisation has been recognised and embodied in the classification of five different Levels of Test in the Handbook published by the DTA [12]. These different levels can be defined as follows:

Level 0: estimation of natural frequencies and damping factors; response levels measured at few points; very short test times.

Level 1: estimation of natural frequencies and damping factors; mode shapes defined qualitatively rather than quantitatively.

Level 2: measurements of all modal parameters suitable for tabulation and mode shape display, albeit un-normalised.

Level 3: measurements of all modal parameters, including normalised mode shapes; full quality checks performed and model usable for model validation.

Level 4: measurements of all modal parameters and residual effects for out-of-range modes; full quality checks performed and model usable for all response-based applications, including modification, coupling and response predictions.

Attention to the appropriate level of test required in each application clearly constitutes an important part of the planning process.

1.8   TERMINOLOGY AND NOTATION

One final word by way of introduction to this important aspect of the work: this subject (like many others but perhaps more than most) has generated a wealth of jargon, not all of it consistent! We have adopted a particular notation and terminology throughout this work but in order to facilitate 'translation' for compatibility with other references, and in particular the manuals of various analysis equipment and software in widespread use, the alternative names will be indicated as the various parameters are introduced.

In the past decade (the 1990s) some attempts have been made to standardise the notation and the terminology used in our subject, [23]. While not succeeding completely, there has been some notable progress, not least by a number of journals, and significant conferences, recommending to their prospective authors a basic notation which was proposed in 1990, [23], and refined over a period of three to four years before being used as the basis of the documentation published by the Dynamic Testing Agency, [12]. This notation is presented here in an Appendix and will be strictly adhered to throughout this text.

It is not only in respect of notation that care is needed in order to ensure clarity in the explanations, and the necessary lack of ambiguity. The terminology that we use should also be carefully considered and there are certain words that are used rather imprecisely, and others that are used by different authors to mean different things. Many of these are 'ordinary' words, and not technical jargon, and so it is worth mentioning a few of these at the outset of our text. We shall be dealing extensively with the imprecision and uncertainty which accompanies all experimental work. Thus we must be clear about the meaning of words such as 'errors', 'uncertainty', 'inaccuracy', 'imprecision'; and 'repeatability' and 'reproducibility'.

- An **Error** is the difference between an obtained value and the true or correct value

- **Uncertainty** refers to the range of values within which we can define a quantity (that we have measured, or otherwise obtained)
- **Inaccuracy** relates to the size of the error which may be associated with a quoted value
- **Imprecision** generally refers to the degree of resolution or detail with which a calculation is performed, and often accompanies the introduction of a truncated or otherwise unrefined element in the computation of a given quantity
- **Repeatability** refers to the extent to which a quantity will be found to have exactly the same (measured) value in a second or subsequent acquisition, using exactly the same process(es)
- **Reproducibility** refers to the extent to which a given result (experiment, measurement) can be reproduced at a later date using similar procedures or equipment

We shall need to consider the difference between **directly** measured data and **indirect** measurements. Usually, we shall make direct measurements of excitation forces and the resulting responses: any response functions which are then derived are indirectly measured, as are the ensuing estimates of natural frequencies and mode shapes. We must always be aware that we do not **measure** these modal properties, but we derive them by some inexact analysis of the quantities that we actually do measure.

Later, we shall be concerned with the reconciliation between predicted properties and measured ones. Indeed, as already mentioned, this process is one of the major reasons for modal tests to be performed. In such applications, there is much talk of 'errors' in the theoretical model and this can lead to unrealistic expectations of the whole test-analysis comparison process. We shall talk then of verification and validation and although these terms will be defined in the relevant chapter, it is perhaps appropriate to introduce the subtle difference intended to be represented by these two words. 'Verification' refers to the process of determining whether something (an algorithm, a calculation, a model) is correct or not. It is black or white: the object is either correct or it is not: the matrix inversion routine is either coded correctly or it is not: the cables are either correctly labelled or they are not. 'Validation', on the other hand, is less black and white and is more concerned with whether the object being described is fit for its intended purpose. 'Valid' is taken here to mean that the object (perhaps a measurement, or a mathematical model) is capable of representing the quantity or behaviour of interest **sufficiently well to serve the needs of that object.** Thus, we have a degree of judgement to exercise in deciding whether something we have obtained or created is valid (i.e. good enough) and this satisfies many of the real life situations in which we are obliged to apply our skills.

These are not the only examples where precise use of the language is necessary, but they serve to illustrate the concern and to justify the comments that will be made from time to time where a similar confusion is to be avoided by careful choice of the words we use.

CHAPTER 2

Theoretical Basis

2.1   INTRODUCTION

It has already been emphasised that the theoretical foundations of modal testing are of paramount importance to its successful implementation. Thus it is appropriate that this first main chapter deals with the various aspects of theory which are used at the different stages of modal analysis and testing.

The majority of this chapter (Sections 2.2 to 2.8) deals with an analysis of the basic vibration properties of general linear structures, including ones which contain rotating components, as these properties form the basis of experimental modal analysis. Later sections extend the theory somewhat to take account of the different ways in which such properties can be measured (Section 2.11) and some of the more complex features which may be encountered, such as complex modes (Section 2.9) and non-linearities, (Section 2.14). Section 2.10 is devoted to the question of different methods of graphical presentation which can be used to enhance both the display and the interpretation of the FRF data which constitute the 'currency' of our methods. There are some topics of which knowledge is assumed in the main text but for which a review is provided in the Appendices in case the current usage is unfamiliar.

Before embarking on the detailed analysis, it is appropriate to put the different stages into context and this can be done by showing what will be called the 'theoretical route' for vibration analysis (Fig. 2.1). This illustrates the three phases through which a typical theoretical vibration analysis progresses. Generally, we start with a description of the structure's physical characteristics, usually in terms of its **mass**, **stiffness** and **damping** properties, and this is referred to as the **Spatial Model**.

Then it is customary to perform a theoretical modal analysis of the spatial model which leads to a description of the structure's behaviour as a set of vibration modes; the **Modal Model**. This model is defined as a set of **natural frequencies** with corresponding **modal damping**

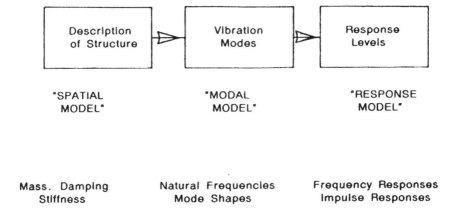

Fig. 2.1   Theoretical route to vibration analysis

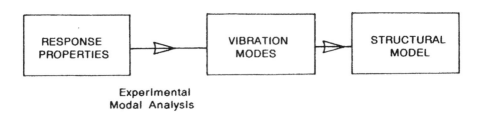

Fig. 2.2   Experimental route to vibration analysis

factors and **vibration mode shapes**. It is important to remember that this solution always describes the various ways in which the structure is capable of vibrating naturally, i.e. without any external forcing or excitation, and so these are called the 'natural' or 'normal' modes of the structure.

The third stage is generally that in which we have the greatest interest; namely, the analysis of exactly how the structure will respond under given excitation conditions and, especially, with what amplitudes. Clearly, this will depend not only upon the structure's inherent properties but also on the nature and magnitude of the imposed excitation and so there will be innumerable solutions of this type. However, it is convenient to present an analysis of the structure's response to a 'standard' excitation (from which the solution for any particular case can be constructed) and to describe this as the **Response Model**. The standard excitation chosen throughout this work will be that of a unit-amplitude sinusoidal force applied to each point on the structure individually, and at every frequency within a specified range. Thus our response model will consist of a set of **frequency response functions** (FRFs) which must be defined over the applicable range of frequency.

Throughout the following analysis we shall be focusing on these three stages and types of model — **Spatial**, **Modal** and **Response** — and it is essential to understand fully their interdependence as it is upon this characteristic that the principles of modal testing are founded. As indicated in Fig. 2.1, it is possible to proceed from the spatial model through to a response analysis. It is also possible to undertake an analysis in the reverse direction — i.e. from a description of the response properties (such as measured frequency response functions) we can deduce modal properties and, in the limit, the spatial properties. This is the 'experimental route' to vibration analysis which is shown in Fig. 2.2 and which will be discussed in detail in Chapter 5.

As a parting comment before we embark on a moderately lengthy voyage through the underlying theory upon which our subject is based, it must be noted that it can seem to be an extreme irony that an experimentally-based technology such as we are describing here demands a richness of theoretical methods that significantly outstrips the corresponding material that would be found in a theoretically-based study of the same general subject. This is simply because in the experimental field we must be prepared to explain and to interpret the most general of circumstances (uncertain damping type; almost inevitable arbitrariness of damping distribution; the distinct possibility of encountering non-linear behaviour, and so on). The luxury of being able to dictate the conditions (or assumptions) at the outset of a study — as we are wont to do in theoretical analyses — is not one that can generally be extended to the experimentalist, and so he or she must be armed with the most general of models.

2.2 SINGLE-DEGREE-OF-FREEDOM (SDOF) SYSTEM THEORY

Although very few practical structures could realistically be modelled by a single-degree-of-freedom (SDOF) system, the properties of such a system are very important because those for a more complex multi-degree-of-freedom (MDOF) system can always be represented as the linear superposition of a number of SDOF characteristics.

Throughout this chapter we shall describe three classes of system model:

(a)   undamped
(b)   viscously-damped
(c)   hysteretically- (or structurally-) damped

and we shall also make extensive use of complex algebra to describe the time-varying quantities of displacement, force etc. (Appendix 1 gives some notes on the use of complex algebra for harmonic quantities.)

The basic model for the SDOF system is shown in Fig. 2.3 where

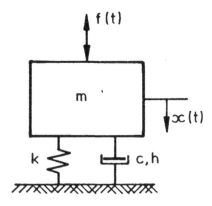

Fig. 2.3   Single-degree-of-freedom (SDOF) system

$f(t)$ and $x(t)$ are general time-varying force and displacement response quantities. The spatial model consists of a mass (m) and a spring (k) plus (when damped) either a viscous dashpot (c) or hysteretic damper (d).

2.2.1 Undamped Systems

As stated, the **spatial model** consists of m and k.

For the modal model, we consider the properties of the system with no external forcing, i.e. $f(t) = 0$ and for this case the governing equation of motion is:

$$m\,\ddot{x} + k\,x = 0 \qquad (2.1)$$

The trial solution, $x(t) = X e^{i\omega t}$ leads to the requirement that:

$$\left(k - \omega^2 m\right) = 0$$

Hence the **modal model** consists of a single solution (mode of vibration) with a natural frequency $\overline{\omega}_0$ given by $(k/m)^{1/2}$.

Turning next to a frequency response analysis, we consider an excitation of the form:

$$f(t) = F e^{i\omega t}$$

and assume a solution of the form

$$x(t) = X e^{i\omega t}$$

where X and F are complex to accommodate both the amplitude and phase information (see Appendix 1). Now the equation of motion is

$$\left(k - \omega^2 m\right) X e^{i\omega t} = F e^{i\omega t} \qquad (2.2)$$

from which we extract the required **response model** in the form of a frequency response function, $\alpha(\omega)$:

$$\boxed{\alpha(\omega) = \frac{X}{F} = \frac{1}{(k - \omega^2 m)} \quad \text{also written as } H(\omega)} \qquad (2.3)$$

This particular form of FRF, where the response parameter is displacement (as opposed to velocity or acceleration), is called a

'Receptance' and is usually written as $\alpha(\omega)$, although sometimes $H(\omega)$ is used as a generic FRF parameter. Note that this function, along with other versions of the FRF, is independent of the excitation.

2.2.2 Viscous Damping
If we add a viscous dashpot, c, the equation of motion for free vibration becomes

$$m\,\ddot{x} + c\,\dot{x} + k\,x = 0 \qquad\qquad (2.4)$$

and we must now use a more general trial solution:

$$x(t) = Xe^{st} \qquad \text{(where s is complex, rather than imaginary, as before)}$$

with which we obtain the condition that must be satisfied for a solution to exist:

$$\left(ms^2 + cs + k\right) = 0 \qquad\qquad (2.5)$$

This, in turn, leads to

$$
\begin{aligned}
s_{1,2} &= -\frac{c}{2m} \pm \frac{\sqrt{c^2 - 4km}}{2m} \\[2mm]
&= -\overline{\omega}_0\,\zeta \pm i\,\overline{\omega}_0\,\sqrt{1-\zeta^2}
\end{aligned}
\qquad\qquad (2.6)
$$

where

$$\overline{\omega}_0^2 = \left(k/m\right) \quad ; \quad \zeta = c/c_0 = \left(c/2\sqrt{km}\right)$$

This implies a modal solution of the form:

$$x(t) = Xe^{-\overline{\omega}_0\zeta t}\,e^{i(\overline{\omega}_0\sqrt{1-\zeta^2})t} = Xe^{-at}\,e^{i\omega_0' t}$$

which is a single mode of vibration with a complex natural frequency having two parts:

- an imaginary or oscillatory part; a frequency of ω_0' $(= \overline{\omega}_0\sqrt{1-\zeta^2}$);
- a real or decay part; a damping rate of $\dot{a}$ $(= \zeta\overline{\omega}_0$).

The physical significance of these two parts of the modal model is illustrated in the typical free response plot, shown in Fig. 2.4.

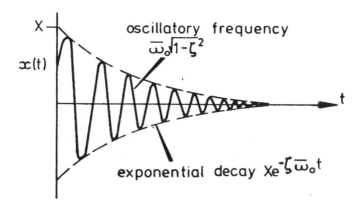

Fig. 2.4   Free vibration characteristic of damped SDOF system

Lastly, we consider the forced response when $f(t) = F e^{i\omega t}$ and, as before, we assume $x(t) = X e^{i\omega t}$. Here, the equation of motion:

$$\left(-\omega^2 m + i \omega c + k\right) X e^{i\omega t} = F e^{i\omega t} \tag{2.7}$$

gives a receptance FRF of the form

$$\boxed{H(\omega) = \alpha(\omega) = \frac{1}{(k - \omega^2 m) + i(\omega c)}} \tag{2.8a}$$

which is now complex, containing both magnitude and phase information.

Note that

$$|\alpha(\omega)| = \frac{|X|}{|F|} = \frac{1}{\sqrt{\left(k - \omega^2 m\right)^2 + (\omega c)^2}} \tag{2.8b}$$

and $\angle\alpha(\omega) = \angle X - \angle F = \operatorname{tg}^{-1}(-\omega c/(k - \omega^2 m)) = \theta_\alpha$.

A corresponding non-dimensional version of the same expression can be developed:

$$\alpha(\omega) = \frac{1/k}{(1 - (\omega/\overline{\omega}_0)^2 + 2i\varsigma(\omega/\overline{\omega}_0))} \tag{2.8c}$$

2.2.3 Structural Damping

Close inspection of the behaviour of real structures suggests that the viscous damping model used above is not very representative when applied to MDOF systems. There is seen to be a frequency-dependence exhibited by real structures which is not described by the standard viscous dashpot and what is required, apparently, is a damper whose rate varies with frequency. This conclusion can be seen by inspection of the underlying physics of typical damping phenomena which exist in most real structures.

All structures exhibit a degree of damping due to the hysteresis properties of the material(s) from which they are made. A typical example of this effect is shown in the force-displacement plot in Fig. 2.5(a) in which the area contained by the loop represents the energy 'lost' in one cycle of vibration between the extremities shown. Interestingly, the area contained by the shaded triangle represents the maximum energy stored in the elastic part of the structure, at the point of greatest deflection. The damping capacity, or effect, of such a component can conveniently be defined by the ratio of these two areas:

damping capacity = energy lost per cycle/maximum energy stored

Another common source of energy dissipation in practical structures, and thus of their damping, is the friction which exist in joints between components of the structure. Whether these effects be macro slip between adjacent parts or, more commonly, micro slip in the areas of connection between them, they may be described very roughly by the simple dry friction model shown in Fig. 2.5(b) with its corresponding force-displacement diagram. There are, of course, other forms of such damping mechanisms, but these two serve to illustrate the essential features of them all.

The mathematical model of the viscous damper which we have used above can be compared with these more physical effects by plotting the corresponding force-displacement diagram for it, and this is shown in Fig. 2.5(c). Because the relationship is linear between force and **velocity**, it is necessary to suppose harmonic motion, at frequency ω, in

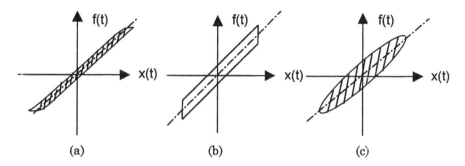

Fig. 2.5   Force-deflection characteristics.
(a) Material hysteresis; (b) Dry friction; (c) Viscous damper

order to construct a force-displacement diagram, and that has been done here. The resulting diagram shows the nature of the approximation provided by the viscous damper model and the concept of the effective or equivalent viscous damping coefficient for any of the actual phenomena as being that which provides the same energy loss per cycle as the real thing. The problem which arises with the viscous damping model is that it has a frequency-dependence in the amount of energy loss per cycle whereas the dry friction mechanism is clearly unaffected by the frequency of loading and experiments suggest that the hysteresis effect is similarly independent of frequency. Thus, we find a problem in obtaining a single equivalent viscous dashpot model which will be valid over a range of frequencies, such as will be necessary to represent the damping of an MDOF system over all, or at least several, of its modes of vibration.

An alternative theoretical damping model is provided by the **hysteretic** or **structural** damper which not only has the advantage mentioned above (in the form of an effective damper rate which varies inversely with frequency, $c_e = (d/\omega)$, and thus satisfies the requirement that the energy lost per cycle is independent of frequency), but also provides a much simpler analysis for MDOF systems, as shown below in Section 2.6. However, it presents difficulties to a rigorous free vibration analysis and its application is generally focused on the forced response analysis. In this case we can write an equation of motion:

$$\left(-\omega^2\, m + k + i\, d\right) X\, e^{i\omega t} = F\, e^{i\omega t} \tag{2.9}$$

giving

$$\frac{X}{F} = \alpha(\omega) = \frac{1}{(k - \omega^2 m) + i(d)} \qquad (2.10)$$

or

$$\boxed{\alpha(\omega) = \frac{1/k}{(1 - (\omega/\overline{\omega}_0)^2 + i\eta)}} \qquad (2.11)$$

where η is the **structural damping loss factor** and replaces the critical damping ratio, ζ, used for the viscous damping model. It can be seen from (2.8c) and (2.11) that an equivalence between the different types of damping can be defined at resonance, at which frequency, $\eta = 2\zeta_e$.

The similarities between the FRF expressions for the different cases are evident from equations (2.3) (2.8) and (2.10).

2.3   PRESENTATION AND PROPERTIES OF FRF DATA FOR SDOF SYSTEM

Having developed expressions for the basic receptance frequency response function of the SDOF system, we now turn our attention to the various ways of presenting or displaying these data. We shall first discuss variations in the basic form of the FRF and then go on to explore different ways of presenting the properties graphically. Finally, we shall examine some useful geometric properties of the resulting plots.

2.3.1   Alternative Forms of FRF

So far we have defined our receptance frequency response function $\alpha(\omega)$ as the ratio between a harmonic displacement response and the harmonic force. This ratio is complex as there is both an amplitude ratio ($|\alpha(\omega)|$) and a phase angle between the two sinusoids (θ_α).

We could equally have selected the response velocity $v(t)$ ($= \dot{x}(t)$) as the 'output' quantity and defined an alternative frequency response function — **Mobility, (or $Y(\omega)$)** — as

$$\boxed{Y(\omega) = \frac{V e^{i\omega t}}{F e^{i\omega t}} = \frac{V}{F}} \qquad (2.12)$$

However, when considering sinusoidal vibration we have a simple relationship between displacement and velocity (and thus between receptance and mobility) because:

$$x(t) = X e^{i\omega t}$$

and

$$v(t) = \dot{x}(t) = V e^{i\omega t} = i\omega X e^{i\omega t}$$

So,

$$Y(\omega) = \frac{V}{F} = i\,\omega\frac{X}{F} = i\,\omega\,\alpha(\omega) \tag{2.13}$$

Thus

$$\left|Y(\omega)\right| = \omega\left|\alpha(\omega)\right|$$

and

$$\theta_Y = \theta_\alpha + 90°$$

so that mobility is closely related to receptance. Similarly, we could use acceleration $(a(t) = x(t))$ as our response parameter (since it is customary to measure acceleration in tests) so we could define a third FRF parameter — **Inertance** or **Accelerance** — as

$$\boxed{A(\omega) = \frac{A}{F} = -\omega^2\,\alpha(\omega)} \tag{2.14}$$

These represent the main formats of FRF although there exist yet more possibilities by defining the functions in an inverse way, namely as:

$$\left(\frac{\text{force}}{\text{displacement}}\right): \quad \text{“\textbf{Dynamic Stiffness}”}$$

$$= \left(k - \omega^2\,m\right) + \left(i\,\omega\,c\right)\text{or}\left(i\,d\right) \tag{2.15}$$

or

$$\left(\frac{\text{force}}{\text{velocity}}\right): \quad \text{“\textbf{Mechanical Impedance}”}$$

36

or

$$\left(\frac{\text{force}}{\text{acceleration}}\right): \quad \textbf{"Apparent Mass"}$$

It should be noted that these latter formats are often discouraged, except in special cases, as they can lead to considerable confusion and error if improperly used in MDOF systems.

Table 2.1 gives details of all six of the FRF parameters and of the names variously used for them.

Table 2.1 Definition of Frequency Response Functions

Response Parameter R	Standard FRF: R/F	Inverse FRF: F/R
DISPLACEMENT	RECEPTANCE ADMITTANCE DYNAMIC COMPLIANCE DYNAMIC FLEXIBILITY	DYNAMIC STIFFNESS
VELOCITY	MOBILITY	MECHANICAL IMPEDANCE
ACCELERATION	ACCELERANCE INERTANCE	APPARENT MASS

Lastly, it should be noted that it is common practice to use $H(\omega)$ to refer to the standard frequency response functions generically, and to use $Z(\omega)$ to describe any of the inverse formats

2.3.2   Graphical Displays of FRF Data
There is an overriding complication to plotting FRF data which derives from the fact that they are complex and thus there are three quantities — frequency plus two parts of the complex function — and these cannot be fully displayed on a standard x-y graph. Because of this, any such simple plot can only show two of the three quantities and so there are different possibilities available, several of which are used from time to time.

The four most common forms of presentation are:

(i)   Modulus (of FRF) vs. Frequency and Phase vs. Frequency (the Bode type of plot, consisting of two graphs);
(ii)   Real Part (of FRF) vs. Frequency and Imaginary Part vs. Frequency (two plots);

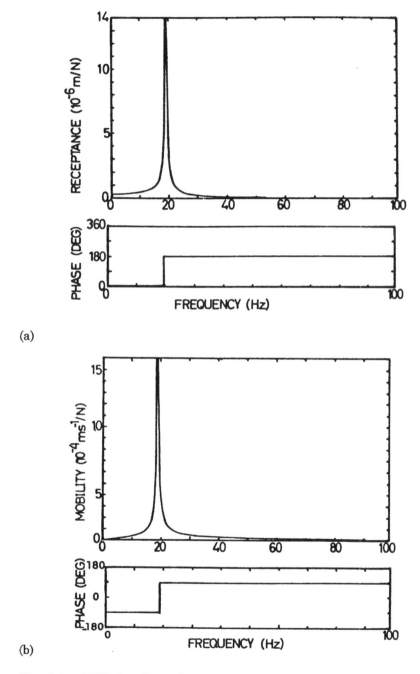

(a)

(b)

Fig. 2.6   FRF plots for undamped SDOF system (linear scales).
(a) Receptance FRF; (b) Mobility FRF

38

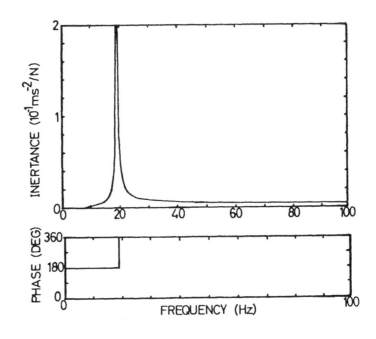

(c)

Fig. 2.6 FRF plots for undamped SDOF system (linear scales).
(c) Accelerance FRF

(iii)   Real Part (of **reciprocal** of FRF) vs. Frequency (or (Frequency)^2)
        and Imaginary Part (of **reciprocal** of FRF) vs. Frequency; and
(iv)   Real Part (of FRF) vs. Imaginary Part (of FRF). (The so-called
        'Nyquist plot': a single graph which does not contain frequency
        information explicitly).

We shall now examine the form and use of these types of graphical
display and identify the particular advantages or features of each.

(i)   A classical Bode plot is shown in Fig. 2.6(a) for the receptance of a
        typical SDOF system without damping. Corresponding plots for
        the mobility and inertance of the same system are shown in Figs.
        2.6(b) and (c), respectively.
            One of the problems with these FRF properties, as with much
        vibration data, is the relatively wide range of values which must
        be encompassed no matter which type of FRF is used. In order to
        cope with this problem, it is often appropriate to make use of
        logarithmic scales and the three functions specified above have
        been replotted in Figs. 2.7(a), (b) and (c) using logarithmic scales

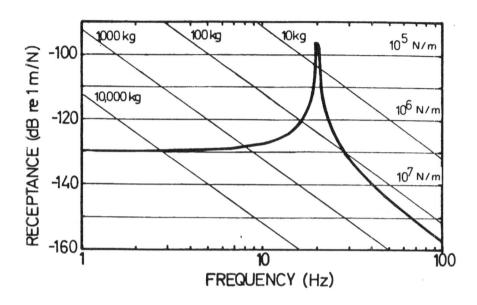

(a)

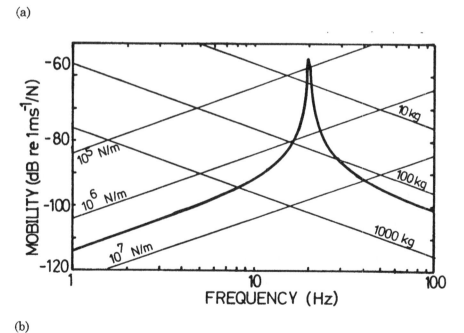

(b)

Fig. 2.7   FRF plots for undamped SDOF system (log-log scales).
(a) Receptance FRF; (b) Mobility FRF

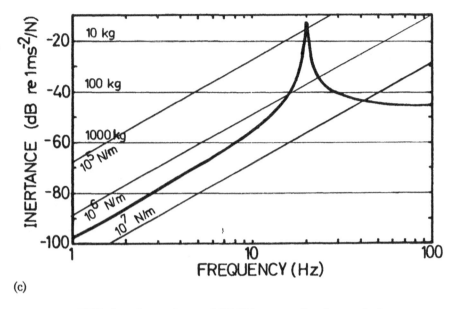

(c)

Fig. 2.7 FRF plots for undamped SDOF system (log-log scales).
(c) Accelerance FRF

for all frequency and modulus axes. The result is something of a transformation in that in each plot can now be divided into three regimes:

- a low-frequency straight-line characteristic;
- a high-frequency straight-line characteristic, and
- the resonant region with its abrupt magnitude and phase variations.

It is helpful and instructive to superimpose on these log-log plots a grid of lines which show the relevant FRF characteristics separately of simple mass elements and simple spring elements. Table 2.2 shows the corresponding expressions for α_m, α_k, Y_m, etc. and from this it is possible to see that mass and stiffness properties will always appear as straight lines on log (modulus) vs. log (frequency) plots, as shown in Fig. 2.8. These have in fact been included in Fig. 2.7 but their significance can be further appreciated by reference to Fig. 2.8 which shows the mobility modulus plots for two different systems. By referring to and interpolating between the mass- and stiffness-lines drawn on the plot, we can deduce that system (a) behaves as would a mass of 1 kg with a spring stiffness of 2.5 kN/m while system (b) has corresponding values of 0.8 kg and 120 kN/m, respectively.

Table 2.2 Frequency Responses of Mass and Stiffness Elements

FRF Parameter	Mass	Stiffness
RECEPTANCE $\alpha(\omega)$ $\log\lvert\alpha(\omega)\rvert$	$-1/\omega^2 m$ $-\log(m)-2\log(\omega)$	$1/k$ $-\log(k)$
MOBILITY $Y(\omega)$ $\log\lvert Y(\omega)\rvert$	$-i/\omega m$ $-\log(m)-\log(\omega)$	$i\omega/k$ $\log(\omega)-\log(k)$
ACCELERANCE $A(\omega)$ $\log\lvert A(\omega)\rvert$	$1/m$ $-\log(m)$	$-\omega^2/k$ $2\log(\omega)-\log(k)$

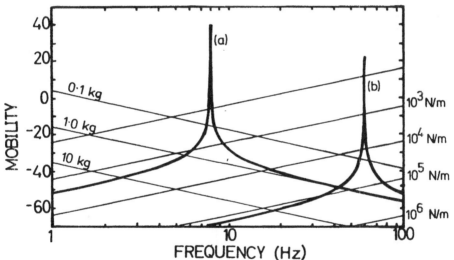

Fig. 2.8   Mobility FRF plots for different undamped SDOF systems

This basic style of displaying FRF data applies to all types of system, whether damped or not, while the other forms are only applicable to damped systems and then tend to be sensitive to the type of damping. Fig. 2.9 shows the basic example system plotted for different levels of viscous damping with a zoomed detail of the narrow band around resonance which is the only region that the damping has any influence on the FRF plot.

(ii)   Companion plots for Real Part (Re.) vs. Frequency and Imaginary Part (Im.) vs. Frequency are shown in Fig. 2.10 for the SDOF system with light viscous damping. All three forms of the FRF are shown and from these we can see how the phase change through

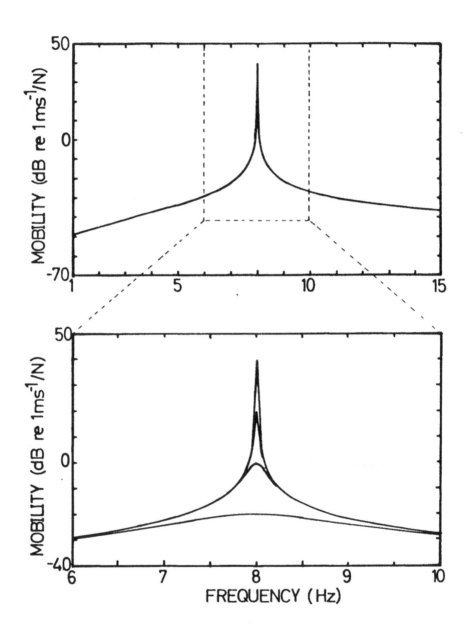

Fig. 2.9   Resonance region detail of FRF plot for damped SDOF system

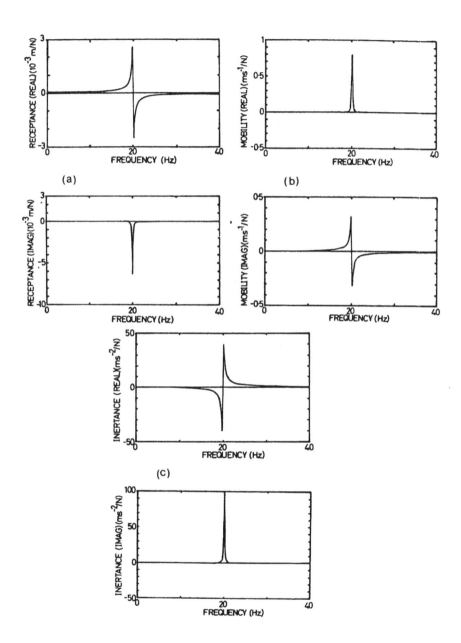

Fig. 2.10   Plots of real and imaginary parts of FRF for damped SDOF
system.
(a) Receptance; (b) Mobility; (c) Accelerance

the resonance region is characterised by a sign change in one part accompanied by a peak (max or min) value in the other part.

It should be noted here that the use of logarithmic scales is not feasible in this case primarily because it is necessary to accommodate both positive and negative values (unlike the modulus plots) and this would be impossible with logarithmic axes. Partly for this reason, and others which become clearer when dealing with MDOF systems, this format of display is not so widely used as the preceding ones.

(iii) The so-called 'inverse' or 'reciprocal' plots are, however, more interesting in that they have the potential of providing rather more insight into the system whose characteristics they represent. First of all, it can be seen from the expression for the inverse receptance (see equation (2.15)) that the Real Part depends entirely on the mass and stiffness properties while the Imaginary Part is a function only of the damping. This separation of the constituent physical properties has not been observed in the other versions of the FRF, or their plots. Fig. 2.11 shows an example of a plot of this form for a system with a combination of both viscous and structural damping. Fig. 2.11(a) shows the Real Part which has a slope of $(-m)$ at the axis crossing point, which is itself at the undamped system natural frequency, $\overline{\omega}_0$. Fig. 2.11(b)

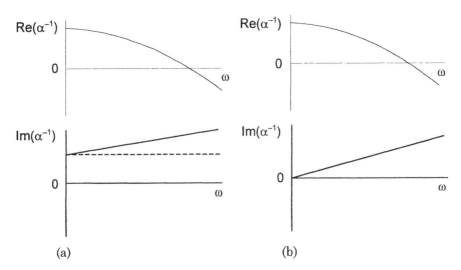

Fig. 2.11 Inverse FRF plot for system with (a) mixed, and (b) viscous damping

shows a straight line whose slope is given by the viscous damping rate, c, and whose intercept at $\omega = 0$ is provided by the structural damping coefficient, d. The potential of this form of presentation, if it carries over to the more general case of MDOF systems, soon becomes apparent.

(iv) The Nyquist or Argand plane plot is widely used and is a very effective way of displaying the important resonance region in some detail.

Fig. 2.12 shows Nyquist-type FRF plots corresponding to the viscously-damped SDOF system previously illustrated in Figs. 2.9 and 2.10. As this style of presentation consists of only a single graph, the missing information (in this case, frequency) must be added by identifying the values of frequency corresponding to particular points on the curve. This is usually done by indicating specific points on the curve at regular increments of frequency. In the examples shown, only those frequency points closest to resonance are clearly identifiable because those away from this area are very close together. Indeed, it is this feature — of distorting the plot so as to focus on the resonance area — that makes the Nyquist plot so effective for modal testing applications.

It is clear from the graphs in Fig. 2.12, and also from the companion set in Fig. 2.13 for hysteretic damping, that each takes the approximate shape of a circle. In fact, as will be shown below, within each set one is an exact circle (marked by *), while the others only approximate to this shape. For viscous damping, it is the **mobility** $Y(\omega)$ which traces out an exact circle while for hysteretic damping, it is the **receptance** $\alpha(\omega)$ and accelerance $A(\omega)$ which do so. In the other cases, the degree of distortion from a circular locus depends heavily on the amount of damping present — becoming negligible as the damping decreases.

Having shown the most common x-y plots used to present FRF data, it is instructive now to provide an illustration of the full three-dimensional quantity which the FRF constitutes. An isometric projection of the Re vs. Im vs. Frequency plot of the Receptance FRF of an SDOF system with viscous damping is shown in Fig. 2.14. From this illustration it is possible to visualise the three projections already shown in (i), (ii) and (iv) above as well as the original three-dimensional curve itself. It is not difficult to understand why this type of presentation is not widely used in practice: its interpretation for more complex systems with many DOFs can become extremely difficult!

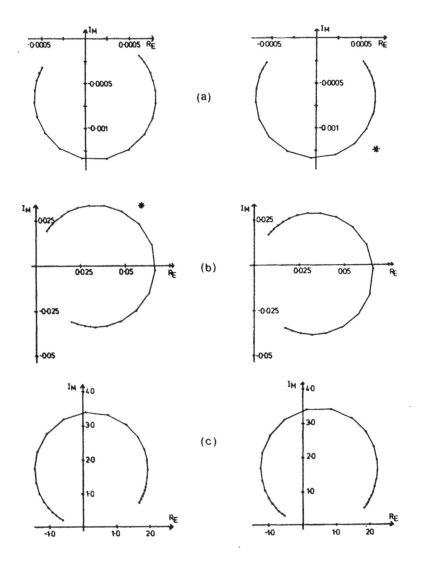

Fig. 2.12 Nyquist FRF plots for
SDOF system with
viscous damping
(a) Receptance;
(b) Mobility;
(c) Accelerance

Fig. 2.13 Nyquist FRF plots for
SDOF system with
structural damping
(a) Receptance;
(b) Mobility;
(c) Accelerance

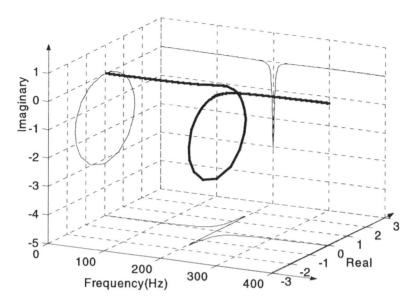

Fig. 2.14   Three-dimensional plot of SDOF system FRF

2.3.3   Properties of SDOF FRF Plots

Lastly, we shall examine some of the basic geometric properties of the various plots we have introduced for the FRF properties of the SDOF system, or oscillator.

We shall consider three specific plots:

(i)   Log mobility modulus versus frequency.
(ii)   Nyquist mobility for viscous damping.
(iii)   Nyquist receptance for hysteretic damping.

It may be observed from Figs. 2.7(b), 2.8 and 2.9, and elsewhere, that for light damping (typically less than 1 per cent) the mobility FRF plot exhibits a degree of symmetry about a vertical line passing through the resonance frequency. As was pointed out by Salter [3], the basic form of this plot can be constructed quite accurately using the reference values indicated on Fig. 2.15.

Turning to the Nyquist plots, we shall show that the two particular cases referred to above, namely:

·   mobility for a viscously-damped system, and
·   receptance for a hysteretically-damped system

48

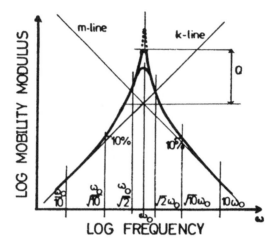

Fig. 2.15 Geometric properties of mobility FRF plot for SDOF system

both trace out exact circles as frequency ω sweeps from 0 to ∞.

Take first the viscous damping case. From equations (2.9) and (2.14) we have that the mobility is

$$Y(\omega) = i\,\omega\,\alpha(\omega) = \frac{i\,\omega}{k - \omega^2\,m + i\,\omega\,c} = \frac{\omega^2\,c + i\,\omega\left(k - \omega^2\,m\right)}{\left(k - \omega^2\,m\right)^2 + \left(\omega\,c\right)^2} \tag{2.16}$$

So,

$$\text{Re}(Y) = \frac{\omega^2\,c}{\left(k - \omega^2\,m\right)^2 + \left(\omega\,c\right)^2} \quad ; \quad \text{Im}(Y) = \frac{\omega\left(k - \omega^2\,m\right)}{\left(k - \omega^2\,m\right)^2 + \left(\omega\,c\right)^2}$$

Let

$$U = \left(\text{Re}(Y) - \frac{1}{2c}\right) \quad \text{and} \quad V = \left(\text{Im}(Y)\right)$$

Then

$$U^2 + V^2 = \frac{\left(\left(k - \omega^2\,m\right)^2 + \left(\omega\,c\right)^2\right)^2}{4c^2\left(\left(k - \omega^2\,m\right)^2 + \left(\omega\,c\right)^2\right)^2} = \left(\frac{1}{2c}\right)^2 \tag{2.17}$$

Hence, a plot of $\text{Re}(Y(\omega))$ vs. $\text{Im}(Y(\omega))$ for $\omega = 0 \to \infty$ will trace out a circle of radius $1/2c$ and with its centre at $(\text{Re} = 1/2c$, $\text{Im} = 0)$, as illustrated clearly in Fig. 2.12(b).

For the hysteretic damping case we have, from equation (2.11), a slightly different expression for the FRF:

$$\alpha(\omega) = \frac{1}{\left(k - \omega^2 m\right) + i d} = \frac{\left(k - \omega^2 m\right) - i(d)}{\left(k - \omega^2 m\right)^2 + (d)^2} \tag{2.18}$$

so that

$$U = \text{Re}(\alpha) = \frac{k - \omega^2 m}{\left(k - \omega^2 m\right)^2 + (d)^2} \quad ; \quad V = \text{Im}(\alpha) = \frac{d}{\left(k - \omega^2 m\right)^2 + (d)^2}$$

Although not the same expressions as those above for viscous damping, it is possible to see that

$$(U)^2 + \left(V + \frac{1}{2d}\right)^2 = \left(\frac{1}{2d}\right)^2 \tag{2.19}$$

demonstrating that a Nyquist plot of receptance for a hysteretically-damped SDOF system will form a circle of radius $1/2d$ and centre at $(0, -1/2d)$, as illustrated in Fig. 2.13(a).

2.4 UNDAMPED MULTI-DEGREE-OF-FREEDOM (MDOF) SYSTEMS

2.4.1 Free Vibration Solution — The Modal Properties

Throughout much of the next seven sections, we shall be discussing multi-degree-of-freedom (MDOF) systems, which might have two degrees of freedom, or 200 or 20000, and in doing so we shall be referring to 'matrices' and 'vectors' of data in a rather abstract and general way. In order to help visualise what some of these generalities mean, a specific 2DOF system, shown in Fig. 2.16, will be used as illustration although the general expressions and solutions will apply to the whole range of MDOF systems.

In this first part, we shall confine our interest to systems which have the feature that their dynamic behaviour is determined predominantly by a combination of inertia and stiffness effects, with damping added in a subsequent section. These systems are typical of those which we would describe as stationary structures, and which relate to the vast majority of situations where it is required to apply modal testing and analysis. Later, in Section 2.8, we shall extend our studies to another class of system which is typified by structures in which one or more components rotate, just as is found in rotating machinery. In this latter class of system, more complicated equations of motion apply, with a consequently greater complication in the modal properties and the response functions which relate. However, that is left until much later: first, we must establish the essential behaviour of the 'standard' type of MDOF system.

For an undamped MDOF system, with N degrees of freedom, the governing equations of motion can be written in matrix form as

$$[M]\{\ddot{x}(t)\} + [K]\{x(t)\} = \{f(t)\} \tag{2.20}$$

where $[M]$ and $[K]$ are $N \times N$ mass and stiffness matrices, respectively, and $\{x(t)\}$ and $\{f(t)\}$ are $N \times 1$ vectors of time-varying displacements and forces.

For our 2DOF example, the equations become

$$m_1 \ddot{x}_1 + (k_1 + k_2)x_1 - (k_2)x_2 = f_1$$
$$m_2 \ddot{x}_2 + (k_2 + k_3)x_2 - (k_2)x_1 = f_2$$

or

$$\begin{bmatrix} m_1 & 0 \\ 0 & m_2 \end{bmatrix}\begin{Bmatrix} \ddot{x}_1 \\ \ddot{x}_2 \end{Bmatrix} + \begin{bmatrix} (k_1 + k_2) & (-k_2) \\ (-k_2) & (k_2 + k_3) \end{bmatrix}\begin{Bmatrix} x_1 \\ x_2 \end{Bmatrix} = \begin{Bmatrix} f_1 \\ f_2 \end{Bmatrix}$$

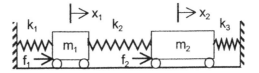

Fig. 2.16   2DOF system used as numerical case study.
$m_1 = 1$ kg; $m_2 = 1$ kg; $k_1 = k_3 = 0.4$ MN/m; $k_2 = 0.8$ MN/m

or, using the numerical data in Fig. 2.16,

$$[M] = \begin{bmatrix} 1 & 0 \\ 0 & 1 \end{bmatrix} \text{(kg)} \quad ; \quad [K] = \begin{bmatrix} 1.2 & -0.8 \\ -0.8 & 1.2 \end{bmatrix} \text{(MN/m)}$$

We shall consider first the free vibration solution (in order to determine the normal or natural modal properties) by taking

$$\{f(t)\} = \{0\}$$

In this case we shall assume that a solution exists of the form

$$\{x(t)\} = \{X\}e^{i\omega t}$$

where $\{X\}$ is an $N \times 1$ vector of time-independent amplitudes for which case it is clear that $\{\ddot{x}\} = -\omega^2\{X\}e^{i\omega t}$.

(NOTE that this assumes that the whole system is capable of vibrating in simple harmonic motion at a single frequency, ω.)

Substitution of this condition and trial solution into the equation of motion (2.20), leads to

$$\left([K] - \omega^2[M]\right)\{X\}e^{i\omega t} = \{0\} \tag{2.21}$$

for which the only non-trivial solutions are those which satisfy

$$\det\left|[K] - \omega^2[M]\right| = 0 \tag{2.22}$$

or

$$a_{2N}\,\omega^{2N} + a_{2N-2}\,\omega^{2N-2} + \dots a_0 = 0$$

from which condition can be found N values of ω^2 : $(\overline{\omega}_1^2, \overline{\omega}_2^2, \dots, \overline{\omega}_r^2, \dots, \overline{\omega}_N^2)$, the undamped system's natural frequencies.

Substituting any one of these back into (2.21) yields a corresponding set of **relative** values for $\{X\}$, i.e. $\{\psi\}_r$, the so-called **mode shape** corresponding to that natural frequency.

The complete solution can be expressed in two $N \times N$ matrices — the eigenmatrices — as

$$\left[\ddots\,\overline{\omega}_r^2\,\ddots\right], [\Psi]$$

where $\overline{\omega}_r^2$ is the rth eigenvalue, or natural frequency squared, and $\{\psi\}_r$ is a description of the corresponding mode shape.

Various numerical procedures are available which take the system matrices $[M]$ and $[K]$ (the **Spatial Model**), and convert them to the two eigenmatrices $[\overline{\omega}_r^2]$ and $[\Psi]$ (which constitute the **Modal Model**).

It is important to realise at this stage that one of these two matrices — the eigenvalue matrix — is unique, while the other — the eigenvector matrix — is not. Whereas the natural frequencies are fixed quantities, the mode shapes are subject to an indeterminate scaling factor which does not affect the **shape** of the vibration mode, only its amplitude. Thus, a mode shape vector of

$$\begin{Bmatrix} 1 \\ 2 \\ 1 \\ 0 \\ \cdot \end{Bmatrix} \quad \text{describes exactly the same vibration mode as} \quad \begin{Bmatrix} 3 \\ 6 \\ 3 \\ 0 \\ \cdot \end{Bmatrix}$$

and so on.

What determines how the eigenvectors are scaled, or 'normalised' is largely governed by the numerical procedures followed by the eigensolution. This topic will be discussed in more detail below.

For our 2DOF example, we find that equation (2.22) becomes

$$\det \begin{vmatrix} \left(k_1 + k_2 - \omega^2 m_1 \right) & \left(-k_2 \right) \\ \left(-k_2 \right) & \left(k_2 + k_3 - \omega^2 m_2 \right) \end{vmatrix}$$

$$= \omega^4 \left(m_1 m_2 \right) - \omega^2 \left(\left(m_1 + m_2 \right) k_2 + m_1 k_3 + m_2 k_1 \right) + \left(k_1 k_2 + k_1 k_3 + k_2 k_3 \right) = 0$$

Numerically:

$$\omega^4 - \omega^2 \left(2.4 \times 10^6 \right) + \left(0.8 \times 10^{12} \right) = 0$$

This condition leads to $\overline{\omega}_1^2 = 4 \times 10^5$ (rad/s)^2 and $\overline{\omega}_2^2 = 2 \times 10^6$ (rad/s)^2. Substituting either value of into the equation of motion, gives

$$\left(k_1 + k_2 - \overline{\omega}_r^2 m_1 \right)_r X_1 = \left(k_2 \right)_r X_2$$

or

$$\{\psi\}_r = \begin{Bmatrix} \psi_{1r} \\ \psi_{2r} \end{Bmatrix} \equiv \begin{Bmatrix} X_1 / X_0 \\ X_2 / X_0 \end{Bmatrix}_r$$

Numerically, we have a solution

$$\left[\overline{\omega}_r^2\right] = \begin{bmatrix} 4 \times 10^5 & 0 \\ 0 & 2 \times 10^6 \end{bmatrix} \quad ; \quad [\Psi] = \begin{bmatrix} 1 & 1 \\ 1 & -1 \end{bmatrix}$$

2.4.2 Orthogonality Properties

Before proceeding with the next phase — the response analysis — it is worthwhile to examine some of the properties of the modal model as these greatly influence the subsequent analysis.

The modal model possesses some very important properties — known as the **Orthogonality** properties — which, concisely stated, are as follows:

$$[\Psi]^T[M][\Psi] = [m_r]$$
$$[\Psi]^T[K][\Psi] = [k_r]$$

(2.23)

from which: $[\overline{\omega}_r^2] = [m_r]^{-1}[k_r]$ where m_r and k_r are often referred to as the **modal mass** and **modal stiffness** of mode r. (See the next subsection — 2.4.3 — for a discussion of modal, generalised, and effective mass and stiffness.) Now, because the eigenvector matrix is subject to an arbitrary scaling factor, the values of m_r and k_r are not unique and so it is inadvisable to refer to 'the' modal mass or stiffness of a particular mode. Many eigenvalue extraction routines scale each vector so that its largest element has unit magnitude (1.0), but this is not universal. In any event, what is found is that the ratio of (k_r / m_r) is unique and is equal to the eigenvalue, $(\overline{\omega}_r^2)$. Among the many scaling or normalisation processes, there is one which has most relevance to modal testing and that is **mass-normalisation**. The mass-normalised eigenvectors are written as $[\Phi]$ and have the particular property that

$$[\Phi]^T[M][\Phi] = [I]$$

and thus

(2.24)

$$[\Phi]^T[K][\Phi] = [\overline{\omega}_r^2]$$

The relationship between the mass-normalised mode shape for mode r, $\{\phi\}_r$, and its more general form, $\{\psi\}_r$, is, simply:

$$\{\phi\}_r = \frac{1}{\sqrt{m_r}}\{\psi\}_r \text{ where } m_r = \{\psi\}_r^T[M]\{\psi\}_r$$

or   (2.25)

$$[\Phi] = [\Psi]\left[m_r^{-1/2}\right]$$

A proof of the orthogonality properties is as follows. The equation of motion for free vibration may be written

$$\left([K]-\omega^2[M]\right)\{X\}e^{i\omega t} = \{0\}$$ (2.26)

For a particular mode, we have

$$\left([K]-\overline{\omega}_r^2[M]\right)\{\psi\}_r = \{0\}$$ (2.27)

Premultiply by a different eigenvector, transposed:

$$\{\psi\}_s^T\left([K]-\overline{\omega}_r^2[M]\right)\{\psi\}_r = 0$$ (2.28)

We can also write

$$\left([K]-\overline{\omega}_s^2[M]\right)\{\psi\}_s = \{0\}$$ (2.29)

which we can transpose, and postmultiply by $\{\psi\}_r$, to give

$$\{\psi\}_s^T\left([K]^T-\overline{\omega}_s^2[M]^T\right)\{\psi\}_r = 0$$ (2.30)

But, since $[M]$ and $[K]$ are generally symmetric*, they are identical to their transposes and equations (2.28) and (2.30) can be combined to give

$$\left(\overline{\omega}_r^2 - \overline{\omega}_s^2\right)\{\psi\}_s^T[M]\{\psi\}_r = 0$$ (2.31)

which, if $\omega_r \neq \omega_s$, can only be satisfied if

$$\{\psi\}_s^T[M]\{\psi\}_r = 0; r \neq s$$ (2.32)

---

*   See later for the analysis in the special cases where $[M]$ and $[K]$ are not symmetric.

Together with either (2.28) or (2.30), this means also that

$$\{\psi\}_s^T [K]\{\psi\}_r = 0; r \neq s \tag{2.33}$$

For the special cases where $r = s$, or if $\bar{\omega}_r = \bar{\omega}_s$, equations (2.32) and (2.33) do not apply, but it is clear from (2.28) that

$$\left(\{\psi\}_r^T [K]\{\psi\}_r\right) = \bar{\omega}_r^2 \left(\{\psi\}_r^T [M]\{\psi\}_r\right) \tag{2.34}$$

so that

$$\{\psi\}_r^T [M]\{\psi\}_r = m_r \quad \text{and} \quad \{\psi\}_r^T [K]\{\psi\}_r = k_r$$

and

$$\bar{\omega}_r^2 = k_r / m_r$$

Putting together all the possible combinations of r and s leads to the full matrix equation (2.23) above.

For our 2DOF example, the numerical results give eigenvectors which are clearly plausible. If we use them to calculate the generalised mass and stiffness, we obtain

$$\begin{bmatrix} 1 & 1 \\ 1 & -1 \end{bmatrix} \begin{bmatrix} 1 & 0 \\ 0 & 1 \end{bmatrix} \begin{bmatrix} 1 & 1 \\ 1 & -1 \end{bmatrix} = \begin{bmatrix} 2 & 0 \\ 0 & 2 \end{bmatrix} = [m_r]$$

$$\begin{bmatrix} 1 & 1 \\ 1 & -1 \end{bmatrix} \begin{bmatrix} 1.2 & -0.8 \\ -0.8 & 1.2 \end{bmatrix} \begin{bmatrix} 1 & 1 \\ 1 & -1 \end{bmatrix} 10^6 = \begin{bmatrix} 0.8 & 0 \\ 0 & 4 \end{bmatrix} 10^6 = [k_r]$$

clearly

$$[m_r]^{-1}[k_r] = \begin{bmatrix} 0.4 & 0 \\ 0 & 2.0 \end{bmatrix} 10^6 = [\omega_r^2]$$

To obtain the mass normalised version of these eigenvectors, we must calculate

$$[\Phi] = \begin{bmatrix} 1 & 1 \\ 1 & -1 \end{bmatrix} \lfloor m_r \rfloor^{-1/2} = \begin{bmatrix} 0.707 & 0.707 \\ 0.707 & -0.707 \end{bmatrix}$$

2.4.3 Modal, Generalised and Effective Mass and Stiffnesses

In the previous section, we introduced a pair of quantities which we called the modal mass and modal stiffness. There is a variety of terminology in this area which is worth mentioning so that at least the different quantities can be identified, even if uniformity of terminology cannot be assured. Three terms are encountered in the literature: modal mass (and stiffness); generalised mass and effective or equivalent mass. In this section we shall seek to explain what these different quantities are and how they may be interpreted or used.

We start with the **modal mass**, already defined above based on the mode shape vector for mode r and the system mass matrix. As mentioned there, there is no unique value for the modal mass as it is directly related to the scaling method which has been used to define the mode shape eigenvector, $\{\psi\}_r$. This scaling is completely arbitrary and so the modal mass could be any value, as also can be the corresponding modal stiffness. However, as already observed, the ratio between any modal stiffness and its associated modal mass **is** unique and is equal to the corresponding eigenvalue. The modal mass is generally used to convert the original mode shape vector, $\{\psi\}_r$, to the more useful mass-normalised mode shape vector, $\{\phi\}_r$. It should be noted that the original vector is dimensionless, while the mass-normalised vector has dimensions of $(\text{mass})^{-0.5}$.

Using the mass-normalised mode shape vectors, we can see how to derive quantities which provide us with information about the **effective mass** (or stiffness) at any point on the structure, such as DOF j. It is helpful to visualise a detail of the point FRF, $H_{jj}(\omega)$, that might be computed and plotted just in the immediate vicinity of the corresponding natural frequency: this would look something similar to the detailed plot shown earlier in Fig 2.15 and would be characterised by a skeleton which is based on an effective mass line, $(m_{jj})_r$, and an effective stiffness line, $(k_{jj})_r$. These quantities can be related to the eigenvector elements by the simple formulae:

Effective mass at DOF_j for mode r,

$$\left(m_{jj}\right)_r = \frac{1}{\left(\phi_{jr}\right)^2} \qquad \text{, which has units of mass,}$$

and effective stiffness at DOF_j for mode r,

$$\left(k_{jj}\right)_r = \frac{\bar{\omega}_r^2}{\left(\phi_{jr}\right)^2}$$

It can be seen that since the mass-normalised eigenvectors are unique, and not subject to any arbitrary scaling factors, these effective mass and stiffness properties are also unique and represent a useful description of the underlying behaviour of the structure point by point, and mode by mode.

The other quantities which are sometimes referred to as unique properties of each mode are the **generalised mass** and **generalised stiffness**. Although there s no universal agreement of the definitions of these properties, that which is adopted in this work is to define the generalised mass (or stiffness) of the rth mode as the effective mass (or stiffness) at the DOF with the largest amplitude of response. This quantity serves to provide a comparison of the relative strength of each mode of the structure.

2.4.4   Repeated Roots or Multiple Modes

There are situations where two (or more) different modes will have the same natural frequency. This is one of the exclusions made above at equations (2.32) and (2.33) but occurs frequently in structures which exhibit a degree of symmetry, especially axisymmetry, as found in most discs, cylinders, rings, etc. In these cases, there is no guarantee that the corresponding two (or more) eigenvectors, $\{\phi\}_r$ and $\{\phi\}_s$, will be orthogonal to each other as required in those equations. However, it can be asserted that two such orthogonal vectors **do** exist and that if this property is not already exhibited by the two vectors available, then two other linear combinations of these two vectors can always be found such that orthogonality is observed between the mode shapes used to describe the motion in each of the two modes which have the same frequency. It should be noted, however, that free vibration at that frequency is possible not only in each of the two vectors thus defined, but also in a deformation pattern which is given by any linear combination of these two vectors. This can be easily demonstrated using the example of a circular-section bar, clamped at one end, as shown in Fig. 2.17(a). There will clearly be two modes which correspond to each bending deflection pattern along the shaft: one in the vertical plane and

58

the other in the horizontal plane, see Figs. 2.17(b) and (c). If the bar itself and the end supports are completely axisymmetric, these two modes will have identical natural frequencies. As a result, any combination of 'the vertical' mode and 'the horizontal' mode will also be a valid mode of vibration at that natural frequency, such as the example shown in Fig. 2.17(d).

As the existence of double modes is commonplace in many structures, we shall return to these features from time to time.

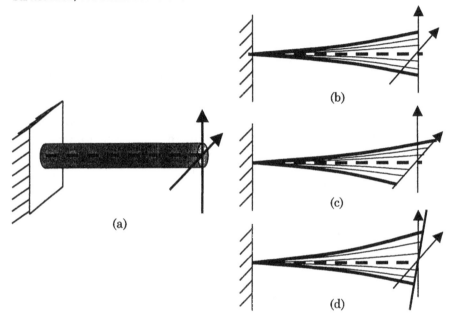

Fig. 2.17 Repeated modes of symmetric structures.
(a) Symmetric structure; (b) Vertical mode; (c) Horizontal mode; (d) Oblique mode

2.4.5   Forced Response Solution — The FRF Characteristics

Turning now to a response analysis, we shall consider the case where the structure is excited sinusoidally by a set of forces all at the same frequency, ω, but with individual amplitudes and phases. Then

$$\{f(t)\} = \{F\}e^{i\omega t}$$

and, as before, we shall assume a solution exists of the form

$$\{x(t)\} = \{X\}e^{i\omega t}$$

where $\{F\}$ and $\{X\}$ are $N \times 1$ vectors of time-independent complex amplitudes.

The equation of motion then becomes

$$\left([K] - \omega^2 [M]\right)\{X\}e^{i\omega t} = \{F\}e^{i\omega t} \tag{2.35}$$

or, rearranging to solve for the unknown responses,

$$\{X\} = \left([K] - \omega^2 [M]\right)^{-1}\{F\} \tag{2.36a}$$

which may be written

$$\{X\} = [\alpha(\omega)]\{F\} \tag{2.36b}$$

where $[\alpha(\omega)]$ is the $N \times N$ receptance FRF matrix for the system and constitutes its **Response Model**. The general element in the receptance FRF matrix, $\alpha_{jk}(\omega)$, is defined as follows:

$$\alpha_{jk}(\omega) = \left(\frac{X_j}{F_k}\right); \quad F_m = 0; \quad m = 1, N; \neq k \tag{2.37}$$

and as such represents an individual receptance FRF expression very similar to that defined earlier for the SDOF system.

It is clearly possible for us to determine values for the elements of $[\alpha(\omega)]$ at any frequency of interest simply by substituting the appropriate values into (2.36). However, this involves the inversion of a system matrix at each frequency and this has several disadvantages, namely:

- it becomes costly for large-order systems (large N);
- it is inefficient if only a few of the individual FRF expressions are required;
- it provides no insight into the form of the various FRF properties.

For these, and other, reasons an alternative means of deriving the various FRF parameters is used which makes use of the modal properties for the system instead of the spatial properties.

Returning to (2.36) we can write

$$\left([K] - \omega^2 [M]\right) = [\alpha(\omega)]^{-1} \tag{2.38}$$

Premultiply both sides by $[\Phi]^T$ and postmultiply both sides by $[\Phi]$ to obtain

$$[\Phi]^T\big([K]-\omega^2[M]\big)[\Phi] = [\Phi]^T[\alpha(\omega)]^{-1}[\Phi]$$

or

$$\big[\big(\overline{\omega}_r^2 - \omega^2\big)\big] = [\Phi]^T[\alpha(\omega)]^{-1}[\Phi] \tag{2.39}$$

which leads to

$$[\alpha(\omega)] = [\Phi]\big[\big(\overline{\omega}_r^2 - \omega^2\big)\big]^{-1}[\Phi]^T \tag{2.40}$$

It is clear from this equation that the receptance matrix $[\alpha(\omega)]$ is symmetric and this will be recognised as the principle of reciprocity which applies to many structural characteristics. Its implications in this situation are that:

$$\alpha_{jk} = \big(X_j/F_k\big) = \alpha_{kj} = \big(X_k/F_j\big)$$

Equation (2.40) permits us to compute any individual FRF parameter, $\alpha_{jk}(\omega)$, using the following formula (noting that the resulting expression is delivered by multiplying the jth row of $[\Phi]$ by the diagonal frequency matrix by the kth column of $[\Phi]^T$:

$$\alpha_{jk}(\omega) = \sum_{r=1}^{N} \frac{\big(\phi_{jr}\big)\big(\phi_{kr}\big)}{\overline{\omega}_r^2 - \omega^2} = \sum_{r=1}^{N} \frac{\big(\psi_{jr}\big)\big(\psi_{kr}\big)}{m_r\big(\overline{\omega}_r^2 - \omega^2\big)} \tag{2.41}$$

or

$$\alpha_{jk}(\omega) = \sum_{r=1}^{N} \frac{{}_r A_{jk}}{\overline{\omega}_r^2 - \omega^2}$$

which is very much simpler and more informative than by means of the direct inverse, equation (2.36a). Here we introduce a new parameter, ${}_r A_{jk}$, which we shall refer to as a **Modal Constant***: in this case, that

---

*   Note that other presentations of the theory sometimes refer to the modal constant as a '**Residue**' together with the use of '**Pole**' instead of our natural frequency.

for mode r for this specific receptance linking coordinates j and k. The above is a most important result and is in fact the central relationship upon which the whole subject is based. From the general equation (2.36a), the typical individual FRF element $\alpha_{jk}(\omega)$, defined in (2.37), would be expected to have the form of a ratio of two polynomials:

$$\alpha_{jk}(\omega) = \frac{b_0 + b_1\omega^2 + b_2\omega^4 + ... + b_{N-1}\omega^{2N-2}}{a_0 + a_1\omega^2 + a_2\omega^4 + a_3\omega^6 ... + a_N\omega^{2N}} \tag{2.42}$$

and in such a format it would be difficult to visualise the nature of the function, $\alpha_{jk}(\omega)$. However, it is clear that an expression such as (2.42) can also be rewritten as

$$\alpha_{jk}(\omega) = B\frac{\left(\Omega_1^2 - \omega^2\right)\left(\Omega_r^2 - \omega^2\right)..\left(\Omega_{N-1}^2 - \omega^2\right)}{\left(\overline{\omega}_1^2 - \omega^2\right)\left(\overline{\omega}_2^2 - \omega^2\right)\left(\overline{\omega}_3^2 - \omega^2\right)..\left(\overline{\omega}_N^2 - \omega^2\right)} \tag{2.43}$$

and by inspection of the form of (2.36a), it is also clear that the factors in the denominator, $\overline{\omega}_1^2$, $\overline{\omega}_2^2$, etc. are indeed the natural frequencies of the system, $\overline{\omega}_r^2$ (this is because the denominator is necessarily formed by the $\det([K] - \omega^2[M])$).

All this means that a forbidding rational fraction expression such as (2.42) can be expected to be reducible to a partial fraction series form, such as

$$\alpha_{jk}(\omega) = \frac{{}_1A_{jk}}{\overline{\omega}_1^2 - \omega_2} + \frac{{}_2A_{jk}}{\overline{\omega}_2^2 - \omega_2} + ... = \sum_{r=1}^{N}\frac{{}_rA_{jk}}{\overline{\omega}_r^2 - \omega^2} \tag{2.44}$$

Thus, the solution we obtain through equations (2.38) to (2.41) is not unexpected, but its significance lies in the very simple and convenient formula it provides for the coefficients, ${}_rA_{jk}$, in the series form.

We can observe some of the above relationships through our 2DOF example. The forced vibration equations of motion give

$$\left(k_1 + k_2 - \omega^2 m_1\right)X_1 + \left(-k_2\right)X_2 = F_1$$
$$\left(-k_2\right)X_1 + \left(k_2 + k_3 - \omega^2 m_2\right)X_2 = F_2$$

which, in turn, give (for example):

$$\left(\frac{X_1}{F_1}\right)_{f_2=0} = \frac{k_2 + k_3 - \omega^2 m_2}{\omega^4 m_1 m_2 - \omega^2\left((m_1 + m_2)k_2 + m_1 k_3 + m_2 k_1\right) + \left(k_1 k_2 + k_2 k_3 + k_1 k_3\right)}$$

or, numerically,

$$= \frac{\left(1.2 \times 10^6 - \omega^2\right)}{\left(\omega^4 - 2.4 \times 10^6 \omega^2 + 0.8 \times 10^{12}\right)}$$

Now, if we use the modal summation formula (2.41) together with the results obtained earlier, we can write

$$\alpha_{11} = \left(\frac{X_1}{F_1}\right) = \frac{(\phi_{11})^2}{\overline{\omega}_1^2 - \omega^2} + \frac{(\phi_{12})^2}{\overline{\omega}_2^2 - \omega^2}$$

or, numerically,

$$= \frac{0.5}{0.4 \times 10^6 - \omega^2} + \frac{0.5}{2 \times 10^6 - \omega^2}$$

which is equal to $\left(1.2 \times 10^6 - \omega^2\right) / \left(0.8 \times 10^{12} - 2.4 \times 10^6 \omega^2 + \omega^4\right)$, as above.

The above characteristics of both the modal and response models of an undamped MDOF system form the basis of the corresponding data for the more general, damped, cases.

The following sections will examine the effects on these models of adding various types of damping, while a discussion of the presentation MDOF frequency response data is given in Section 2.10.

2.5   MDOF SYSTEMS WITH PROPORTIONAL DAMPING
2.5.1   General Concept and Features of Proportional Damping
In approaching the more general case of damped systems, it is convenient to consider first a special type of damping which has the advantage of being particularly easy to include in our analysis. This type of damping is usually referred to as 'proportional' damping (for reasons which will be clear later) although this is a somewhat restrictive title. The particular advantage of using a proportional damping model in the analysis of structures is that the modes of such a

structure are almost identical to those of the undamped version of the model. Specifically, the mode shapes **are** identical and the natural frequencies are very similar to those of the simpler undamped system. In fact, it is possible to derive the modal properties of a proportionally-damped system by analysing the undamped version in full and then making a correction for the presence of the damping. While this procedure is often used in the theoretical analysis of structures, it should be noted that it is only valid in the case of this special type or distribution of damping, which may not generally apply to the real structures studied in modal tests.

If we return to the general equation of motion for an MDOF system, equation (2.20), and add a viscous damping matrix $[C]$, we obtain:

$$[M]\{\ddot{x}\}+[C]\{\dot{x}\}+[K]\{x\}=\{f\} \tag{2.45}$$

which is not so amenable to the type of solution followed in Section 2.4. A general solution will be presented in the next section, but here we shall examine the properties of this equation for the case where the damping matrix is directly proportional to the stiffness matrix; i.e. where

$$[C]=\beta[K] \tag{2.46}$$

(NOTE — It should be noted that this is not the only type of proportional damping — see below.)

In this case, it is clear that if we pre- and post-multiply the damping matrix by the eigenvector matrix for the undamped system, $[\Psi]$, in just the same way as was done in equation (2.23) for the mass and stiffness matrices, then we shall find:

$$[\Psi]^{T}[C][\Psi]=\beta[k_r]=[c_r] \tag{2.47}$$

where the diagonal elements, c_r, represent the modal damping of the various modes of the system. The fact that this matrix is also diagonal means that the undamped system mode shapes are also those of the damped system, and this is a particular feature of this type of damping. This statement can easily be demonstrated by taking the general equation of motion above (2.45) and, for the case of no excitation, pre- and post-multiplying the whole equation by the eigenvector matrix, $[\Psi]$. We shall then find:

$$[m_r]\{\ddot{p}\}+[c_r]\{\dot{p}\}+[k_r]\{p\}=\{0\} \quad ; \quad \{p\}=[\Psi]^{-1}\{x\} \tag{2.48}$$

64

from which the rth individual equation is:

$$m_r \ddot{p}_r + c_r \dot{p}_r + k_r p_r = 0 \qquad (2.49)$$

which is clearly that of a single-degree-of-freedom system, or of a single mode of the system. This mode has a complex natural frequency with an imaginary (oscillatory) part of

$$\omega_r' = \bar{\omega}_r \sqrt{1 - \zeta_r^2} \quad ; \quad \bar{\omega}_r^2 = \frac{k_r}{m_r} \quad ; \quad \zeta_r = \frac{c_r}{2\sqrt{k_r m_r}} = \frac{1}{2}\beta\bar{\omega}_r$$

and a real (decay) part of

$$a_r = \zeta_r \bar{\omega}_r = \frac{\beta}{2}$$

(using the notation introduced above for the SDOF analysis).

These characteristics carry over to the forced response analysis in which a simple extension of the steps detailed between equations (2.35) and (2.41) leads to the definition for the general receptance FRF as:

$$[\alpha(\omega)] = \left[K + i\omega C - \omega^2 M\right]^{-1}$$

or

$$\alpha_{jk}(\omega) = \sum_{r=1}^{N} \frac{(\psi_{jr})(\psi_{kr})}{(k_r - \omega^2 m_r) + i(\omega c_r)} \qquad (2.50)$$

which has a very similar form to that for the undamped system except that now it becomes complex in the denominator as a result of the inclusion of damping.

2.5.2 General Forms of Proportional Damping

It may be seen from the above that other distributions of damping will bring about the same type of result and these are collectively included in the classification 'proportional damping'. In particular, if the damping matrix is proportional to the mass matrix, then exactly the same type of result ensues and, indeed, the usual definition of proportional damping is that the damping matrix [C] should be of the form:

$$[C] = \beta[K] + \gamma[M] \tag{2.51}$$

In this case, the damped system will have eigenvalues and eigenvectors as follows:

$$\omega'_r = \overline{\omega}_r \sqrt{1 - \zeta_r^2} \quad ; \quad \zeta_r = \beta \overline{\omega}_r / 2 + \gamma / 2 \overline{\omega}_r$$

and

$$[\Psi_{damped}] = [\Psi_{undamped}]$$

Distributions of damping of the type described above are sometimes, though not always, found to be plausible from a practical standpoint: the actual damping mechanisms are usually to be found in parallel with stiffness elements (for internal material or hysteresis damping) or with mass elements (for friction damping). There is a more general definition of the condition required for the damped system to possess the same mode shapes as its undamped counterpart, and that is:

$$\left([M]^{-1}[K]\right)\left([M]^{-1}[C]\right) = \left([M]^{-1}[C]\right)\left([M]^{-1}[K]\right) \tag{2.52}$$

although it is more difficult to make a direct physical interpretation of its form.

Finally, it can be noted that an identical treatment can be made of an MDOF system with proprtional hysteretic damping, producing the same essential results. If the general system equations of motion are expressed as:

$$[M]\{\ddot{x}\} + \left([K + iD]\right)\{x\} = \{f\} \tag{2.53}$$

and the hysteretic damping matrix $[D]$ is 'proportional', typically;

$$[D] = \beta[K] + \gamma[M] \tag{2.54}$$

then we find that the mode shapes for the damped system are again identical to those of the undamped system and that the eigenvalues take the complex form:

$$\lambda_r^2 = \overline{\omega}_r^2 (1 + i\eta_r) \quad ; \quad \overline{\omega}_r^2 = k_r / m_r \quad ; \quad \eta_r = \beta + \gamma / \overline{\omega}_r^2 \tag{2.55}$$

Also, the general FRF expression is written:

$$\alpha_{jk}(\omega) = \sum_{r=1}^{N} \frac{\left(\psi_{jr}\right)\left(\psi_{kr}\right)}{\left(k_r - \omega^2 m_r\right) + i\,\eta_r\,k_r} \tag{2.56}$$

2.6 MDOF SYSTEMS WITH STRUCTURAL (HYSTERETIC) DAMPING — GENERAL CASE

2.6.1 Free Vibration Solution — Complex Modal Properties

The analysis in the previous section for proportionally-damped systems gives some insight into the characteristics of this more general description of practical structures. However, as was stated there, the case of proportional damping is a particular one which, although often justified in a theoretical analysis because it is realistic and also because of a lack of any more accurate model, does not apply to all cases. In our studies here, it is very important that we consider the most general case if we are to be able to interpret and analyse correctly the data we observe on real structures. These, after all, know nothing of our predilection for assuming proportionality in the distribution of damping. Thus, in the next two sections we consider the properties of systems with general damping elements, first of the hysteretic type, then viscous.

We start by writing the general equation of motion for an MDOF system with hysteretic damping and harmonic excitation (as it is this that we are working towards):

$$[M]\{\ddot{x}\} + [K]\{x\} + i[D]\{x\} = \{F\}e^{i\omega t} \tag{2.57}$$

Now, consider first the case where there is no excitation and assume a solution of the form:

$$\{x\} = \{X\}e^{i\lambda t} \tag{2.58}$$

where λ is allowed to be complex. Substituted into (2.57), this trial solution leads to a complex eigenproblem whose solution is in the form of two matrices (as for the earlier undamped case), containing the eigenvalues and eigenvectors. In this case, however, these matrices are both complex, meaning that each natural frequency and each mode shape is described in terms of complex quantities. We choose to write the rth eigenvalue as

$$\lambda_r^2 = \omega_r^2\left(1 + i\eta_r\right) \tag{2.59}$$

where ω_r is the natural frequency and η_r is the damping loss factor for that mode. It is important to note that the natural frequency ω_r is not (necessarily) equal to the natural frequency of the undamped system, $\overline{\omega}_r$, as was the case for proportional hysteretic damping, although the two values will generally be very close in practice.

The complex mode shapes are at first more difficult to interpret but in fact what we find is that the amplitude of each DOF can be considered as having both a magnitude and a phase angle. This is only very slightly different from the undamped case as there we effectively have a magnitude at each point plus a phase angle which is either 0° or 180°, both of which can be completely described using real numbers. What the inclusion of general damping effects does is to generalise this particular feature of the mode shape data to a situation in which the phase may take any value, not only 0° and 180°. Further discussion of this feature is given in Section 2.9.

This eigensolution can be seen to possess the same type of orthogonality properties as those demonstrated earlier for the undamped system and may be defined by the equations:

$$[\Psi]^T [M][\Psi] = [m_r] \quad ; \quad [\Psi]^T [K + iD][\Psi] = [k_r] \tag{2.60}$$

Again, the modal mass and stiffness parameters (now complex) depend upon the normalisation of the mode shape vectors for their magnitudes but always obey the relationship:

$$\lambda_r^2 = \frac{k_r}{m_r} \tag{2.61}$$

and here again we may define a set of mass-normalised eigenvectors as:

$$\{\phi\}_r = (m_r)^{-1/2} \{\psi\}_r \tag{2.62}$$

68

Numerical examples with structural damping

Some further numerical examples are included to illustrate the characteristics of more general damped systems, based on the following 3DOF model:

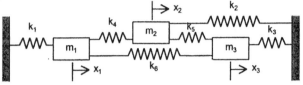

Model 1

$$m_1 = 0.5 \text{ kg} \qquad m_3 = 1.5 \text{ kg}$$

$$m_2 = 1.0 \text{ kg} \qquad k_1 = k_2 = k_3 = k_4 = k_5 = k_6 = 1.0 \times 10^3 \text{ N/m}$$

Case 1(a) — Undamped

$$\left[\overline{\omega}_r^2\right] = \begin{bmatrix} 950 & 0 & 0 \\ 0 & 3352 & 0 \\ 0 & 0 & 6698 \end{bmatrix} \quad ; \quad [\Phi] = \begin{bmatrix} 0.464 & -0.218 & -1.318 \\ 0.536 & -0.782 & 0.318 \\ 0.635 & 0.493 & 0.142 \end{bmatrix}$$

Case 1(b) — Proportional structural damping ($d_j = 0.05k_j; j = 1,6$)

$$\left[\lambda_r^2\right] = \begin{bmatrix} 950(1+0.05i) & 0 & 0 \\ 0 & 3352(1+0.05i) & 0 \\ 0 & 0 & 6698(1+0.05i) \end{bmatrix}$$

$$[\Phi] = \begin{bmatrix} 0.464(0°) & 0.218(180°) & 1.318(180°) \\ 0.536(0°) & 0.782(180°) & 0.318(0°) \\ 0.635(0°) & 0.493(0°) & 0.142(0°) \end{bmatrix}$$

Case 1(c) — Non-proportional structural damping

($d_1 = 0.3k_1$, $d_{2-6} = 0$, i.e. a single damper between m_1 and ground)

$$\left[\lambda_r^2\right] = \begin{bmatrix} 957(1+0.067i) & 0 & 0 \\ 0 & 3354(1+0.042i) & 0 \\ 0 & 0 & 6690(1+0.078i) \end{bmatrix}$$

$$[\Phi] = \begin{bmatrix} 0.463(-5.5°) & 0.217(173°) & 1.321(181°) \\ 0.537(0°) & 0.784(181°) & 0.316(-6.7°) \\ 0.636(1.0°) & 0.492(-1.3°) & 0.142(-3.1°) \end{bmatrix}$$

NOTES:
(i)   Each mode has a different damping factor.
(ii) All eigenvector arguments within 10° of 0° or 180° (i.e. the modes are almost 'real').

Model 2

$m_1 = 1.0\,\text{kg}$ $m_3 = 1.05\,\text{kg}$

$m_2 = 0.95\,\text{kg}$ $k_1 = k_2 = k_3 = k_4 = k_5 = k_6 = 1.0\times10^3\,\text{N/m}$

Case 2(a) — Undamped

$$\left[\overline{\omega}_r^2\right]=\begin{bmatrix}999 & 0 & 0 \\ 0 & 3892 & 0 \\ 0 & 0 & 4124\end{bmatrix} \quad ; \quad \left[\Phi\right]=\begin{bmatrix}0.577 & -0.602 & 0.552 \\ 0.567 & -0.215 & -0.827 \\ 0.587 & 0.752 & 0.207\end{bmatrix}$$

NOTE: this system has two close natural frequencies.

Case 2(b) — Proportional structural damping ($d_j = 0.05k_j$)

$$\left[\lambda_r^2\right]=\begin{bmatrix}999(1+0.05i) & 0 & 0 \\ 0 & 3892(1+0.05i) & 0 \\ 0 & 0 & 4124(1+0.05i)\end{bmatrix}$$

$$\left[\Phi\right]=\begin{bmatrix}0.577(0°) & 0.602(180°) & 0.552(0°) \\ 0.567(0°) & 0.215(180°) & 0.827(180°) \\ 0.587(0°) & 0.752(0°) & 0.207(0°)\end{bmatrix}$$

Case 2(c) — Non-proportional structural damping
($d_1 = 0.3k_1$, $d_{2-6} = 0$)

$$\left[\lambda_r^2\right]=\begin{bmatrix}1006(1+0.10i) & 0 & 0 \\ 0 & 3942(1+0.031i) & 0 \\ 0 & 0 & 4067(1+0.019i)\end{bmatrix}$$

$$\left[\Phi\right]=\begin{bmatrix}0.578(-4°) & 0.851(162°) & 0.685(40°) \\ 0.569(2°) & 0.570(101°) & 1.019(176°) \\ 0.588(2°) & 0.848(12°) & 0.560(-50°)\end{bmatrix}$$

2.6.2   Forced Response Solution — FRF Characteristics

We turn next to the analysis of forced vibration for the particular case of harmonic excitation and response, for which the governing equation of motion is:

$$\left([K]+i[D]-\omega^2[M]\right)\{X\}e^{i\omega t} = \{F\}e^{i\omega t} \tag{2.63}$$

As before, a direct solution to this problem may be obtained by using the equations of motion to give:

$$\{X\} = \left([K]+i[D]-\omega^2[M]\right)^{-1}\{F\} = [\alpha(\omega)]\{F\} \tag{2.64}$$

but again this is very inefficient for numerical application and we shall make use of the same procedure as before by multiplying both sides of the equation by the eigenvectors. Starting with (2.64), and following the same procedure as between equations (2.38) and (2.40), we can write:

$$[\alpha(\omega)] = [\Phi]\left[\left(\lambda_r^2-\omega^2\right)\right]^{-1}[\Phi]^T \tag{2.65}$$

and from this full matrix equation we can extract any one FRF element, such as $\alpha_{jk}(\omega)$, and express it explicitly in a series form:

$$\alpha_{jk}(\omega) = \sum_{r=1}^{N} \frac{(\phi_{jr})(\phi_{kr})}{\omega_r^2-\omega^2+i\eta_r\omega_r^2} \tag{2.66}$$

which may also be rewritten in various alternative ways, such as:

$$\alpha_{jk}(\omega) = \sum_{r=1}^{N} \frac{(\psi_{jr})(\psi_{kr})}{m_r\left(\omega_r^2-\omega^2+i\eta_r\omega_r^2\right)}$$

or

$$\alpha_{jk}(\omega) = \sum_{r=1}^{N} \frac{{}_rA_{jk}}{\omega_r^2-\omega^2+i\eta_r\omega_r^2}$$

In these expressions, the numerator (as well as the denominator) is now complex as a result of the complexity of the eigenvectors. It is in this respect that the general damping case differs from that for proportional damping.

2.6.3   Excitation by a General Force Vector

2.6.3.1   Operating deflection shape (ODS)

Having derived an expression for the general term in the frequency response function matrix, $\alpha_{jk}(\omega)$, it is appropriate to consider next the analysis of a situation where the system is excited simultaneously at several points (rather than at just one, as is the case for the individual FRF expressions).

The general behaviour for this case is governed by equation (2.63) and the solution in (2.64). However, a more explicit (and perhaps useful) form of this solution may be derived from (2.64) — although not very easily! — as:

$$\{X\} = \sum_{r=1}^{N} \frac{\{\phi\}_r^T \{F\}\{\phi\}_r}{\omega_r^2 - \omega^2 + i\eta_r \omega_r^2} \tag{2.67}$$

This equation permits the calculation of one or more individual responses to an excitation of several simultaneous harmonic forces (all of which must have the same frequency but may vary in magnitude and phase) and it may be seen that the special case of one single response to a single force (a frequency response function) is clearly that quoted in (2.66). The resulting vector of responses is sometimes referred to as **forced vibration mode** or, more commonly, as an **operating deflection shape** (ODS). When the excitation frequency is close to one of the system's natural frequencies, the ODS will usually reflect the shape of the nearby mode because one term in the series of (2.67) will dominate, but will not be identical to it because of the contributions, albeit small, of all the other modes.

2.6.3.2   Pure mode excitation 1 — damped system normal modes

There are a number of cases of multi-point harmonic excitation of special interest which are worth mentioning here. These are generally associated with the notion that by carefully choosing or tuning the vector of individual forces it is possible to set up a response of the structure which is entirely controlled by a single normal mode of the structure. When we enter this domain, we sometimes run into difficulties of nomenclature, especially in respect of the meaning of the term 'normal mode'. As explained earlier in this work, the normal modes are the characteristic modes of the structure in its actual, damped, state. While it is possible to talk of the modes 'that the structure would have if the damping could, by some magic, be removed', these are not the 'normal' modes of the structure in any strict sense. They may be referred to as the 'normal modes of the associated undamped structure' and it is true that they are properties of some interest because in most cases of test-analysis comparison, the analytical model will be undamped and so there is a desire to be able to extract the test structures 'undamped' modes from the test data in order to effect a direct comparison between prediction and measurement.

So, we find ourselves seeking procedures which would enable us to identify the normal modes of the structure directly, one by one, by generation of a suitable excitation force vector, $\{F\}$.

The first of these cases that we shall describe is the genuine normal mode excitation, in which an excitation vector $\{F\}$ is sought such that the response, $\{X\}$, shall consist of a single modal component so that all terms but one in the summation of (2.67) shall be zero. It can be seen that this situation can be attained if the excitation vector $\{F\}_s$ is chosen such that the product $\{\phi_r\}^T\{F\}_s = 0$ for all terms except $r = s$. If this can be achieved, then the said excitation vector generates a response in just mode s. Depending upon the exact damping conditions, this exclusive excitation vector may be more or less easy to define, and indeed, its elements may well be complex (i.e. they will each have different phases) but it will always exist.

2.6.3.3   Pure mode excitation 2 — associated undamped system normal modes

It is also worth mentioning another special case of some interest: namely, that where the harmonic excitation is described by a vector of mono-phased forces. Here, the complete generality admitted in the previous paragraph is restricted somewhat by insisting that all forces have the same frequency **and** phase, although their magnitudes may vary. What is of interest in this case is to see that there exist conditions under which it is possible to obtain a similarly mono-phased response (the whole system responding with a single phase angle). This is not generally the case in the solution to equation (2.67) above.

Thus, let the force and response vectors be represented by

$$\{f\} = \left\{\hat{F}\right\}e^{i\omega t}$$
$$\{x\} = \left\{\hat{X}\right\}e^{i(\omega t - \theta)} \tag{2.68}$$

where both $\{\hat{F}\}$ and $\{\hat{X}\}$ are vectors of real quantities, and substitute these into the equation of motion, (2.63). This leads to a complex equation which can be split into real and imaginary parts to give:

$$\left(\left(-\omega^2[M]+[K]\right)\cos\theta + [D]\sin\theta\right)\left\{\hat{X}\right\} = \left\{\hat{F}\right\}$$
$$\left(\left(-\omega^2[M]+[K]\right)\sin\theta - [D]\cos\theta\right)\left\{\hat{X}\right\} = \{0\} \tag{2.69}$$

The second of this pair of equations can be treated as an eigenvalue problem which has 'roots' θ_s and corresponding 'vectors' $\{\kappa\}_s$. These may be inserted back into the first of the pair of equations (2.69) in order to establish the form of the (mono-phased) force vector necessary to bring about the (mono-phased) response vector described by $\{\kappa\}_s$. Thus we find that there exist a set of N mono-phased force vectors each of which, when applied as excitation to the system, results in a mono-phased response characteristic.

It must be noted that this analysis is even more complicated than it appears at first, mainly because the equations used to obtain the above mentioned solution are functions of frequency. Thus, each solution obtained as described above applies only at one specific frequency, ω_s. However, it is particularly interesting to determine what frequencies must be considered in order that the characteristic phase lag (θ) between (all) the forces and (all) the responses is exactly 90°. Inspection of equation (2.69) shows that if θ is to be 90°, then that equation reduces to:

$$\left(-\omega^2 [M] + [K]\right)\{\hat{x}\} = 0 \tag{2.70}$$

which is clearly the equation to be solved to find the undamped system natural frequencies and mode shapes. Thus, we have the important result that it is always possible to find a set of mono-phased forces which will cause a mono-phased set of responses and, moreover, if these two sets of mono-phased parameters are separated by exactly 90°, then the frequency at which the system is vibrating is identical to one of its undamped natural frequencies and the displacement 'shape' is the corresponding undamped mode shape.

This most important result is the basis for many of the multi-shaker test procedures used (particularly in the aircraft industry) to isolate the undamped modes of structures for comparison with theoretical predictions. It is also noteworthy that this is one of the few methods for obtaining directly the undamped modes as almost all other methods extract the actual damped modes of the system under test. The physics of the technique are quite simple: the force vector is chosen so that it exactly balances all the damping forces, whatever these may be, and so the principle applies equally to other types of damping.

2.6.4 Postscript

It is often observed that the analysis for hysteretic damping is less than rigorous when applied to the free vibration situation, as we have done above. However, it is an admissible model of damping for describing harmonic forced vibration and this is the objective of most of our studies. Moreover, it is always possible to express each of the receptance (or other FRF) expressions either as a ratio of two polynomials (as explained in Section 2.3) or as a series of simple terms such as those we have used above. Each of the terms in the series may be identified with one of the 'modes' we have defined in the earlier free vibration analysis for the system. Thus, whether or not the solution is strictly valid for a free vibration analysis, we can usefully and confidently consider each of the uncoupled terms or modes as being a genuine characteristic of the system. As will be seen in the next section, the analysis required for the general case of viscous damping — which is more rigorous — is considerably more complicated than that used here which is, in effect, a very simple extension of the undamped case.

2.7   MDOF SYSTEMS WITH VISCOUS DAMPING — GENERAL CASE

2.7.1   Free Vibration Solution — Complex Modal Properties

We turn now to a corresponding treatment for the case of general viscous damping. Exactly the same introductory comments apply in this case as were made at the beginning of Section 2.6 and the only difference is in the specific model chosen to represent the damping behaviour.

The general equation of motion for an MDOF system with viscous damping is:

$$[M]\{\ddot{x}\} + [C]\{\dot{x}\} + [K]\{x\} = \{f\} \tag{2.71}$$

As before, we consider first the case where there is zero excitation in order to determine the natural modes of the system and to this end we assume a solution to the equations of motion which has the form:

$$\{x\} = \{X\}e^{st} \tag{2.72}$$

Substituting this into the appropriate equation of motion gives:

$$\left(s^2[M] + s[C] + [K]\right)\{X\} = \{0\} \tag{2.73}$$

the solution of which constitutes a complex eigenproblem, although one with a somewhat different solution to that of the corresponding stage for the previous case with hysteretic damping. In this case, there are $2N$ eigenvalues, s_r (as opposed to N values of λ_r^2 before) but these now occur in complex conjugate pairs. (This is an inevitable result of the fact that all the coefficients in the matrices are real and thus any characteristic values, or roots, must either be real or occur in complex conjugate pairs.) As before, there is an eigenvector corresponding to each of these eigenvalues, but these also occur as complex conjugates. Hence, we can describe the eigensolution as:

$$
\text{and} \quad \left. \begin{array}{c} s_r, s_r^* \\ \{\psi\}_r, \{\psi\}_r^* \end{array} \right\} \quad r = 1, N \tag{2.74}
$$

It is customary to express each eigenvalue s_r in the form:

$$
s_r = \omega_r \left(-\zeta_r + i\sqrt{1 - \zeta_r^2} \right)
$$

where ω_r is the 'natural frequency' and ζ_r is the critical damping ratio for that mode. Sometimes, the quantity ω_r is referred to as the 'undamped natural frequency' but this is not strictly correct except in the case of proportional damping (or, of course, of a single degree of freedom system).

This eigensolution possesses orthogonality properties although these, also, are different to those of the earlier cases. In order to examine these properties, we first note that any eigenvalue/eigenvector pair satisfies the equation

$$
\left(s_r^2 [M] + s_r [C] + [K] \right) \{\psi\}_r = \{0\} \tag{2.75}
$$

and then we pre-multiply this equation by $\{\psi\}_q^T$ so that we have:

$$
\{\psi\}_q^T \left(s_r^2 [M] + s_r [C] + [K] \right) \{\psi\}_r = 0 \tag{2.76}
$$

A similar expression to (2.75) can be produced by using s_q and $\{\psi\}_q$:

$$
\left(s_q^2 [M] + s_q [C] + [K] \right) \{\psi\}_q = \{0\} \tag{2.77}
$$

which can be transposed, taking account of the symmetry of the system matrices, to give:

$$
\{\psi\}_q^T \left(s_q^2 [M] + s_q [C] + [K] \right) = \{0\}^T \tag{2.78}
$$

If we now postmultiply this expression by $\{\psi\}_r$ and subtract the result from that in equation (2.76), we obtain:

$$\left(s_r^2 - s_q^2\right)\{\psi\}_q^T[M]\{\psi\}_r + \left(s_r - s_q\right)\{\psi\}_q^T[C]\{\psi\}_r = 0 \tag{2.79}$$

and, provided s_r and s_q are different, this leads to the first of a pair of orthogonality equations:

$$\left(s_r + s_q\right)\{\psi\}_q^T[M]\{\psi\}_r + \{\psi\}_q^T[C]\{\psi\}_r = 0 \tag{2.80a}$$

A second equation can be derived from the above expressions as follows: multiply (2.76) by s_q and (2.78) by s_r and subtract one from the other to obtain:

$$s_r s_q \{\psi\}_q^T[M]\{\psi\}_r - \{\psi\}_q^T[K]\{\psi\}_r = 0 \tag{2.80b}$$

These two equations — (2.80a) and (2.80b) — constitute the orthogonality conditions of the system and it is immediately clear that they are far less simple than those we have encountered previously. However, it is interesting to examine the form they take when the modes r and q are a complex conjugate pair. In this case, we have that

$$s_q = \omega_r\left(-\zeta_r - i\sqrt{1-\zeta_r^2}\right) \tag{2.81}$$

and also that

$$\{\psi\}_q = \{\psi\}_r^* \tag{2.82}$$

Inserting these into equation (2.80a) gives

$$-2\omega_r\zeta_r\{\psi\}_r^H[M]\{\psi\}_r + \{\psi\}_r^H[C]\{\psi\}_r = 0 \tag{2.83}$$

where $\{\ \}^H$ denotes the complex conjugate (Hermitian) transpose, from which we obtain:

$$2\omega_r\zeta_r = \frac{\{\psi\}_r^H[C]\{\psi\}_r}{\{\psi\}_r^H[M]\{\psi\}_r} = \frac{c_r}{m_r} \tag{2.84}$$

Similarly, inserting (2.81) and (2.82) into (2.80b) gives

$$\omega_r^2 \{\psi\}_r^H [M]\{\psi\}_r + \{\psi\}_r^H [K]\{\psi\}_r = 0 \tag{2.85}$$

from which

$$\omega_r^2 = \frac{\{\psi\}_r^H [K]\{\psi\}_r}{\{\psi\}_r^H [M]\{\psi\}_r} = \frac{k_r}{m_r} \tag{2.86}$$

In these expressions, m_r, k_r and c_r may be described as modal mass, stiffness and damping parameters, respectively, although the meaning is slightly different to that used in the other systems.

2.7.2   Forced Response Solution

In this stage of the analysis, this case of general viscous damping again presents a more complex task. Returning to the basic equation, (2.71), and assuming a harmonic response:

$$\{x(t)\} = \{X\}e^{i\omega t} \tag{2.87}$$

we can write the forced response solution directly as

$$\{X\} = \left([K] - \omega^2 [M] + i\omega [C]\right)^{-1} \{F\} \tag{2.88}$$

but as in previous cases, this expression is not particularly convenient for numerical application. We shall seek a similar series expansion to that which has been used in the earlier cases of undamped, proportionally-damped and hysteretically-damped systems but now we find that the eigenvalue solution presented in the above equations is not directly amenable to this task. In fact, it is necessary to recast the equations into the state-space form in order to achieve our goal.

Define a new coordinate vector $\{u\}$ which is of order $2N$ and which contains both the displacements $\{x\}$ and the velocities $\{\dot{x}\}$:

$$\{u\} = \begin{Bmatrix} x \\ \dot{x} \end{Bmatrix}_{(2N \times 1)} \tag{2.89}$$

Equation (2.71) can then be written as:

$$[C : M]_{N \times 2N} \{\dot{u}\}_{2N \times 1} + [K : 0]\{u\} = \{0\}_{N \times 1} \tag{2.90}$$

However, in this form we have N equations and $2N$ unknowns and so we add an identity equation of the type:

$$[M:0]\{\dot{u}\}+[0:-M]\{u\}=\{0\} \tag{2.91}$$

which can be combined to form a set of $2N$ equations

$$\begin{bmatrix} C & M \\ M & 0 \end{bmatrix}\{\dot{u}\}+\begin{bmatrix} K & 0 \\ 0 & -M \end{bmatrix}\{u\}=\{0\} \tag{2.92a}$$

which can be simplified to:

$$[A]\{\dot{u}\}+[B]\{u\}=\{0\} \tag{2.92b}$$

These equations are now in a standard eigenvalue form and by assuming a trial solution of the form $\{u\}=\{U\}e^{st}$, we can obtain the $2N$ eigenvalues and eigenvectors of the system, s_r and $\{\theta\}_r$, which together satisfy the general equation:

$$(s_r[A]+[B])\{\theta\}_r=\{0\} \quad ; \quad r=1,2N \tag{2.93}$$

These eigenproperties will, in general, be complex although for the same reasons as previously they will always occur in conjugate pairs. They possess orthogonality properties which are simply stated as

$$[\Theta]^T[A][\Theta]=[a_r]$$
$$[\Theta]^T[B][\Theta]=[b_r] \tag{2.94}$$

and which have the usual characteristic that

$$s_r=-\frac{b_r}{a_r} \quad ; \quad r=1,2N \tag{2.95}$$

Now we may express the forcing vector in terms of the new coordinate system as:

$$\{P\}_{2N\times1}=\begin{Bmatrix} F \\ 0 \end{Bmatrix} \tag{2.96}$$

and, assuming a similarly harmonic response and making use of the previous development of a series form expression of the response (equations (2.35) to (2.41)), we may write:

$$
\left\{ \begin{array}{c} X \\ \cdots \\ i\omega X \end{array} \right\}_{2N\times 1} = \sum_{r=1}^{2N} \frac{\{\theta\}_r^T \{P\}\{\theta\}_r}{a_r(i\omega - s_r)}
\tag{2.97}
$$

However, because the eigenvalues and vectors occur in complex conjugate pairs, this last equation may be rewritten as:

$$
\left\{ \begin{array}{c} X \\ \cdots \\ i\omega X \end{array} \right\} = \sum_{r=1}^{N} \left(\frac{\{\theta\}_r^T \{P\}\{\theta\}_r}{a_r(i\omega - s_r)} + \frac{\{\theta\}_r^H \{P\}\{\theta^*\}_r}{a_r^*(i\omega - s_r^*)} \right)
\tag{2.98}
$$

At this stage, it is convenient to extract a single response parameter, say X_j, resulting from a single force such as F_k — the receptance frequency response function, α_{jk} — and in this case, equation (2.98) leads to:

$$
\alpha_{jk}(\omega) = \sum_{r=1}^{N} \left(\frac{(\theta_{jr})(\theta_{kr})}{a_r\left(\omega_r\zeta_r + i\left(\omega - \omega_r\sqrt{1-\zeta_r^2} \right) \right)} \right.
$$
$$
\left. + \frac{(\theta_{jr})^*(\theta_{kr})^*}{a_r^*\left(\omega_r\zeta_r + i\left(\omega + \omega_r\sqrt{1-\zeta_r^2} \right) \right)} \right)
\tag{2.99}
$$

Using the fact that $s_r = \omega_r(-\zeta_r + i\sqrt{1-\zeta_r^2})$, this expression can be further reduced to the form:

$$
\alpha_{jk}(\omega) = \sum_{r=1}^{N} \frac{(_rR_{jk}) + i(\omega/\omega_r)(_rS_{jk})}{\omega_r^2 - \omega^2 + 2i\omega\omega_r\zeta_r}
\tag{2.100}
$$

where the coefficients R and S are obtained from:

$$\{_rR_k\} = 2\left(\zeta_r \mathrm{Re}\{_rG_k\} - \mathrm{Im}\{_rG_k\}\sqrt{1-\zeta_r^2}\right);$$

$$\{_rS_k\} = 2\mathrm{Re}\{_rG_k\}; \tag{2.101}$$

$$\{_rG_k\} = (\theta_{kr}/a_r)\{\theta\}_r$$

The similarity between this expression and that derived in Section 2.5 (equation (2.66)) is evident, the main difference being in the frequency-dependence of the numerator in the case of viscous damping. If we confine our interest to a small range of frequency in the vicinity of one of the natural frequencies (i.e. $\omega \approx \omega_r$), then it is clear that (2.100) and (2.66) are very similar indeed.

2.7.3 Excitation by General Force Vector

Although we have only fully developed the analysis for the case of a single force, the ingredients already exist for the more general case of multi-point excitation in the equations (2.96) to (2.98). The particular case of excitation by mono-phased forces has effectively been dealt with in Section 2.6 because it was there shown that the results obtained would apply to any type of damping.

2.8 MODAL ANALYSIS OF ROTATING STRUCTURES
2.8.1 General Features of Rotating Structure Dynamics
2.8.1.1 Nonsymmetry in system matrices

All the cases considered until now have contained certain properties which make them less than the most general type of linear systems that can be encountered. There is another class of system which we should consider in this work, and that is the class that includes components in the structure under consideration which are rotating and which are therefore subject to forces in addition to those already considered in our analysis so far. These additional forces include, for example:

- gyroscopic forces;
- rotor-stator rub forces;
- electromagnetic forces;
- unsteady aerodynamic forces;
- time-varying fluid forces;

and any or all of these can have the effect of destroying the symmetry of the system matrices used to define the equations which govern the motion of the system. Symmetry of the mass, stiffness and damping matrices is a feature which has been present in all cases we have dealt with so far. As a consequence of this feature, the solutions we have made for the simpler systems have themselves possessed certain simplicities and symmetries, such as reciprocity, which will no longer apply to this new class of system, examples of which are referred to as 'non-self-adjoint' (NSA) systems.

We shall now extend our studies of the underlying theory of linear systems into this new area, with the specific interest focused on structures which contain rotating components. Although it can be noted that other systems may well exist that exhibit similar features, we shall not consider these in detail here.

In the most general case treated to date, we have already seen that the eigenvalues and eigenvectors (natural frequency, damping factor and mode shape for each mode) will all be complex: the complexity of the eigenvalues resulting from the damped nature of the modes and the complexity of the mode shapes signifying an arbitrary distribution of the damping within the structure. However, in all those earlier cases the system matrices were symmetric, as were the frequency response function matrices. Further generalisation of the systems being considered can lead to non-symmetric system matrices, particularly in the stiffness and so-called damping matrices. This, in turn, leads to a situation where there are two sets of eigenvectors, referred to as the 'left-hand' and the 'right-hand' vectors. The right-hand vectors are those generally associated with the mode shapes, while the left-hand vectors are the set which are obtained by analysing the transformed equations and are associated with patterns of forces (rather than of responses) associated with a single mode. Both sets of vectors are required for a full reconstruction or description of the system's dynamic behaviour.

Two special cases of non-symmetric matrices are worth mentioning at this stage because they frequently occur in rotating structure analysis: the first is where there is a skew-symmetric component in the 'damping' matrix and the second where there is a skew-symmetric component in the stiffness matrix. The first of these conditions arises directly due to the existence of gyroscopic forces which have a velocity-dependent characteristic (hence their appearance in the 'damping' matrix) although they do not constitute dissipative effects as do true damping terms. The second case can arise from the behaviour of hydrodynamic or other bearings, or from the presence of 'internal' damping* in the rotor.

In the first of these two cases, if the mass and stiffness matrices are symmetric, and the 'damping' matrix (or, more correctly, the velocity-dependent matrix) is skew-symmetric — which happens if there are only gyroscopic terms present — then the eigenvalues of the system will be real (i.e. the system will be undamped) while the mode shapes will be complex, and heavily so. Furthermore, the left-hand eigenvectors are found to be the complex conjugates of the right-hand vectors.

In the second case, if the mass matrix is symmetric, and the damping matrix is zero, then skew-symmetry in the stiffness matrix results in complex eigenvectors **and** complex eigenvalues, the latter indicating a system which is effectively damped, although it will be seen that such modal damping can be either positive (indicating stability) or negative (indicating an unstable system), depending upon the particular conditions which apply in each case. Once again, the left-hand eigenvectors are the complex conjugates of the right-hand eigenvectors.

2.8.1.2 Stationary and rotating frames of reference

As a further complication, in addition to the loss of symmetry in the system matrices, some types of rotating structure are found to have equations of motion with time-varying coefficients in place of the constant coefficients we have had previously. For these cases, there are no fixed modes as such, and we shall need to extend our methods of analysis, as well as those of measurement, to be able to apply the techniques of modal analysis and testing to this important class of structure. It will be seen below how geometrical symmetry in the form of either the rotating or the stationary structures (**not** related to the symmetry in the describing system matrices), and the choice of coordinate system used to describe the motion, have a direct effect on this feature.

We have hitherto written the equations of motion for our various systems in terms of coordinates which are referred to fixed axes but now there exists the choice of using these same coordinates **or** of using coordinates which are set in axes that rotate with the structure. In practical terms, this is equivalent to measuring vibration with transducers that are fixed in space (e.g. inductance displacement transducers) or with those which are fixed to the rotating component (e.g. strain gauges).

---

*   'Internal' and 'external' damping refers to damping elements which are, respectively: (a) internal to and (b) external to the rotating component. These derive from (a) joints, splines etc in the rotor assembly, as well as to its inherent material damping properties, and (b) bearings and other externally-attached devices.

Fig. 2.18 illustrates the choice. Coordinates in the fixed axis set are denoted by x_S, y_S and z_S (usually simplified to x, y and z) while those in the rotating axes frame are referred to as: x_R, y_R and z_R. Transformation from one set to the other is relatively straightforward, as will be shown in the following paragraphs, but can result in significantly different equations of motion.

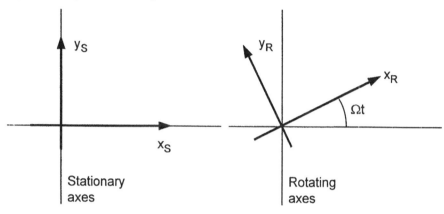

Fig. 2.18   Coordinate systems in stationary and rotating axes sets

2.8.2   Dynamic Analysis of Symmetric Rotor with Gyroscopic Effects

2.8.2.1   Non-rotating system properties

As in earlier sections of the book, it will be helpful to approach the more general types of rotating structures via a simple example and that can be provided by a simple model of a symmetrical rotor supported in bearings, as illustrated in Fig. 2.19. This system consists of a rigid disc mounted on the free end of a rigid shaft of length L, the other end of which is effectively pin-jointed (by the bearings which hold it there). At the 'free' end, the rotor is supported by a flexible bearing, represented by vertical and horizontal stiffnesses, k_y and k_x, respectively. The polar moment of inertia of the disc and shaft is J, while its moment of inertia about a lateral axis through 0 is I_0. The rotor, when spinning, has an angular velocity of Ω_z, anticlockwise viewed from the free end of the shaft.

The two equations of motion for this simple structure can be written as follows, starting with the simplest case possible where the rotation speed (Ω_z) is zero and there is no damping or external forcing:

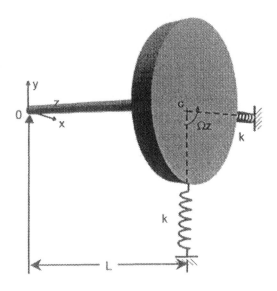

Fig. 2.19 Simple 2DOF rotor-stator system

$$(I_0 / L)\ddot{x} + k_x L \; x \; = 0$$

$$(I_0 / L)\ddot{y} + k_y L \; y \; = 0$$

(2.102)

Solution of these uncoupled equations of motion reveals the expected two modes of vibration, one in the vertical plane and the other in the horizontal plane, both of which will have the same natural frequency if k_x and k_y are identical at k, as will be the case for the fully-symmetrical system. (Comments made earlier in Section 2.4.3 concerning systems with two or more identical natural frequencies should be noted.)

Extension of the model to include 'external' dampers, c_x and c_y, alongside the two springs and vertical and horizontal external forces applied at the bearing location, $f_x(t)$ and $f_y(t)$, leads to the following equations for forced vibration and, in turn to the derivation of expressions for the four FRF properties for this 2DOF system:

$$(I_0 / L)\ddot{x} \; + c_x L \; \dot{x} \; + k_x L \; x \; = f_x(t)L$$

$$(I_0 / L)\ddot{y} \; + c_y L \; \dot{y} \; + k_y L \; y \; = f_y(t)L$$

(2.103)

which, for harmonic vibration where $f_x(t) = F_x e^{i\omega t}; x(t) = X e^{i\omega t}$; etc. lead to:

$$\alpha_{xx}(\omega) = X / F_x = \frac{L^2}{(k_x L^2 + i\omega c_x L^2 - \omega^2 I_0)}$$

$$\alpha_{yy}(\omega) = Y / F_y = \frac{L^2}{(k_y L^2 + i\omega c_y L^2 - \omega^2 I_0)} \qquad (2.104)$$

$$\alpha_{xy}(\omega) = \alpha_{yx}(\omega) = 0$$

2.8.2.2 Modes of the undamped rotating system

If we now return to the basic system, with no damping and no external excitation, and allow the disc to spin at speed Ω_z, then there will be some additional forces generated by the fact that, when vibrating in the x and/or y directions, the disc is rotating simultaneously about more than one axis. These forces are due to the Coriolis accelerations set up by this complex motion and are usually referred to as 'gyroscopic' forces. Essentially, simultaneous rotation about the z-axis (at angular speed Ω_z) and about the y-axis (with angular speed $\dot{\theta}_y = \dot{x}/L$) can only exist if there is a moment applied to the system about the third (x-) axis with a magnitude $M_x = J\Omega_z \dot{x}/L$. When added to the equations of motion, this moment, and its counterpart for the other combination of rotations, has the effect of coupling the two equations, which now take the form:

$$(I_0 / L)\ddot{x} + (J\Omega_z / L)\dot{y} + k_x L \; x \; = 0$$

$$(I_0 / L)\ddot{y} - (J\Omega_z / L)\dot{x} + k_y L \; y \; = 0$$

or

$$\begin{bmatrix} (I_0/L) & 0 \\ 0 & (I_0/L) \end{bmatrix} \begin{Bmatrix} \ddot{x} \\ \ddot{y} \end{Bmatrix} + \begin{bmatrix} 0 & J\Omega_z/L \\ -J\Omega_z/L & 0 \end{bmatrix} \begin{Bmatrix} \dot{x} \\ \dot{y} \end{Bmatrix} + \begin{bmatrix} k_x L & 0 \\ 0 & k_y L \end{bmatrix} \begin{Bmatrix} x \\ y \end{Bmatrix} = \begin{Bmatrix} 0 \\ 0 \end{Bmatrix}$$

This equation is more conveniently written as:

$$
\begin{bmatrix} (I_0/L^2) & 0 \\ 0 & (I_0/L^2) \end{bmatrix} \begin{Bmatrix} \ddot{x} \\ \ddot{y} \end{Bmatrix} + \begin{bmatrix} 0 & J\Omega_z/L^2 \\ -J\Omega_z/L^2 & 0 \end{bmatrix} \begin{Bmatrix} \dot{x} \\ \dot{y} \end{Bmatrix}
$$

$$
+ \begin{bmatrix} k_x & 0 \\ 0 & k_y \end{bmatrix} \begin{Bmatrix} x \\ y \end{Bmatrix} = \begin{Bmatrix} 0 \\ 0 \end{Bmatrix} \tag{2.105}
$$

Note that these equations include velocity-dependent terms (although these are not damping effects) but that they are not symmetric: indeed, the velocity-dependent matrix is skew-symmetric. The free vibration solution of these equations reveals that there are still two modes of vibration and that now they have different natural frequencies, even when the two bearing stiffnesses are identical. Furthermore, it can be seen that these are still the natural frequencies of an undamped system, even though there are velocity-dependent forces present. It will also be noted that this type of equation has two different sets of eigenvectors — the so-called 'left-hand' and 'right-hand' sets, the latter representing the mode shapes themselves while the former are associated with preferred excitation patterns.

(a)   Symmetric stator

The essential solution for the case where the vertical and horizontal stiffnesses are identical, and both equal to k, is as follows. Assume, as before, a simple harmonic solution of the type:

$$
x = Xe^{i\omega t}
$$
$$
y = Ye^{i\omega t} \tag{2.106}
$$

which, applying the symmetry of the stator so that $k_x = k_y = k$, leads to:

$$
\begin{bmatrix} \left(k - \omega^2 I_0/L^2\right) & \left(i\omega J\Omega_z/L^2\right) \\ \left(-i\omega J\Omega_z/L^2\right) & \left(k - \omega^2 I_0/L^2\right) \end{bmatrix} \begin{Bmatrix} X \\ Y \end{Bmatrix} = \begin{Bmatrix} 0 \\ 0 \end{Bmatrix} \tag{2.107}
$$

and this, in turn, to the following characteristic equation:

$$
(kL^2)^2 - \omega^2(2I_0kL^2 + J^2\Omega_z^2) + \omega^4 I_0^2 = 0 \tag{2.108}
$$

This equation can be solved to find the natural frequencies, ω_1 and ω_2, using the following notation, to give:

$$\omega_1 = \omega_\Omega - \tfrac{1}{2}(\gamma\Omega); \omega_2 = \omega_\Omega + \tfrac{1}{2}(\gamma\Omega)$$

where   (2.109)

$$\omega_0^2 = \frac{kL^2}{I_0} \quad ; \quad \gamma = \frac{J}{I_0} \quad ; \quad \omega_\Omega^2 = \omega_0^2 + \tfrac{1}{4}(\gamma\Omega)^2$$

Fig. 2.20(a) shows the dependence of the two natural frequencies on the speed of rotation, Ω_z, indicating the two relevant asymptotes, $\omega = \gamma\Omega$ and $\omega = 0.5\gamma\Omega$, respectively. Also shown is the once-per-rev line, $\omega = \Omega$ (sometimes referred to as the 'first engine order' (1EO) line), and the two critical speeds where this line intersects the two natural frequency lines, at

$$\Omega_1 = \frac{\omega_0}{\sqrt{(1+\gamma)}} \quad ; \quad \text{and} \quad \Omega_2 = \frac{\omega_0}{\sqrt{(1-\gamma)}}$$ (2.110)

respectively.

Completion of the free vibration solution reveals that the mode shapes corresponding to the two natural frequencies are complex — entirely complex, in fact, in that the two elements are exactly in quadrature with each other. The right-hand eigenvector for the lower-frequency of the two modes, $r = 1$, can be shown to take the form:

$$\{\phi_{RH}\}_1 = \begin{Bmatrix} 1 \\ i \end{Bmatrix}$$

while that for the second mode, $r = 2$, is:

$$\{\phi_{RH}\}_2 = \begin{Bmatrix} i \\ 1 \end{Bmatrix} \equiv \begin{Bmatrix} 1 \\ -i \end{Bmatrix}$$

The interpretation of these two complex mode shapes is straightforward: the first one (which corresponds to the lower natural frequency) represents a motion which constitutes a circular orbit of the disc centre which is *backwards* with respect to the spinning motion of the rotor, while the second mode shape, corresponding to the higher natural frequency, represents a *forward* circular orbiting motion, in the same direction as the spinning of the disc itself. Fig. 2.20(b) seeks to illustrate these mode shapes and further explanation of the

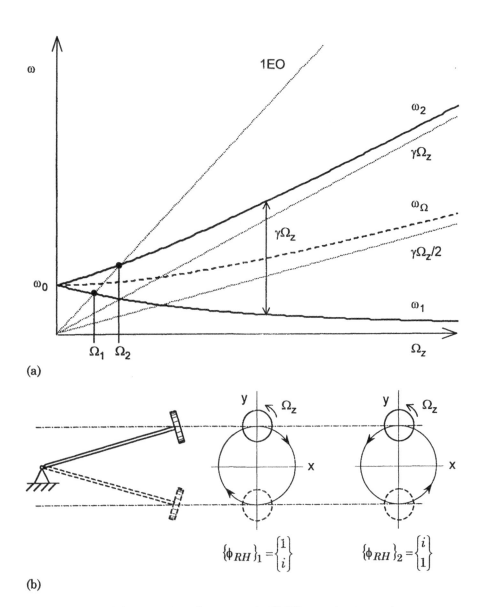

Fig. 2.20 Modal properties of symmetric 2DOF rotor/stator system.
(a) Natural frequencies; (b) Mode shapes

interpretation of complex modes in terms of stationary and rotating components is provided in later sections of the book.

These two mode shapes combine to form the matrix of the RH eigenvectors, $[\Phi_{RH}]$, which can be compared with its left-hand counterpart, $[\Phi_{LH}]$, and shown to observe the orthogonality conditions for this type of system, which are of the rather complicated form previously encountered in the case of general viscously-damped systems (see Section 2.7, equations (2.80)).

(b) Non-symmetric stator

It is relatively straightforward to extend the above analysis to include the case where the two bearing (stator) stiffnesses are not identical, but have the relationship:

$$k_x = a\,k_y$$

In this case, the two system natural frequencies differ from those quoted previously, as illustrated in Fig. 2.21(a), but the main difference is seen in the two mode shapes, which once again reflect backward and forward orbits but this time of an elliptical rather than circular form, as shown by the relevant right-hand eigenvectors:

$$\{\phi_{RH}\}_1 = \begin{Bmatrix} A \\ i \end{Bmatrix} \quad ; \quad \{\phi_{RH}\}_2 = \begin{Bmatrix} B \\ -i \end{Bmatrix}$$

and which are illustrated alongside the natural frequency plot in Fig. 2.21(b).

2.8.2.3 FRFs of the rotating structure with external damping

If we extend our analysis further, to include external damping and excitation forces, then we can — as before — derive expressions for the FRF characteristics of this system. There are again four FRFs applicable at the bearing support point, but this time they contain important differences from those developed earlier for the non-spinning case. Now, we find that the equations of motion for damped forced vibration of the symmetric system can be written as:

$$\begin{bmatrix} \left(\dfrac{I_0}{L^2}\right) & 0 \\[2mm] 0 & \left(\dfrac{I_0}{L^2}\right) \end{bmatrix}\begin{Bmatrix} \ddot{x} \\ \ddot{y} \end{Bmatrix} + \begin{bmatrix} c & \left(\dfrac{J\Omega_z}{L^2}\right) \\[2mm] -\left(\dfrac{J\Omega_z}{L^2}\right) & c \end{bmatrix}\begin{Bmatrix} \dot{x} \\ \dot{y} \end{Bmatrix} + \begin{bmatrix} k & 0 \\ 0 & k \end{bmatrix}\begin{Bmatrix} x \\ y \end{Bmatrix} = \begin{Bmatrix} f_x \\ f_y \end{Bmatrix} \quad (2.111)$$

and the corresponding expressions for the same four FRFs as shown previously are found by setting f_x and f_y to be $F_x e^{i\omega t}$ and $F_y e^{i\omega t}$, respectively, to yield:

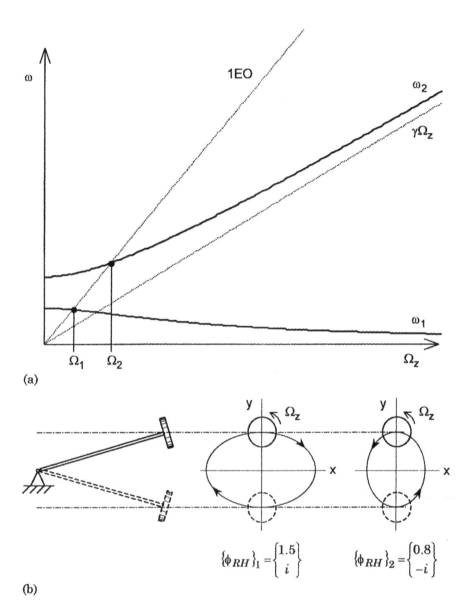

$$\{\phi_{RH}\}_1 = \begin{Bmatrix} 1.5 \\ i \end{Bmatrix}$$

$$\{\phi_{RH}\}_2 = \begin{Bmatrix} 0.8 \\ -i \end{Bmatrix}$$

(b)

Fig. 2.21 Modal properties of non-symmetric 2DOF rotor/stator system.
(a) Natural frequencies; (b) Mode shapes

$$\alpha_{xx}(\omega) = X/F_x$$

$$= \frac{(kL^2 + i\omega c L^2 - \omega^2 I_0)}{(k^2 L^2 + 2i\omega c k L^2 - \omega^2 (c^2 L^2 + 2I_0 k + (J\Omega_z/L)^2) - 2i\omega^3 c I_0 + \omega^4 (I_0/L)^2)}$$

$$\alpha_{yy}(\omega) = Y/F_y = \alpha_{xx}(\omega) \qquad\qquad (2.112)$$

$$\alpha_{xy}(\omega) = -\alpha_{yx}(\omega)$$

$$= \frac{i\omega J\Omega_z}{(k^2 L^2 + 2i\omega c k L^2 - \omega^2 (c^2 L^2 + 2I_0 k + (J\Omega_z/L)^2) - 2i\omega^3 c I_0 + \omega^4 (I_0/L)^2)}$$

Plots of these FRFs are shown in Fig. 2.22. The loss of reciprocity between $\alpha_{xy}(\omega)$ and $\alpha_{yx}(\omega)$ is visible in these plots and is evident from the formula, as is the fact that a force in the vertical (y-) direction now generates a response in the orthogonal horizontal (x-) direction, and vice versa, in direct contrast to the results which apply without any spinning of the rotor. Also clear from the plots is the high degree of complexity of the mode shapes, as witnessed by the significant imaginary part of the transfer FRF curves.

Similar expressions apply to the more general case in which the stator stiffnesses are not identical in the two planes, although the differences are of detail only, and bring no additional features.

Of course, the response of the system to the simultaneous application of several forces at the same frequency can be derived by appropriate superposition of the relevant components. Such a situation could be envisaged if there were simultaneous horizontal and vertical forces applied to the bearing in the current example. It also applies in the important case of internally-generated out-of-balance forces, which will be dealt with in the next section.

2.8.2.4 Response of externally-damped rotating structure to synchronous and non-synchronous out-of-balance excitations

In addition to the classical forced response analysis which yields the FRF properties of the system, it is also of interest to undertake a similar analysis for the particular case of an excitation which is provided by an out-of-balance force. Such forces may be of a *synchronous* nature, when they result from an out-of-balance mass on the rotating component itself, or of a *non-synchronous* nature, when derived from a similar out-of-balance on a co- or counter-rotating shaft.

92

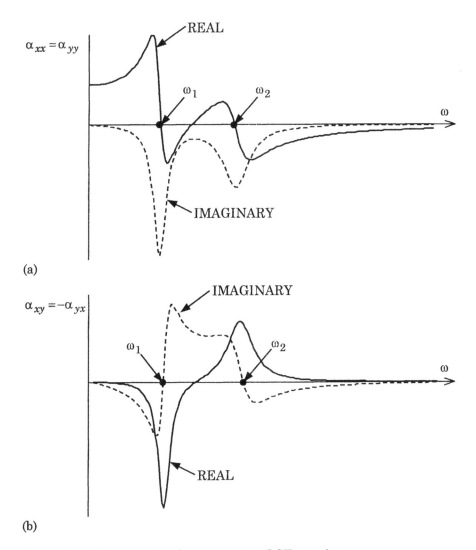

(a)

(b)

Fig. 2.22 FRF properties for symmetric 2DOF rotor/stator system

The analysis is first presented for the simpler and more common case (synchronous out-of-balance) and then extended to the more general non-synchronous case.

If we suppose there to be an out-of-balance of magnitude (mr), which rotates with the rotor, then this will develop forces in the fixed-axis x and y directions of the form:

$$F_x = F_{OOB}\cos(\Omega t) \quad ; \quad F_y = F_{OOB}\sin(\Omega t)$$

where   (2.113a)

$$F_{OOB} = mr\Omega^2$$

This multiple excitation case can be defined as an harmonic force vector, $\{F\}$, which has the form:

$$\{F\} = \begin{Bmatrix} F_x \\ F_y \end{Bmatrix} = F_{OOB} \begin{Bmatrix} 1 \\ -i \end{Bmatrix} e^{i\Omega t} \qquad (2.113b)$$

(a)   Synchronous excitation: symmetric stator

It can be shown that such an excitation vector will generate a similarly harmonic response which can be expressed as $\{X\ Y\}^T$ which, in the case of the symmetric stator system, is:

$$\begin{Bmatrix} X \\ Y \end{Bmatrix} e^{i\Omega t} = \begin{Bmatrix} A \\ -iA \end{Bmatrix} F_{OOB}\, e^{i\Omega t}$$

where

$$A = \frac{L^2}{I_0 \left(\omega_0^2 - \Omega^2 (1-\gamma)\right)} \qquad (2.114)$$

From this expression we can see the interesting result that only one of the two modes of the system is excited into resonance at the appropriate speed of rotation. As the rotation speed (and thus the excitation frequency) increases and passes through the lower natural frequency, ω_1 (with the corresponding critical speed, Ω_1, as given in (2.110)) it is found that no resonant response is generated at all. In contrast, when the speed/frequency reaches the second natural frequency, a clear resonance condition is achieved in this case, see Fig. 2.23(a). This reflects the well-known physical phenomenon that out-of-balance excitation can excite resonance only in a forward whirl mode for a symmetrical rotor/stator system, and not at all in a backward whirl motion, which is just what the first mode of vibration of this system is.

Closer inspection of this result shows it to be entirely predictable. The force vector in this case can be seen to be orthogonal to the mode shape vector of the first mode and, under such conditions, no response would be expected from that mode, even when the excitation and natural frequencies coincide. The second mode shape vector, however, is

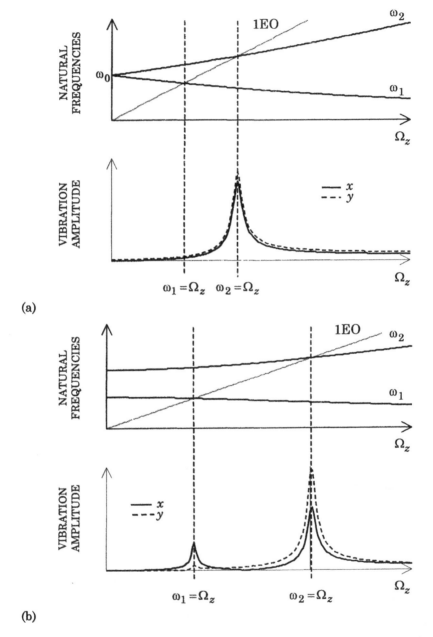

Fig. 2.23 Response to out-of-balance excitation of 2DOF rotor/stator system.
(a) Symmetric stator; (b) Non-symmetric stator

not orthogonal to the excitation force vector and so a response is found in that case.

(b)   Synchronous excitation: non-symmetric stator

The situation is different, in the more general case where the stator is not symmetric: i.e. when $k_x \neq k_y$. Here, the excitation force vector (which is unchanged from the previous case) is orthogonal to neither mode shape vector and so a resonance is expected at both critical speeds. This prediction is confirmed by the typical results shown in Fig. 2.23(b).

(c)   Non-synchronous excitation

It is possible to envisage a situation where an out-of-balance excitation force is generated by a second rotor or disc which is spinning at a different speed to the test rotor: say, at a speed of $\beta\Omega$. In this case, the only difference lies in the frequency and magnitude of the excitation: the essential results are the same as for the synchronous case, as illustrated on the example shown in equation (2.115):

$$\begin{Bmatrix} X \\ Y \end{Bmatrix} e^{i\beta\Omega t} = \begin{Bmatrix} A \\ -iA \end{Bmatrix} F_{OOB}\, e^{i\beta\Omega t}$$

where

$$A = \frac{L^2}{I_0\left[\omega_0^2 - \beta\Omega^2\,(\beta - \gamma)\right]} \tag{2.115}$$

2.8.2.5   Analysis using coordinates in rotating frame of reference

At the outset of this analysis, we elected to write the equations of motion for the above system using 'stationary' coordinates, x_S and y_S, these being deflections of the disc centre in the fixed-in-space horizontal and vertical directions, respectively. In rotating structures there is a natural alternative to this choice of coordinates: namely, to use the coordinates which are fixed in the rotor, and which rotate with it. These are identified here as x_R and y_R, as opposed to x_S and y_S (which are generally abbreviated to plain x and y, as here).

Fig. 2.18 shows the two sets of coordinates and from this diagram it is a simple matter to define the transformation matrices which allow us to convert the equations of motion from one set to the other:

$$\begin{Bmatrix} x \\ y \end{Bmatrix} = \begin{bmatrix} c & -s \\ s & c \end{bmatrix} \begin{Bmatrix} x_R \\ y_R \end{Bmatrix} \quad \text{and} \quad \begin{Bmatrix} x_R \\ y_R \end{Bmatrix} = \begin{bmatrix} c & s \\ -s & c \end{bmatrix} \begin{Bmatrix} x \\ y \end{Bmatrix}$$

where

$$c = \cos(\Omega t) \quad ; \quad s = \sin(\Omega t) \tag{2.116}$$

Using the normal transformation process for equations of motion, we can rewrite the equations shown above in (2.105) to an alternative version in terms of the rotating coordinates, x_R and y_R, to obtain*:

$$\begin{bmatrix} \dfrac{I_0}{L^2} & 0 \\ 0 & \dfrac{I_0}{L^2} \end{bmatrix} \begin{Bmatrix} \ddot{x}_R \\ \ddot{y}_R \end{Bmatrix} + \begin{bmatrix} 0 & -\dfrac{2\Omega I_0}{L^2} + \dfrac{J\Omega}{L^2} \\ \dfrac{2\Omega I_0}{L^2} - \dfrac{J\Omega}{L^2} & 0 \end{bmatrix} \begin{Bmatrix} \dot{x}_R \\ \dot{y}_R \end{Bmatrix}$$

$$+ \begin{bmatrix} -\dfrac{I_0\Omega^2}{L^2} + \dfrac{J\Omega^2}{L^2} + k_x c^2 + k_y s^2 & cs(k_y - k_x) \\ cs(k_y - k_x) & -\dfrac{I_0\Omega^2}{L^2} + \dfrac{J\Omega^2}{L^2} + k_y c^2 + k_x s^2 \end{bmatrix} \begin{Bmatrix} x_R \\ y_R \end{Bmatrix} = \{0\}$$

$$\tag{2.117}$$

Although these equations simplify further for the case of a symmetric stator to the form shown below, they have been written in this way in order to introduce a feature which will take on greater importance in the next section. It can be seen in equation (2.117) that some of the coefficients in the equations of motion written in terms of these rotating variables are influenced by the time-varying terms introduced by the rotation of the system. In the particular case where the stator is symmetric with $k_x = k_y = k$, all the coefficients are constant and equation (2.117) simplifies to:

---

* See Appendix 4 for details of the transformations between stationary and rotating coordinates.

$$\begin{bmatrix} \dfrac{I_0}{L^2} & 0 \\ 0 & \dfrac{I_0}{L^2} \end{bmatrix} \begin{Bmatrix} \ddot{x}_R \\ \ddot{y}_R \end{Bmatrix} + \begin{bmatrix} 0 & -\dfrac{2\Omega I_0}{L^2} + \dfrac{J\Omega}{L^2} \\ \dfrac{2\Omega I_0}{L^2} - \dfrac{J\Omega}{L^2} & 0 \end{bmatrix} \begin{Bmatrix} \dot{x}_R \\ \dot{y}_R \end{Bmatrix}$$

$$+ \begin{bmatrix} -\dfrac{I_0\Omega^2}{L^2} + \dfrac{J\Omega^2}{L^2} + k & 0 \\ 0 & -\dfrac{I_0\Omega^2}{L^2} + \dfrac{J\Omega^2}{L^2} + k \end{bmatrix} \begin{Bmatrix} x_R \\ y_R \end{Bmatrix} = \{0\} \qquad (2.118)$$

This equation shows the system matrices to have constant coefficients in this symmetric case but we have been alerted to the possibility that this will not always be the case, particularly if the stator does not have symmetric properties. Nevertheless, there are yet more features to be teased out of this simple example.

Clearly, these equations describe the dynamic behaviour of the same system as do the equations in (2.105), yet they are different and we might suppose that the eigensolution of (2.118) will differ from that of (2.105). Indeed, this is the case and it can be shown that the two eigenvalues of the last equations are:

$$\omega_1 = \omega_\Omega - \tfrac{1}{2}(\gamma\Omega) + \Omega \quad ; \quad \omega_2 = \omega_\Omega + \tfrac{1}{2}(\gamma\Omega) - \Omega \qquad (2.119a)$$

which can be compared with the corresponding values obtained earlier:

$$\omega_1 = \omega_\Omega - \tfrac{1}{2}(\gamma\Omega) \quad ; \quad \omega_2 = \omega_\Omega + \tfrac{1}{2}(\gamma\Omega) \qquad (2.119b)$$

the primary difference being the addition or subtraction of the rotation speed, Ω, which reflects the difference between the two different axes sets. The eigenvectors (mode shapes) remain unchanged.

Finally, it is interesting to examine the effect of transformation to the rotating frame of reference of the excitation forces that are applied in the stationary frame. In other words, to determine how the rotating structure 'sees' an excitation which is applied externally. If we suppose that there is a single point excitation force, $F_{xS} = F_0 \cos\omega t$ and that $F_{yS} = 0$, then we can transform the excitation vector from the stationary frame of reference to the rotating frame, as follows:

$$\begin{Bmatrix} F_{xR} \\ F_{yR} \end{Bmatrix} = \begin{bmatrix} c & s \\ -s & c \end{bmatrix} \begin{Bmatrix} F_{xS} \\ F_{yS} \end{Bmatrix} = \begin{bmatrix} c & s \\ -s & c \end{bmatrix} \begin{Bmatrix} F_0 \\ 0 \end{Bmatrix} \cos \omega t$$

$$= 0.5 F_0 \begin{Bmatrix} \cos(\omega - \Omega)t + \cos(\omega + \Omega)t \\ \sin(\omega - \Omega)t - \sin(\omega + \Omega)t \end{Bmatrix} \tag{2.120}$$

From this we note the interesting, but obvious, result that a stationary harmonic excitation at frequency ω is seen by a structure rotating at speed, Ω, as an excitation with two frequency components: $(\omega - \Omega)$ and $(\omega + \Omega)$. Accordingly, the structure will respond at these two apparent frequencies so that it will be difficult to derive a conventional frequency response function (FRF) under these circumstances, where the response is measured on the structure and the excitation is measured fixed in space. Clearly, both response and excitation must be measured in the same frame of reference for such response functions to be obtained. Later, we shall develop a generalisation of this feature (see Section 2.8.4.3).

2.8.2.6 Effects of damping in both stationary and rotating components

In the preceding sections, we referred to two types of damping — 'internal' and 'external' — but have only included the latter type in our discussion so far. In the more general case where the rotor and the stator are both flexible components, it is clearly possible for either or both to possess damping as well as elasticity and inertia. We can see the differing effects of internal and external damping by reverting to our simple 2DOF example, and to the case of the equations of motion for the symmetric stator case, described in terms of coordinates referred to the rotating frame of reference, equation (2.118). If we take those equations and introduce the extra terms which are generated by internal damping elements, c_I, assumed to be identical in the x and y directions and which are effective in the rotating frame of reference (i.e. rotating with the rotor), we obtain the following equation:

$$
\begin{bmatrix} \dfrac{I_0}{L^2} & 0 \\[2mm] 0 & \dfrac{I_0}{L^2} \end{bmatrix} \begin{Bmatrix} \ddot{x}_R \\ \ddot{y}_R \end{Bmatrix} + \begin{bmatrix} c_I & -\dfrac{2\Omega I_0}{L^2} + \dfrac{J\Omega}{L^2} \\[2mm] \dfrac{2\Omega I_0}{L^2} - \dfrac{J\Omega}{L^2} & c_I \end{bmatrix} \begin{Bmatrix} \dot{x}_R \\ \dot{y}_R \end{Bmatrix}
$$

$$
+ \begin{bmatrix} -\dfrac{I_0\Omega^2}{L^2} + \dfrac{J\Omega^2}{L^2} + k & 0 \\[3mm] 0 & -\dfrac{I_0\Omega^2}{L^2} + \dfrac{J\Omega^2}{L^2} + k \end{bmatrix} \begin{Bmatrix} x_R \\ y_R \end{Bmatrix} = \{0\} \qquad (2.121a)
$$

We may now transform this equation of motion back to the coordinates referred to the stationary frame of reference using the transformation in (2.116) (see Appendix 4 for details of this process) to obtain:

$$
\begin{bmatrix} (I_0/L^2) & 0 \\ 0 & (I_0/L^2) \end{bmatrix} \begin{Bmatrix} \ddot{x} \\ \ddot{y} \end{Bmatrix} + \begin{bmatrix} c_I & J\Omega_z/L^2 \\ -J\Omega_z/L^2 & c_I \end{bmatrix} \begin{Bmatrix} \dot{x} \\ \dot{y} \end{Bmatrix}
$$

$$
+ \begin{bmatrix} k & \Omega c_I \\ -\Omega c_I & k \end{bmatrix} \begin{Bmatrix} x \\ y \end{Bmatrix} = \begin{Bmatrix} 0 \\ 0 \end{Bmatrix} \qquad (2.121b)
$$

and if we then add some symmetrical external damping as well, c_E, we construct the most general case for this simple 2DOF system:

$$
\begin{bmatrix} (I_0/L^2) & 0 \\ 0 & (I_0/L^2) \end{bmatrix} \begin{Bmatrix} \ddot{x} \\ \ddot{y} \end{Bmatrix} + \begin{bmatrix} c_E + c_I & J\Omega_z/L^2 \\ -J\Omega_z/L^2 & c_E + c_I \end{bmatrix} \begin{Bmatrix} \dot{x} \\ \dot{y} \end{Bmatrix}
$$

$$
+ \begin{bmatrix} k & \Omega_z c_I \\ -\Omega_z c_I & k \end{bmatrix} \begin{Bmatrix} x \\ y \end{Bmatrix} = \begin{Bmatrix} 0 \\ 0 \end{Bmatrix} \qquad (2.121c)
$$

It can be seen in this equation that the internal and external damping features have quite different effects, most notably by the presence of c_I in the stiffness matrix, and in a skew-symmetric format. In the next Section 2.8.3.2, we shall explore numerically the consequences of the loss of matrix symmetry brought about by these two principle effects — gyroscopic forces in the velocity-dependent matrix and internal damping in the displacement-dependent matrix, but it can be noted here that the former causes the eigenvectors to become complex (as we have already seen) and the latter causes the eigenvalues to become complex, sometimes with a negative real part (stable system) but sometimes with a positive real part, indicating an unstable system. The deciding factor in determining the stability/instability condition relates directly to the relative values of c_I, c_E, ω_r and Ω_z, with the general feature that instability is approached and encountered as Ω_z increases to a critical value. This shows the theoretical origins of the well-known practical phenomenon that internal damping can cause instability in rotors running at supercritical speeds (Ω_z greater than ω_r).

2.8.3   Dynamic Analysis of General Rotor-Stator Systems
2.8.3.1   Equations of motion for general rotor-stator systems
By now it is becoming evident that structures in rotating machines, and their modal testing, may be much more complex than the relatively simple cases described in the earlier sections of the book and it is necessary to extend our analysis far beyond that stage in order to deal with these cases. The complications which can arise include:

*   non-symmetric bearing supports;
*   observations in fixed- or rotating-coordinate systems;
*   non-axisymmetric rotors;
*   internal (rotating) and external (non-rotating) damping.

These, in turn, can give rise to a number of effects not found in the simplest non-rotating structure, including:

*   time-varying modal properties;
*   response components at frequencies not present in the excitation signal;
*   instabilities (negative modal damping).

All of these need to be considered if modal analysis and testing is to be applied to practical structures which contain rotating elements.

We have already seen in the preceding paragraphs how the equations of motion for systems which include a rotating component are prone (a) to lose the symmetry in the describing matrices (that we have come to expect in conventional stationary structures), (b) to generate complex eigenvalues and/or eigenvectors from nonsymmetry in the displacement- and/or velocity-related system matrices, respectively, and (c) to include some coefficients in these matrices which are time-varying (instead of the constant coefficients that we have encountered elsewhere). All of these features result in equations of motion which are more complex in their solution and properties than we have described in the earlier sections. It is appropriate here to seek to define the general forms of the equation of motion to be encountered in systems with rotating components, and of the solutions which will follow from these different types of equation.

We have seen that, when the most immediate feature which is unique to rotating systems — the gyroscopic effect — is included in the analysis, at least one of the matrices which describe the equations of motion incurs a skew-symmetric component. We have also seen that a similar result is found when including internal damping in the analysis: then, the stiffness matrix incurs a skew-symmetric component. Thus, it is an inherent feature of even the simplest systems with rotating components that there will be a violation of the previously-standard 'rule' that the system matrices should be symmetric and so we must expect in all cases of dynamic analysis of rotating structures that at least one, if not more, of the system matrices will be non-symmetric.

A second complication was encountered in Section 2.8.2.5 where it was found that, when the equations of motion of the simple 2DOF system are expressed in coordinates which refer to the rotating frame of reference, some of the coefficients in the system equations are time-varying. Such a feature presents particular difficulties for a modal analysis as the conventional eigensolution procedure is based on the premise that the matrices to be thus decomposed are populated by constant coefficients — in other words that they are 'linear, time-invariant' or LTI equations or matrices. These latest matrices represent 'linear, time-dependent' or L(t) equations and their eigensolution is non-standard. Indeed, the analysis of such equations is beyond the scope of this work and the interested reader is referred to one of a small number of specialised texts which do address this case [24]. Suffice it now to observe here that the type of time-dependent system matrices just encountered are those in which some of the coefficients are periodic, with the rotation speed of the rotor, Ω, being the basis of the periodicity. The result of analysing such matrices is to establish a series of time-dependent (but specifically, periodic) eigenvalues and eigenvevctors. However, even when obtained, these characteristic properties fall outside the scope of what can be described by conventional structural analysis and so such systems will not be amenable to treatment using the standard tools of modal analysis or modal testing.

It is useful to identify the circumstances in which these two types of equation might be encountered and this is summarised in the accompanying table, which shows that the only case in which LTI equations cannot be produced is for the extreme case of a non-symmetric rotor supported in a non-symmetric stator. The other three combinations can all be described by LTI equations, by adopting a suitable choice of coordinate system.

System Type	Stationary Coordinates	Rotating Coordinates
R-symm; S-symm	LTI	LTI
R-symm; S-nonsymm	LTI	L(t)
R-nonsymm; S-symm	L(t)	LTI
R-nonsymm; S-nonsymm	L(t)	L(t)

If we confine our interest now to the LTI type of equations, we can see that the equations of motion for the more general type of system might be expressed in matrix form as follows:

$$[M]\{\ddot{x}\} + [[C] + [G(\Omega)]]\{\dot{x}\} + [[K] + i[D] + [E(\Omega)]]\{x\} = \{f(t)\} \qquad (2.122)$$

In this equation, the two speed-dependent matrices, $[G]$ and $[E]$ are both skew-symmetric, while all other matrices are symmetric. Solution of the equations will follow different routes depending upon the specific features in each case but in the classical example of viscous damping only, we shall need to convert these $N \times N$ second-order differential equations into $2N \times 2N$ first-order equations, as already seen in Section 2.7, by creating two matrices, $[A]$ and $[B]$:

$$[A] = \begin{bmatrix} [[C] + [G(\Omega)]] & [M] \\ -[M] & [0] \end{bmatrix}$$

$$[B] = \begin{bmatrix} [[K] + [E(\Omega)]] & [0] \\ [0] & [M] \end{bmatrix} \qquad (2.123a)$$

$$\{u\} = \begin{Bmatrix} \{x\} \\ \{\dot{x}\} \end{Bmatrix}$$

which combine to provide the equation of motion:

$$[A]\{\dot{u}\} + [B]\{u\} = \{0\} \qquad (2.123b)$$

In this equation, one or both of the system matrices is non-symmetric and, as a result, the eigensolution will yield a single eigenvalue matrix as usual, $[s_r]$, and two sets of eigenvectors, $[\Theta_{LH}]$ and $[\Theta_{RH}]$, both of which will be complex. This is the general case eigensolution for a system with constant coefficients. In this general case, each complex eigenvalue comprises the oscillation and decay rates (frequency and damping) for one mode of vibration and the two corresponding eigenvectors (RH and LH) will describe the mode shape and a normal excitation 'shape', respectively. In the particular case where there is no damping, the only non-symmetric matrix is [A] and that is skew-symmetric with the result that the left-hand (LH) vectors are the complex conjugates of the right-hand (RH) vectors. However, this is not generally the case.

A numerical example for this type of system (symmetric rotor, but non-symmetric support) is shown in Fig. 2.24 which displays the natural frequencies of the two modes as a function of speed of rotation. It is clearly seen that there are two modes, as before, and that one represents forward whirl — the higher natural frequency — while the other represent backwards whirl. In these cases, the mode shapes reveal that the orbit of the whirl is not circular, but elliptical.

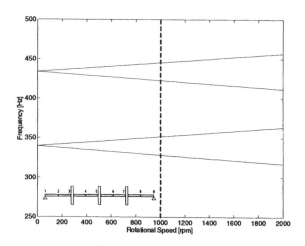

Fig. 2.24   Natural frequencies of MDOF rotor/stator system

Further study of the most general LTI system reveals that the FRF properties of such a system may be derived from the modal properties and that the receptance FRF matrix, $[\alpha(\omega)]$, can be expressed as follows:

$$[\alpha(\omega)] = [\Theta_{RH}][(s_r - i\omega)]^{-1}[\Theta_{LH}]^{H} \tag{2.124}$$

from which it can be seen that the property of reciprocity which we have observed in earlier cases no longer necessarily applies. Plots of some of the FRF properties for the same case study as used for Fig. 2.24 are included in Fig. 2.25.

104

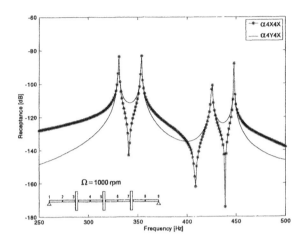

Fig. 2.25 FRF properties of MDOF rotor/stator system

2.8.3.2 Eigen-properties of general LTI system matrices

The previous subsections have revealed some trends that are to be expected in the matrices which are generated for the equations of motion of system with rotating components and it is appropriate here to illustrate the properties of such matrices in a more general form than has been shown previously.

The essential features of interest are those which result in a loss of symmetry of the velocity-dependent or displacement-dependent matrices (referred to loosely as the 'damping' and 'stiffness' matrices, respectively). It may be readily seen that any matrix can be expressed as the linear combination of a symmetric matrix and a skew-symmetric matrix and it is convenient to think of these dynamic system matrices in this way because most of the reasons for a loss of symmetry are, in fact, the introduction of a degree of skew-symmetry — the gyroscopic forces in the first case and the internal damping effects in the second. Thus it is useful to examine first the effect of a skew-symmetric damping matrix and then of a skew-symmetric stiffness matrix. These two effects will be described and illustrated by representative numerical examples based on the 2DOF system already studied.

(a) Skew-symmetry in the damping matrix

As we have already seen, the effect of introducing a skew-symmetric damping matrix to otherwise symmetric system matrices is to generate complex eigenvectors (mode shapes) but to retain real eigenvalues (i.e. undamped natural frequencies with no decay component). A set of numerical examples based on the 2DOF system are shown in the Table below. For the case where the three system matrices are described by:

$$[M] = \begin{bmatrix} 1 & 0 \\ 0 & 1 \end{bmatrix} \quad ; \quad [K] = \begin{bmatrix} 3 & -1 \\ -1 & 3 \end{bmatrix}$$

$$[C] = \Delta C \cdot \begin{bmatrix} 0 & 0.5 \\ -0.5 & 0 \end{bmatrix} + (1 - \Delta C) \cdot \begin{bmatrix} 1 & -0.5 \\ -0.5 & 1 \end{bmatrix}$$

we obtain the following results for the second mode as ΔC (the extent of skew symmetry) is varied from 0 to 1:

ΔC	s_2	X_1	X_2
0.0	$-0.75 + 1.85\,i$	1	-1.00
0.1	$-0.68 + 1.88\,i$	1	$-1.05 + 0.08\,i$
0.3	$-0.52 + 1.94\,i$	1	$-1.08 + 0.28\,i$
0.5	$-0.37 + 1.99\,i$	1	$-1.03 + 0.49\,i$
0.7	$-0.23 + 2.04\,i$	1	$-0.90 + 0.63\,i$
0.9	$-0.07 + 2.08\,i$	1	$-0.76 + 0.71\,i$
1.0	$2.11\,i$	1	$-0.69 + 0.73\,i$

(b) Skew-symmetry in the stiffness matrix
We can conduct a similar exercise for the second example, in which the stiffness matrix is found to have a skew-symmetric element due to the effect of internal damping. Thus, for the same basic 2DOF system, we can introduce the following variable elements in the stiffness matrix:

$$M = \begin{bmatrix} 1 & 0 \\ 0 & 1 \end{bmatrix} \quad ; \quad C = [0] \quad ; \quad K = \Delta K \cdot \begin{bmatrix} 0 & 1 \\ -1 & 0 \end{bmatrix} + (1 - \Delta K) \cdot \begin{bmatrix} 3 & -1 \\ -1 & 3 \end{bmatrix}$$

which yields the results shown in the following table, again focusing on the second mode of vibration. In many ways, these results are more interesting than those from the previous set because they show a transition from an undamped system to an unstable damped one. In this example, the extent of the skew-symmetry could be adjusted by varying the speed of rotation of the rotor since this parameter appears directly in the off-diagonal terms that relate to the internal damping.

ΔK	s_2	X_1	X_2
0.0	$2.00\,i$	1	-1.00
0.1	$1.90\,i$	1	-1.12
0.3	$1.65\,i$	1	-1.58
0.5	$1.23\,i$	1	Infinity
0.7	$0.32 + 1.00\,i$	1	$1.58\,i$
0.9	$0.57 + 0.79\,i$	1	$1.12\,i$
1.0	$0.70 + 0.70\,i$	1	i

Thus we arrive at the conclusion that as the speed of rotation increases, the system becomes less and less stable until there comes a point when the boundary of stability is crossed so that both the eigenvalues and eigenvectors are complex and the system is unstable. In real structures, this boundary of stability will be affected by all the sources of damping which are present (only the effect of internal damping on the stiffness matrix has been included here) but this analysis demonstrates in a modal analysis context and, using a simple example, the well-known phenomenon that 'internal damping has a destabilising effect on a rotor that is running at a super-critical speed'.

Before leaving this topic, it is worth noting that there is another common phenomenon which comes into the same category in that it gives rise to a non-symmetric stiffness matrix, and that is friction-induced rub between rotating and a stationary components. When forces created by rubs are included in the equations of motion, they will frequently have the effect of contributing a non-symmetric component to the displacement-dependent (stiffness) matrix. Such a feature will have the same possible consequences as those we have seen above: in some cases, depending on the specific numerical values of all the relevant parameters, this effect can give rise to an unstable mode of vibration. Although it seems instinctively to be unlikely, friction forces developed by a rotor rubbing on the stator can cause instability, and this property has been witnessed in countless situations, often to the distress of the relevant components.

2.8.4 Dynamic Analysis of Rotating Flexible Disc-like Structures

2.8.4.1 Classification and modal properties of flexible disc-like structures

It is found that many of the most critical (i.e. vibration-prone) components in rotating machines fall into a class of structure which is

generally described as 'quasi-axisymmetric': components which are
'disc-like' (wheels, gears, discs, impellers,...). The various types of
structure which concern us here are:

(i)   axisymmetric;
(ii)   cyclically-periodic, or
(iii)   slightly asymmetric, or aperiodic (which means slightly imperfect
        structures of the first and second types), referred to as 'quasi-
        periodic'.

We shall not describe the vibration properties of these structures in
detail here, but simply note that the essential modal properties are as
follows:

*   **axisymmetric** structures (plain discs, wheels, etc.) have modes
    whose shapes are described in the circumferential direction, θ , by
    variations of the form:

    $$\phi(\theta) = \cos(n\theta + \alpha_n)$$

    and are therefore known as 'n-nodal diameter' or nND modes;

*   **cyclically-periodic** structures (such as bladed discs, impellers,
    gear wheels) have modes whose shapes are described in the
    circumferential direction, θ , by expressions of the form:

    $$\phi(\theta) = \{A_n \cos(n\theta + \alpha_n) + A_{N-n} \cos((N-n)\theta + \alpha_{N-n})$$
    $$+ A_{N+n} \cos((N+n)\theta + \alpha_{N+n}) + ...\}$$

    where N is the number of blades, vanes, teeth etc. These are also
    referred to as 'n-nodal diameter modes' although this description
    is less precise in this case than for the preceding one because the
    mode shape clearly includes more components than just the cos
    $n\theta$ one. It is possible to find modes in which the largest component
    is cos $n\theta$, and such a mode is well described as one with nND,
    but, equally, modes exist in which the most significant component
    in the mode shape is the second one $((N-n)\theta)$ or an even higher
    one, and in these cases, the nND description is not the most
    appropriate. (It should be noted that if such mode shapes are
    defined by determining the amplitude ratios only at the N
    discrete points around the rim which carry the blades/vanes/etc.,
    then the discrete Fourier description which results will be
    incapable of discriminating above the first term in this series —

$(N-n)$ would be indistinguishable from n — hence the classification simply as 'n-nodal diameter' modes.);

• **quasi-periodic** structures, in which the loss of axisymmetry is small, usually due to manufacturing tolerances, have mode shapes which are described (in the circumferential direction) by:

$$\phi(\theta) = \sum A_r \cos(r\theta + \alpha_r)$$

where $r = 1, 2, 3, ...$, but is generally — although not always — dominated by n, $N-n$, $N+n$, etc. When these modes are so dominated by a single, or few, terms of this type, the nodal diameter label is still used, although it is rather less precise than for the other cases. However, it should be noted that there are situations where small deviations from true axisymmetry, or cyclic symmetry, can lead to mode shapes which contain many significant diametral components and which cannot then be realistically described at all by the 'nND' label.

A second feature of the modes of these disc-like structures concerns their natural frequencies. In the case of type (i) and (ii) structures, most of the modes exist in pairs of 'double' modes: two modes with identical natural frequencies and mode shapes which differ only in the angular orientation of the nodal lines, i.e. in α_n. As is the case generally, when there are two or more modes with identical natural frequencies, any combination of the individual two mode shapes is also a mode shape. This can lead to some unexpected features in the case of these axisymmetric structures where, for example, a valid mode shape can be produced by a combination of $1.0 \, x \cos(n\theta + \alpha_n)$ plus $i \, x \sin(n\theta + \alpha_n)$, the result of which is a $\cos n\theta$ mode shape rotating around the structure in a travelling wave motion. Type (iii) structures also possess these double modes but in this case the natural frequencies of each pair of modes are slightly separated, or 'split', resulting in two distinct modes (i.e. not repeated roots, in the mathematical sense) but which may be very difficult to distinguish from each other in observed response characteristics because of the inevitable 'coupling' effects of the structure's damping properties.

One consequence of these features is that the structures which display such modal properties can be much more difficult to test than are structures with single modes. As a result, measured modal properties of axisymmetric, or quasi-axisymmetric, structures are often in error, sometimes through ignorance on the part of the analyst and sometimes because of the inherent difficulties in making the measurements.

2.8.4.2 Response properties of stationary disc-like structures

As intimated above, the response properties of the family of structures included in this 'disc' category can be very complicated as a direct result of the special 'symmetry' they possess, even when they are stationary. (Note that this 'symmetry' is unconnected with the symmetry of the system matrices which has been discussed at length earlier in this section.) While it is not appropriate to provide a full exposition of these vibration properties here, it **is** useful to illustrate the basic phenomena using a simple example. The simplest illustration of the effects of interest can be provided by considering a plain uniform disc which is excited at a single point on its rim with an harmonic force tuned to the exact natural frequency of the fundamental mode with two nodal diameters. The resulting response can be summarised by the sketch in Fig. 2.26(a) which shows the two nodal diameter lines symmetrically disposed about the excitation point. If we were to relocate the point of application of the excitation force around the disc rim by 45 degrees, we might then expect to obtain no resonant response, because excitation at a nodal point has that particular result. However, what we find in practice is that the nodal lines 'move' around to follow the excitation point so that the response pattern obtained in the second case is as shown in Fig. 2.26(b). This result, which is both intuitive and correct, can only be explained if there is more than one mode with the two nodal diameter shape: indeed — two, each with the same natural frequency but with mode shapes that are orthogonal to each other — one as sin 2θ and the other as cos 2θ — and this is exactly what happens.

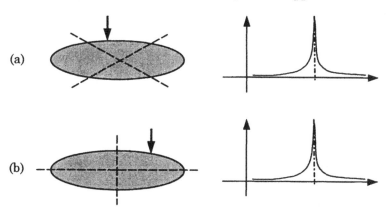

Fig. 2.26 Forced response of disc in 2ND resonance

Each double mode of a 'disc' structure can be described as two modes, with identical or very close (depending upon the exact degree of symmetry which obtains) natural frequencies and mode shapes which are essentially of the form:

$$\phi_1(\theta) = \sin n\theta \quad \text{and} \quad \phi_2(\theta) = \cos n\theta$$

(in which we have dispensed with the arbitrary offset, α_n, for simplicity). We shall now suppose that there is a point harmonic excitation $F_0 \cos(\omega t)$ applied at circumferential location $\theta = \theta_j$, and note that this excitation condition can also be described more completely as:

$$f(\theta,t) = \left(\frac{F_0}{\pi}\right) \sum_{n=0}^{\infty} \cos\left(n(\theta - \theta_j)\right) \cos(\omega t) \tag{2.125a}$$

It can be seen that such an excitation can also be described as:

$$f(\theta,t) = \left(\frac{F_0}{\pi}\right) \left\{ \sum_{n=0}^{\infty} a_j(\cos n\theta) + b_j(\sin n\theta) \right\} \cos(\omega t) \tag{2.125b}$$

$$a_j = \cos n\theta_j \quad ; \quad b_j = \sin n\theta_j$$

and that the first term will excite the first of the pair of nND modes while the second term will excite the second one, so that the response will be of the form:

$$x(\theta,t) = \left\{ A_j(\cos n\theta) + B_j(\sin n\theta) \right\} \cos(\omega t)$$

where A_j, B_j depend on the proximity of the excitation frequency, ω, to the natural frequency of each mode, and the damping, but which will be proportional to a_j, b_j when the two modes have identical natural frequencies, in which case:

$$x(\theta,t) = \left\{ C\left(\cos n\left(\theta - \theta_j\right)\right) \right\} \cos(\omega t)$$

This brief analysis serves to explain the above example of the 2ND modes on the simple disc by showing that in the case of a perfectly tuned (i.e. axisymmetric) disc the nodal lines will align themselves to be symmetric with respect to the excitation, no matter where that is applied.

Implicit in the above analysis is the fact that a mode with a $\cos n\theta$ mode shape will only be excited by an excitation force pattern which has a corresponding spatial distribution in it: i.e. the excitation force must contain a component of $F_n \cos n\theta$ in it. Clearly, a single point excitation force can be seen to contain components of all orders, from $n = 0$ to infinity. From this result there follows the property that response in certain modes can be suppressed by arranging for an excitation pattern that eliminates certain orders from its circumferential decomposition. Thus, if we apply two single-point excitation forces which are spaced 90 degrees apart around the disc rim, then we shall find that this excitation has a zero component of 2θ excitation and, as a result, modes with 2ND will not be excited.

The above comments relate primarily to the reference case where the structure is tuned, or perfectly axi- or cyclically symmetric. This is not always the case, and it is necessary to extend the foregoing analysis to the more general case where the symmetry is imperfect, in which case the two modes of each pair (or double mode) have slightly different natural frequencies. In this condition, the two modes will resonate at different frequencies and, strictly, the nodal lines will be fixed in the disc and will not move around to follow the excitation. Instead, two resonances will be observed, one at the first natural frequency with nodal lines in their fixed orientation, independent of the location of the excitation, and a second resonance at the second natural frequency with its nodal lines likewise fixed in the disc. The one caveat to this relatively straightforward description concerns the transition from the former, tuned case, to the latter, mistuned situation. How much natural frequency separation or 'split' is necessary before we can observe the above two-resonance phenomenon? The answer depends upon the level of damping which prevails in the structure because if the frequency separation or split between the two modes is very small, and the damping is moderate, then the two resonances will not be individually discernable and the structure will still appear to be tuned. As a rough guide, we can expect that if the natural frequency split of the pair of modes (as a percentage of the mean) is greater than the prevailing modal damping (as a percentage of critical) then we should be able to see the two modes individually but, if not, then the structure will effectively be tuned.

2.8.4.3 Excitation and response of rotating disc-like structures

We shall now extend the above discussion of excitation forces applied to the disc-like structures of interest in many rotating machines to the case where these structures are rotating. We shall start by considering again a single point harmonic force, $F_0 \cos(\omega t)$, which is applied at a fixed point in space, $\theta_S = 0$ and which can be defined by:

$$f(\theta_S,t) = \left(\frac{F_0}{\pi}\right)\sum_{n=0}^{\infty}\cos(n\theta_S)\cos(\omega t) \qquad (2.126a)$$

We next consider the effect of this force pattern as 'seen' in the rotating axes frame by setting $\theta_R = \theta_S + \Omega t$ (where Ω is the speed of rotation), so that we obtain:

$$f(\theta_R,t) = \left(\frac{F_0}{2\pi}\right)\sum_{n=1}^{\infty}\left(\begin{array}{c}\cos(\omega+n\Omega)t\cos(n\theta_R)+\sin(\omega+n\Omega)t\sin(n\theta_R)\\+\cos(\omega-n\Omega)t\cos(n\theta_R)-\sin(\omega-n\Omega)t\sin(n\theta_R)\end{array}\right) \qquad (2.126b)$$

This expression shows that the single-harmonic fixed excitation, $F_0 e^{i\omega t}$, will generate a series of excitations applied to the rotating component with the particular feature that all modes (in the rotating disc) which have a $\cos(n\theta_R)$ or $\sin(n\theta_R)$ form (i.e. modes with n nodal diameters) will be excited at two frequencies, $\omega_{1,2} = \omega \pm n\Omega$. Thus, a single harmonic excitation fixed in space will generate vibrations in the rotating components at a large number of different frequencies, two for every n, although the strength of each of the different response components will vary according to the proximity of its frequency to a natural frequency of one of the corresponding modes of the disc. This result means that although it is possible to understand and to explain the reason for these complicated excitation/response characteristics, it will nevertheless be very difficult to derive FRF data in the usual format if excitation is to be applied and measured in the stationary axes set (for example) and the response is to be measured in the rotating axes set.

A special case of widespread interest can be noted here: when the excitation force is a static force (i.e. when $\omega = 0$), there will still be an effective dynamic excitation experienced by the rotating disc and this will be experienced at a frequency of $n\Omega$ for a mode with nND.

It can now be noted that the previously-reported example of this effect with the rigid rotor (in equation (2.120)) constitutes a special case of this more general analysis. Such a rotor only has the possibility of vibrating in modes for which $n = 0$ (axial and torsional motion) or $n = 1$ (lateral motion) and in the example quoted only the second of these two groups was active, hence the two response frequencies of $\omega \pm \Omega$, as described in the earlier sections.

2.9 COMPLEX MODES

2.9.1 Real and Complex Modes, Stationary and Travelling Waves

Earlier in this chapter we have encountered not only complex eigenvalues — whose real and imaginary parts can be interpreted as representing both decay and oscillatory components in the natural frequencies — but also complex eigenvectors. The significance of these complex eigenvectors is that the mode shapes are complex and we need to understand what this means in practice.

In effect, a complex mode is one in which each part of the structure has not only its own amplitude of vibration but also its own phase. As a result, each part of a structure which is vibrating in a complex mode will reach its own maximum deflection at a different instant in the vibration cycle to that of its neighbours which all have different phases. A real mode is one in which the phase angles are all identically 0° or 180° and which therefore has the usual property that all parts of the structure do reach their own maxima all at the same instant in the vibration cycle. Equally, in a real mode, all parts of the structure pass through their zero deflection position at the same instant so that there are two moments in each vibration cycle when the structure is completely undeformed. This is not a property of a complex mode because, by the same token that results in the maxima being attained at different times, the zero positions are reached at different times also. Thus, while the real mode has the appearance of a standing wave, the complex mode is better described as exhibiting travelling waves. An attempt to illustrate these two types of mode pictorially is offered in Fig. 2.27 where a succession of frames are overplotted. The standing and travelling wave effects can be seen here although the best demonstrations are to be obtained by using the animation facilities on a computer.

Another method of displaying modal complexity is by plotting the elements of the eigenvector on an Argand diagram, such as the ones shown in Fig. 2.28, which include examples of both highly-complex (a) and almost-real (b) mode shapes. (Note that the almost-real mode shape does not necessarily have vector elements with near-0° or near-180°: what matters are the relative phases between the different elements.)

2.9.2 Measurement of Modal Complexity

In the next section we shall be considering the provenance of modal complexity and, in the course of that discussion, the question of the degree of complexity will arise. It is thus necessary to have a means of measuring modal complexity and there are one or two parameters which have been proposed for this task, although none of these has yet been established as the universally-accepted indicator. Two will be

114

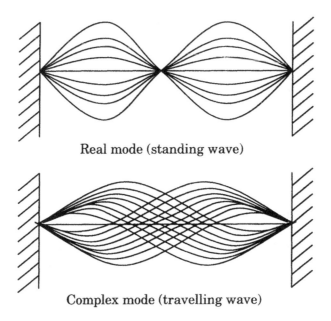

Real mode (standing wave)

Complex mode (travelling wave)

Fig. 2.27   Real and complex mode shapes displays

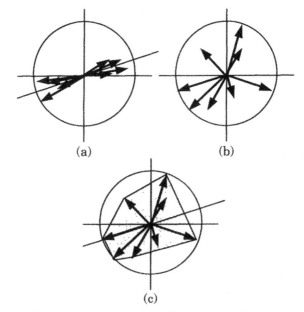

(a)   (b)

(c)

Fig. 2.28   Complex mode shapes plotted on Argand diagrams.
(a) Almost-real mode; (b) Complex mode; (c) Measure of
complexity

presented here: a simple and crude one (MCF1) which simply measures the phase differences between all pairs of mode shape vector elements, regardless of the magnitude of those elements, and a more subtle (and realistic) one (MCF2) which reflects the magnitude as well as the phase of each of the elements.

The first of these indicators, MCF1, is computed simply by summing all the phase differences between every combination of two eigenvector elements. Thus:

$$MCF1 = \sum_{j=1}^{N} \sum_{k=1; \neq j}^{N} \left(\theta_{rj} - \theta_{rk} \right)$$

The second measure can be explained with reference to the complex-plane plots shown in Fig. 2.28. If a polygon is drawn around the extremities of the individual vectors, as shown in Fig. 2.28(c), but without permitting re-entrant parts, then it encloses an area which can be compared with the area of the circle which is based on the length of the largest vector element. The resulting ratio of these two areas is used as an indication of the complexity of the mode in question, and is defined as MCF2.

Following the discussion concerning the origins of complex modes, a series of case studies will be summarised in Section 2.9.4, illustrating the essential features of complex modes and reinforcing the explanations of their origins in conventional structures.

2.9.3   Origins of Complex Modes

Complex modes occur in practice for a variety of reasons and it is important to know what conditions are necessary for them to exist so that it can be established whether ones obtained in a modal test are genuine or the result of poor measurement or analysis (a situation which can happen rather too easily).

The types of modes which are referred to as 'operating deflection shapes' (and which are not normal modes, in any form) will frequently exhibit the relative phases differences between responses of adjacent parts of the structure which indicate a complex mode. This is to be expected and is quite normal behaviour.

We have seen in the preceding section that complex normal modes can exist in even the simplest of structures which contain rotating components that are prone to gyroscopic forces. Thus, for this special class of structure, complex modes are the norm and are to be expected in most cases.

However, normal modes of conventional (i.e. non-rotating) linear structures can be complex only if the damping is distributed in a non-proportional way. This situation can arise quite readily in practice because while the internal (hysteresis) damping of most structural elements is distributed in a way which is essentially proportional to the stiffness distribution, the majority of the damping in real structures is generally found to be concentrated at the joints between components of a structural assembly and this does not usually result in a proportional distribution. Thus, in most structures, this basic ingredient of non-proportionality for complex modes is likely to exist. However, it is found that while non-proportionality is a necessary condition for complex modes to exist, it is not sufficient, at least not if the degree of complexity is to be other than trivial. Another ingredient is found to be necessary to generate significant complexity in a structure's modes and that is the requirement that two or more of its modes are 'close'. 'Close' modes are those whose natural frequencies are separated by an amount which is less than the prevailing damping in either or both modes. (In simple terms, two modes with natural frequencies of 105 Hz and 110 Hz are 'close' if the modal dampings are of the order of 5 per cent or more, but are not 'close' if their damping factors are typically 1 per cent or less.) Some specific case studies will be shown at the end of this section.

One further example in which complex modes may be encountered is worth mentioning: that concerns structures with repeated roots, or two or more modes with identical natural frequencies. As previously mentioned, such circumstances are not rare in practical structures and so the special features which they can exhibit are certainly worth including. In the case of a simple undamped structure which has a double symmetry, such as the circular bar discussed in Section 2.4.3, there are two modes with identical natural frequencies and thus mode shapes with a degree of arbitrariness in their form. In fact, possible mode shapes for this structure include any linear combination of the two reference mode shapes that would be computed from a theoretical modal analysis, or measured in a modal test, which would normally be bending in the vertical plane and bending in the horizontal plane. Any combination of these two mode shapes includes: bending in any plane, when the two modes' contributions are in phase, or an orbiting motion, when the components of the two modes are out of phase by 90°. This latter case is clearly a complex mode, and can be exhibited by such a symmetrical structure, even in the absence of damping and the gyroscopic effects of rotation.

2.9.4   Case Studies of Complex Modes

In order to illustrate the effects of close natural frequencies and damping levels on modal complexity, a series of case studies based on a 3DOF system are shown below. The system is a variant of that already used in Section 2.6, and simply involves a systematic changing of the values of the three individual masses, m_1, m_2 and m_3, in such a way that the two of the three natural frequencies get closer and closer, eventually becoming identical. For each case in this series, the modal complexity is recorded, as is the modal damping for each of the modes.

The results are presented in Fig. 2.29, which shows both these parameters as a function of the separation between the natural frequencies of modes 2 and 3, $(\omega_3 - \omega_2)$. From this example, it is seen how the level of complexity of the two close modes is directly related to the proximity of their natural frequencies and the prevailing level of damping in the two modes. At the extreme case, where the two modes have identical natural frequencies, we find that one of them becomes entirely real, but this is a peculiarity of the particular model used here, and corresponds to one of the modes becoming undamped as a result of the geometric symmetry of the system and the ineffectiveness of the single damper to provide damping to that mode as a result.

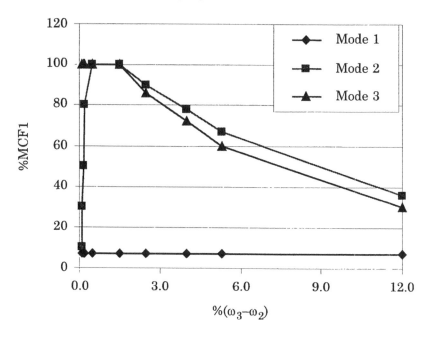

Fig. 2.29   Modal complexity for system with close modes

2.10   CHARACTERISTICS AND PRESENTATION OF MDOF FRF DATA

2.10.1   A Note About Natural Frequencies

Having now presented all the basic theory, it is appropriate to comment on the various definitions which have been introduced for 'natural' frequencies. The basic definition derives from the undamped system's eigenvalues which yield the frequencies at which free vibration of the system can take place. These undamped system natural frequencies are given by the square roots of the eigenvalues and are identified by the symbol ω_r, and this occurs in expressions for both free vibration response:

$$x(t) = \sum_{r=1}^{N} x_r e^{i\overline{\omega}_r t} \qquad (2.127\text{a})$$

and for forced vibration, the FRF:

$$\alpha(\omega) = \sum_{r=1}^{N} \frac{A_r}{\overline{\omega}_r^2 - \omega^2} \qquad (2.127\text{b})$$

For damped systems, the situation is more complicated and leads to two alternative characteristic frequency parameters being defined — both called 'natural' frequencies — one for free vibration (ω_r') and the other for forced vibration (ω_r). The former constitutes the oscillatory part of the free vibration characteristic which, being complex, contains an exponential decay term as well. Thus we have:

$$x(t) = \sum_{r=1}^{N} x_r e^{-a_r t} e^{i\omega_r' t} \qquad (2.127\text{c})$$

where ω_r' may or may not be identical to $\overline{\omega}_r$, depending on the type and distribution of the damping (see Table 2.3). The second definition comes from the general form of the FRF expression which, combining all the previous cases, may be written in the form:

$$\alpha(\omega) = \sum_{r=1}^{N} \frac{C_r}{\omega_r^2 - \omega^2 + iD_r} \qquad (2.127\text{d})$$

Here, C_r may be real or complex and D_r will be real, both may be constant or frequency-dependent and ω_r will, in general, be different to both $\overline{\omega}_r$ and ω_r', except in some special cases. Table 2.3 summarises all the different cases which have been included. In these cases, the precise relationship between the **eigenvalues** (the roots of the characteristic equation) and the **natural frequencies** is also more complex. In the cases of hysteretic damping, each complex eigenvalue yields an oscillatory frequency component in the square root of the Real Part and the decay part (the loss factor) from the Imaginary Part. In the case of viscous damping, each complex eigenvalue yields an oscillatory frequency component in its Imaginary Part and the damping or decay rate in the Real Part.

Table 2.3 FRF Formulae and Natural Frequencies

Case	Eqn. for FRF	C	D	Natural Frequency	
				Free ω_r'	Forced ω_r
UNDAMPED	2.41	REAL, CONST.	0	$\overline{\omega}_r$	$\overline{\omega}_r$
PROP. HYST.	2.56	REAL, CONST.	REAL, CONST.	$\overline{\omega}_r$	$\overline{\omega}_r$
PROP. VISC.	2.50	REAL, CONST.	REAL (ω)	$\omega_r\sqrt{1-\zeta_r^2}$	$\overline{\omega}_r$
GEN. HYST.	2.66	COMPLEX, CONST.	REAL, CONST.	ω_r	ω_r
GEN. VISC.	2.100	COMPLEX (ω)	REAL (ω)	$\omega_r\sqrt{1-\zeta_r^2}$	ω_r

2.10.2 Mobility and Impedance FRF Parameters

In every case, the most important feature of the general expression in (2.127d) is its close relationship with the FRF expression for the much simpler SDOF system, studied in detail in Section 2.2. We shall now consider the properties of this type of function and then examine the various means used to display the information it contains. It should be emphasised that a thorough understanding of the form of the different plots of FRF data is invaluable to an understanding of the modal analysis processes which are described in Chapter 4.

First, we consider the various forms of FRF. As before, there are three main alternatives, using displacement, velocity or acceleration response to produce respectively receptance, mobility or inertance (or acceleration). These three forms are interrelated in just the same way as described earlier, so that we may write:

$$[Y(\omega)] = i\omega[\alpha(\omega)]$$
$$[A(\omega)] = i\omega[Y(\omega)]$$
$$= -\omega^2[\alpha(\omega)]$$

(2.128)

However, the FRF data of multi-degree-of-freedom systems have a more complex form than their SDOF counterparts and this may be seen from the strict definition of the general receptance, α_{jk}, which is:

$$\alpha_{jk}(\omega) = \left(\frac{X_j}{F_k}\right) \quad ; \quad F_l = 0, l = 1, N; l \neq k \tag{2.129}$$

and it is the footnote qualification which is particularly important.

We saw in Sections 2.2 and 2.3 that there exist a further three formats for FRF data, these being the inverses of the standard receptance, mobility and inertance and generally known as 'dynamic stiffness', 'mechanical impedance' and 'apparent mass', respectively. Whereas with a SDOF system there is no difficulty in using either one or its inverse, the same cannot be said in the case of MDOF systems. It is true to say that for MDOF systems we can define a complete set of dynamic stiffness or impedance data (and indeed such data are used in some types of analysis), but it is not a simple matter to derive these inverse properties from the standard mobility type as the following explanation demonstrates.

In general, we can determine the response of a structure to an excitation using the equation:

$$\{\dot{X}\} = \{V\} = [Y(\omega)]\{F\} \tag{2.130a}$$

Equally, we can write the inverse equation using impedances instead of mobilities, as:

$$\{F\} = [Z(\omega)]\{V\} \tag{2.130b}$$

The problem arises because the general element in the mobility matrix, $Y_{ik}(\omega)$, is not simply related to its counterpart in the impedance matrix, $Z_{ik}(\omega)$, as was the case for a SDOF system. Stated simply:

$$Y_{jk}(\omega) \equiv Y_{kj}(\omega) \neq \frac{1}{Z_{jk}(\omega)} \tag{2.131}$$

and the reason for this unfortunate fact derives from the respective definitions, which are:

$$Y_{kj}(\omega) = \left(\frac{V_k}{F_j}\right)_{F_l=0} \quad \text{and} \quad Z_{jk}(\omega) = \left(\frac{F_j}{V_k}\right)_{V_l=0} \quad ; \quad l=1,N; l \neq k \tag{2.132}$$

It is clear from these expressions that while it is entirely feasible to measure the mobility type of FRF by applying just a single excitation force and ensuring that no others are generated, it is far less straightforward to measure an impedance property which demands that all DOFs except one are grounded. Such a condition is almost impossible to achieve in a practical situation.

Thus, we find that the only types of FRF which we can expect to measure directly are those of the mobility or receptance type. Further, it is necessary to guard against the temptation to derive impedance-type data by measuring mobility functions and then computing their reciprocals: these are **not** the same as the elements in the matrix inverse. We can also see from this discussion that if one changes the number of coordinates considered in a particular case (in practice we will probably only measure at a small fraction of the total number of degrees of freedom of a practical structure), then the mobility functions involved remain exactly the same but the impedances will vary.

Lastly, we should just note some definitions used to distinguish the various types of FRF.

A **point mobility** (or receptance, etc.) is one where the response DOF and the excitation coordinate are identical.

A **transfer mobility** is one where the response and excitation DOFs are different.

Sometimes, these are further subdivided into **direct** and **cross mobilities**, which describe whether the types of the DOFs for response and excitation are identical — for example, whether they are both x-direction translations (direct), or one is x-direction and the other is y-direction (cross), etc.

2.10.3 Display for Undamped System FRF Data
2.10.3.1 Construction of FRF plots for 2DOF system
As in the earlier section, it is helpful to examine the form which FRF data take when presented in various graphical formats. This knowledge can be invaluable in assessing the validity of and interpreting measured data.

We shall start with the simplest case of undamped systems, for which the receptance expression is given in equation (2.41). Using the type of log-log plot described in Section 2.3, we can plot the individual terms in the FRF series as separate curves, as shown in Fig. 2.30 (which is actually a plot of mobility). In this way, we can envisage the form which the total FRF curve will take as it is simply the summation of all the individual terms, or curves. However, the exact shape of the curve is not quite so simple to deduce as first appears because part of the information (the phase) is not shown. In fact, in some sections of each curve, the receptance is actually positive in sign and in others, it is

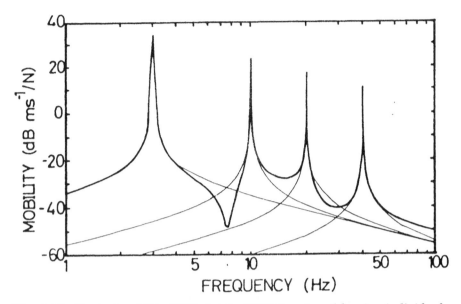

Fig. 2.30 Typical mobility FRF plot for MDOF system (showing individual modal contributions)

negative but there is no indication of this on the logarithmic plot which only shows the modulus. However, when the addition of the various components is made to determine the complete receptance expression, the signs of the various terms are obviously of considerable importance. We shall examine some of the important features using a simple example with just two modes — in fact, based on the system used in Section 2.3 — and we shall develop the FRF plots for two parameters, the point (receptance or mobility) FRF, α_{11}, and the transfer FRF, α_{21}. Fig. 2.30 is a mobility plot showing both the individual terms in the series and applies to both the above-mentioned FRFs. If we look at the expressions for the receptances we have:

$$\alpha_{11}(\omega) = 0.5/\left(\omega_1^2 - \omega^2\right) + 0.5/\left(\omega_2^2 - \omega^2\right)$$

and

$$\alpha_{21}(\omega) = 0.5/\left(\omega_1^2 - \omega^2\right) - 0.5/\left(\omega_2^2 - \omega^2\right) \tag{2.133}$$

from which it can be seen that the only difference between the point and the transfer receptances is in the sign of the modal constant (the numerator) of the second mode and as the plots only show the modulus, they are apparently insensitive to this difference. However, if we consider what happens when the two terms are added to produce the actual FRF for the MDOF system, we find the following characteristics.

Consider first the point mobility, Fig. 2.31(a). We see from (2.133) that at frequencies below the first natural frequency, both terms have the same sign and are thus additive, making the total FRF curve higher than either component, but we must note that the plot uses a logarithmic scale so that the contribution of the second mode at these low frequencies is relatively insignificant. Hence, the total FRF curve is only slightly above that for the first term. A similar argument and result apply at the high frequency end, above the second natural frequency, where the total plot is just above that for the second term alone. However, in the region between the two resonances, we have a situation where the two components have opposite signs to each other so that they are subtractive, rather than additive, and indeed at the point where they cross, their sum is zero since there they are of equal magnitude but opposite sign. On a logarithmic plot of this type, this produces the antiresonance characteristic which reflects that of the resonance. In the immediate vicinity of either resonance, the contribution of the term whose natural frequency is nearby is so much greater than the other one that the total is, in effect, the same as that one term. Physically, the response of the MDOF system just at one of its natural frequencies is dominated by that mode and the other modes have very little influence. (Remember that at this stage we are still concerned with undamped, or effectively undamped, systems.)

Now consider the transfer FRF plot, Fig. 2.31(b). We can apply similar reasoning as we progress along the frequency range with the sole difference that the signs of the two terms in this case are opposite. Thus, at very low frequencies and at very high frequencies, the total FRF curve lies just below that of the nearest individual component while in the region between the resonances, the two components now have the same sign and so we do not encounter the cancelling-out feature which gave rise to the antiresonance in the point mobility. In fact, at the frequency where the two terms intersect, the total curve has a magnitude of exactly twice that at the intersection.

2.10.3.2 FRF modulus plots for MDOF systems

The principles illustrated here may be extended to any number of degrees of freedom and there is a fundamental rule which has great value and this is that if two consecutive modes have the same sign for the modal constants, then there will be an antiresonance at some frequency between the natural frequencies of those two modes. If they have opposite signs, there will not be an antiresonance, but just a minimum. (The most important feature of the antiresonance is perhaps the fact that there is a phase' change associated with it, as well as a

124

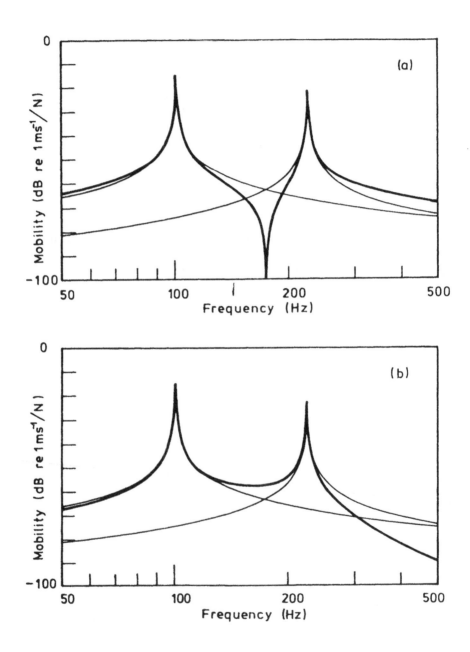

Fig. 2.31 Mobility FRF plot for undamped 2DOF system.
(a) Point FRF; (b) Transfer FRF

very low magnitude.)

It is also very interesting to determine what controls whether a particular FRF will have positive or negative modal constants, and thus whether it will exhibit antiresonances or not. Considerable insight may be gained by considering the origin of the modal constant: it is the product of two eigenvector elements, one at the response point and the other at the excitation point. Clearly, if we are considering a point mobility, then the modal constant for every mode must be positive, it being the square of a number. This means that for a point FRF, there must be an antiresonance following every resonance, without exception.

The situation for transfer FRFs is less categorical because clearly the modal constant will sometimes be positive and sometimes negative, depending upon whether the excitation and response move in phase with each other, or not. Thus, we expect transfer FRF measurements to show a mixture of antiresonances and minima. However, this mixture can be anticipated to some extent because it can be shown that, in general, the further apart are the two points in question, the more likely are the two eigenvector elements to alternate in sign as one progresses through the modes. Thus, we might expect a transfer FRF between two positions widely separated on the structure to exhibit fewer antiresonances than one for two points relatively close together. A clear example of this is given in Fig. 2.32 for a 6DOF system, showing a complete set of FRFs for excitation at one extremity.

Finally, it should be noted that if either the excitation or the response coordinates happen to coincide with a node for one of the modes (i.e. $(\phi_{jr}),(\phi_{kr}) = 0$), then that mode will not appear as a resonance on the FRF plot. This arises since, for such a case, we shall have $_r A_{jk} = 0$ and so the only response which will be encountered at or near $\omega = \overline{\omega}_r$ will be due to the off-resonant contribution of all the other modes.

2.10.3.3 FRF phase plots
Before we leave this topic, it is instructive to consider also the phase of the FRF data. All the plots referred to in this section have displayed only the modulus part of the response function (and this, indeed, is usually the most important part) but there is some value in examining the phase, also. If the phase plots for the six FRFs shown in Fig. 2.32 are presented as well, we can see certain features which, if not expected, are of some interest. Consider first the point FRF, α_{11}, whose phase is shown in Fig. 2.32(b): we see that this function remains within the range $-180° < \theta_\alpha < 0°$, switching from $0°$ to $180°$ on passing through each resonance, and then returning to $0°$ at each antiresonance. However, the plots for the other, transfer, FRFs are the more interesting because we see essentially the same properties (on passing

126

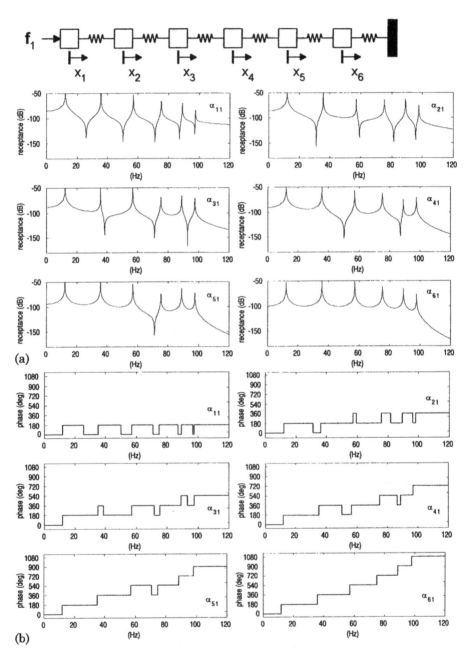

Fig. 2.32 Point and transfer mobility FRF data for 6DOF system
(a) Modulus plots; (b) Phase plots

through resonances and antiresonances) but as the incidence of antiresonances decreases as the separation between excitation and response points grows, we find a progressive increase in the phase angle of the transfer FRFs as the frequency gets higher. In the last case, for α_{61} — Fig. 2.32(b) — the phase angle above the last resonance is 1080°. However, the phase is often plotted within the restricted range of just 360°, whether that be between −180° and +180° or 0° and 360°, and so this unwrapping feature is often not visible, although it is there and has some practical significance.

2.10.3.4 Simple mode shape analysis from undamped FRF plots

There is a simple technique for deducing the essential form of each of the mode shapes of an MDOF system by visual inspection of an appropriate set of FRFs, such as those given in Fig. 2.32 for the 6DOF system shown there. The technique requires the labelling of the sign of each modal constant for each resonance on each curve, as shown in Fig. 2.32 for the first two resonances (the sequence can be continued by the reader to encompass all six modes on the plots). Then, the essential shape of each mode can be identified by noting the signs of the modal constants for one mode as follows: for mode 1, this sequence is:

$$\{+ + + + + +\}^T,$$

while for mode 2 it is

$$\{+ + - - - -\}^T.$$

This means that for mode 1, all the six masses will be moving in phase with each other, while for mode 2, the first and second masses will move in one direction while masses 3, 4, 5 and 6 will be moving in the opposite direction.

The simple explanation of this technique is based on the fact that the sequence of modal constants which make up the rth 'column' consist of the following quantities:

$$\{(\phi_{1r}\,\phi_{1r}), (\phi_{2r}\,\phi_{1r}), (\phi_{3r}\,\phi_{1r}), (\phi_{4r}\,\phi_{1r}), (\phi_{5r}\,\phi_{1r}), (\phi_{6r}\,\phi_{1r})\}^T$$

and these are simply a constant (ϕ_{1r}) multiplied by the rth mode shape vector:

$$\{(\phi_{1r}), (\phi_{2r}), (\phi_{3r}), (\phi_{4r}), (\phi_{5r}), (\phi_{6r})\}^T (\phi_{1r})$$

128

from which it is seen that the essential form of the mode shape, in terms of elements moving in- or out-of- phase, can be readily deduced.

2.10.4   Display of FRF Data for Damped Systems
2.10.4.1 Bode plots

If we turn our attention now to damped systems, we find that the form of the FRF plot of the type just discussed is rather similar to that for the undamped case. The resonances and antiresonances are blunted by the inclusion of damping, and the phase angles (not shown) are no longer exactly 0° or 180°, but the general appearance of the plot is a natural extension of that for the system without damping. Fig. 2.33 shows a plot for the same mobility parameter as appears in Fig. 2.31(a) but here for a system with damping added. Most mobility plots have this general form as long as the modes are relatively well-separated. This condition is satisfied unless the separation between adjacent natural frequencies (expressed as a percentage of their mean) is of the same order as, or less than, the modal damping factors, in which case it becomes difficult to distinguish the individual modes.

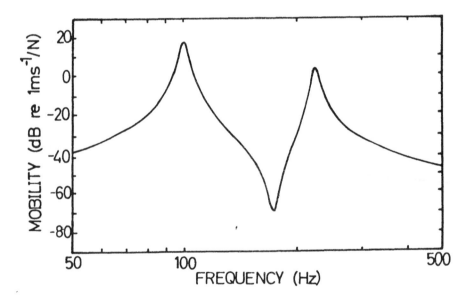

2.10.4.2 Nyquist diagrams

As for the SDOF case, it is interesting to examine the nature of the other types of plot for FRF data, and the most profitable alternative version is again the Nyquist or Argand diagram plot. We saw earlier how one of the standard FRF parameters of a SDOF system produced an approximate circle when plotted in the Nyquist format (and that by choosing the appropriate FRF parameter, an exact circle can be formed for either type of damping). This also applies to the MDOF system in that each of its frequency responses is composed of a number of SDOF components.

Fig. 2.34(a) shows the result of plotting the point receptance, α_{11}, for the 2DOF system described above. Proportional hysteretic damping has been added with damping loss factors of 0.02 and 0.04 for the first and second modes, respectively. It should be noted that it is not always as easy to visualise the total curve from the individual components as is the case here (with well-separated modes) although it is generally found that the basic characteristics of each (mainly the diameter and the frequency at which the local maximum amplitude is reached) are carried through into the complete expression.

A corresponding plot for the transfer receptance, α_{21}, is presented in Fig. 2.34(b) where it may be seen that the opposing signs of the modal constants (remember that these are still real quantities because for a proportionally-damped system the eigenvectors are identical to those for the undamped version) of the two modes have caused one of the modal circles to be in the upper half of the complex plane.

The examples given in Figs. 2.34 were for a proportionally-damped system. In the next two figures, 2.35(a) and (b), we show corresponding data for an example of non-proportional damping. In this case a relative phase has been introduced between the first and second elements of the eigenvectors: of 30° in mode 1 (previously it was 0°) and of 150° in mode 2 (where previously it was 180°). Now we find that the individual modal circles are no longer 'upright' but are rotated by an amount dictated by the complexity of the modal constants. The general shape of the resulting Nyquist plot is similar to that for the proportionally-damped system although the resonance points are no longer at the 'bottom' (or 'top') of the corresponding circles. The properties of an isolated modal circle are described in Chapter 4.

2.10.4.3 Real and imaginary plots

Next, we present in Fig. 2.36 plots of the Real and Imaginary Parts of the FRF vs. Frequency to illustrate the general form of this type of display.

The plots shown in Figs. 2.33 to 2.36 all refer to systems with structural or hysteretic damping. A similar set of results would be obtained for the case of viscous damping with the difference that the exact modal circles will be produced for mobility FRF data, rather than receptances, as has been the case here.

130

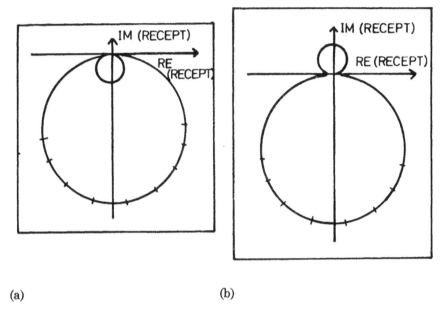

(a) (b)

Fig. 2.34 Nyquist FRF plot for proportionally-damped system.
(a) Point receptance; (b) Transfer receptance

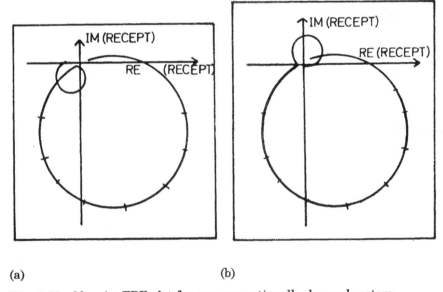

(a) (b)

Fig. 2.35 Nyquist FRF plot for non-proportionally-damped system.
(a) Point receptance; (b) Transfer receptance

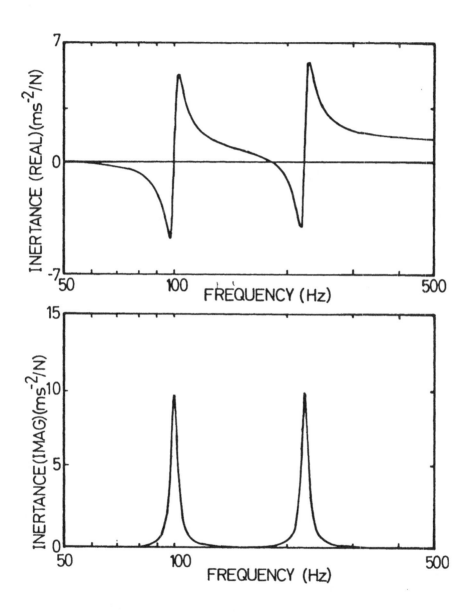

Fig. 2.36 Real and imaginary plots of accelerance FRF for damped 2DOF system

132

2.10.4.4 Three-dimensional plots

Lastly, we show in Fig. 2.37, a three-dimensional plot of an MDOF system FRF, using the same basic format as shown earlier for an SDOF system. Once again, this form of presentation has more curiosity value than practical, it being very difficult to interpret.

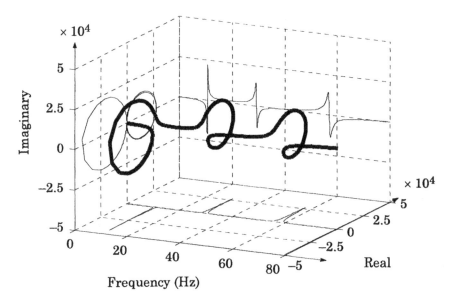

Fig. 2.37   Three-dimensional plot of FRF for MDOF system

2.10.5   Summary

The purpose of this section has been to predict the form which will be taken by plots of FRF data using the different display formats which are in current use. Although we may have a working familiarity with measured FRF plots, the results shown above have been derived entirely from consideration of the theoretical basis of structural vibration theory and the exercise in so doing proves to be invaluable when trying to understand and interpret actual measured data.

2.11   NON-SINUSOIDAL VIBRATION AND FRF PROPERTIES

With receptance and other FRF data we have a means of computing the response of a MDOF system to an excitation which consists of a set of harmonic forces of different amplitudes and phases but all of the same frequency. In the general case, we can simply write

$$\{X\}e^{i\omega t} = [\alpha(\omega)]\{F\}e^{i\omega t} \qquad\qquad (2.134)$$

We shall now turn our attention to a range of other excitation/response situations which exist in practice and which can be analysed using the same frequency response functions. Also, we shall show the mathematical basis of how the FRF properties can be obtained from measurements made during non-sinusoidal vibration tests.

2.11.1 Periodic Vibration
2.11.1.1 Periodic signals as Fourier series
The first of these cases is that of periodic vibration, in which the excitation (and thus the response) is not simply sinusoidal although it does retain the property of periodicity. Such a case is illustrated in the sketch of Fig. 2.38(a) which shows a sawtooth type of excitation and two of the responses it produces from a system. Clearly, in this case there is no longer a simple relationship between the input and the outputs, such as exists for harmonic vibration where we simply need to define the amplitude and phase of each parameter. As a result, any function relating input and output in this case will necessarily be quite complicated.

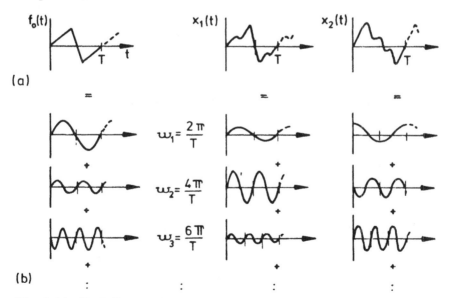

Fig. 2.38 Periodic signals and their sinusoidal (Fourier) components

It transpires that the easiest way of computing the responses in such a case as this is by means of the Fourier Series. The basic principle of Fourier analysis is that any periodic function (such as $f_0(t)$, $x_1(t)$ or

$x_2(t)$ in Fig. 2.38(a)) can be represented by a series of sinusoids of suitable frequencies, amplitudes and phases, as illustrated in Fig. 2.38(b), based on the fundamental period, T. (A more detailed discussion of Fourier Series is given in Appendix 5.) So

$$f_0(t) = \sum_{n=1}^{\infty} {}_0 F_n e^{i\omega_n t} \quad ; \quad \omega_n = \frac{2\pi n}{T} \qquad (2.135)$$

Once a frequency decomposition of the forcing function has been obtained, we may use the corresponding FRF data, computed at the specific frequencies present in the forcing spectrum, in order to compute the corresponding frequency components of the responses of interest:

$$x_j(t) = \sum_{n=1}^{\infty} \alpha_{j0}(\omega_n) {}_0 F_n e^{i\omega_n t} \quad ; \quad \omega_n = \frac{2\pi n}{T} \qquad (2.136)$$

What must be noted here is that although the response contains exactly the same frequencies in its spectrum as does the forcing, the relative magnitudes of the various components are different in the two cases because the FRF data vary considerably with frequency. Thus, we obtain response time histories which are periodic with the same period as the excitation, T, but which have quite different shapes to it and, incidentally, to each other (see Fig. 2.38).

2.11.1.2 To derive FRF from periodic vibration signals

It is possible to determine a system's FRF properties from excitation and response measurements when the vibration is periodic. To do this, it is necessary to determine the Fourier Series components of both the input force signal and of the relevant output response signal(s). Both of these series will contain components at the same set of discrete frequencies; these being integer multiples of $2\pi/T$, where T is the fundamental period.

Once these two series are available, the frequency response function can be defined at the same set of frequency points by computing the ratio of the response component to the input component. For both data sets, there will be two parts to each component — magnitude and phase (or sine and cosine).

2.11.2 Transient Vibration

We shall turn our attention next to the case of transient vibration which, strictly speaking, cannot be treated by the same means as above because the signals of excitation and response are not periodic.

However, as discussed in the Appendix on Fourier analysis, it is sometimes possible to extend the Fourier Series approach to a Fourier Transform for the case of an infinitely long period. Here, it is generally possible to treat both the transient input and response in this way and to obtain an input/output equation in the frequency domain. It is also possible to derive an alternative relationship working in the time domain directly and we shall see that these two approaches arrive at the same solution.

2.11.2.1 Analysis via Fourier Transform
For most transient cases, the input function $f(t)$ will satisfy the Dirichlet condition and so its Fourier Transform, $F(\omega)$, can be computed from:

$$F(\omega) = (1/2\pi) \int_{-\infty}^{\infty} f(t)e^{-i\omega t} dt \tag{2.137}$$

Now, at any frequency ω, the corresponding Fourier Transform of the response, $X(\omega)$, can be determined from:

$$X(\omega) = H(\omega)F(\omega) \tag{2.138}$$

where $H(\omega)$ represents the appropriate version of the FRF for the particular input and output parameters considered. We may then derive an expression for the response itself, $x(t)$, from the Inverse Fourier Transform of $X(\omega)$

$$x(t) = \int_{-\infty}^{\infty} (H(\omega)F(\omega))e^{i\omega t} d\omega \tag{2.139}$$

2.11.2.2 Response via time domain (superposition)
This alternative analysis is sometimes referred to as convolution, or 'Duhamel's Method', and is based on the ability to compute the response of a system to a simple (unit) impulse. Fig. 2.39(a) shows a typical unit impulse excitation applied at time $t = t'$ which has the property that although the function has infinite magnitude and lasts for an infinitesimal period of time, the area underneath it (or the integral $f(t)dt$) is equal to unity. The response of a system at time t (after t') is defined as the system's unit Impulse Response Function (IRF) and has a direct relationship to the Frequency Response Function (FRF) — one

136

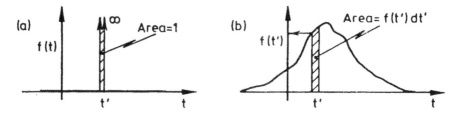

Fig. 2.39 Transient signals.
(a) Delta function; (b) Arbitrary time function

being in the time domain and the other in the frequency domain. The IRF is written as:

$$h(t - t')$$

If we now consider a more general transient excitation or input function, as shown in Fig. 2.39(b), we see that it is possible to represent this as the superposition of several impulses, each of magnitude $(f(t')dt')$ and occurring at different instants in time. The response of a system (at time, t) to just one of these incremental impulses (at time t') can be written as:

$$\delta x(t) = h(t - t') f(t') \, dt' \tag{2.140}$$

and the total response of the system will be given by superimposing or integrating all the incremental responses as follows:

$$x(t) = \int_{-\infty}^{\infty} h(t - t') f(t') \, dt' \quad ; \quad h(t - t') = 0; \ t \le t' \tag{2.141a}$$

This input/output relationship appears somewhat different to that obtained via the Fourier Transform method, equation (2.139), but we shall find that there is in fact a very close relationship between them. This we can do by using the Fourier Transform approach to compute the response of a system to a unit impulse. Thus, let $f(t) = \delta(0)$ and determine its Fourier Transform, $F(\omega)$. In this case, application of equation (2.137) is relatively easy and yields:

$$F(\omega) = 1/2\pi$$

If we now insert this expression into the general response equation, (2.139), and note that, by definition, this must be identical to the impulse response function, we obtain:

$$x(t) = (1/2\pi) \int_{-\infty}^{\infty} H(\omega)e^{i\omega t}\, d\omega \equiv h(t) \qquad (2.141b)$$

Thus we arrive at a most important result in finding the Impulse and Frequency Response Functions to constitute a Fourier Transform pair.

Following from this derivation, we can also see that the IRF can also be expressed as a modal series, just as is possible for the FRF. Thus, if we write

$$H(\omega) = \sum H_r(\omega)$$

where $H_r(\omega)$ is the contribution for a single mode, we can likewise write:

$$h(t) = \sum h_r(t) \qquad (2.142)$$

where, for a viscously-damped system, $h_r(t) = A_r e^{s_r t}$.

2.11.2.3 To derive FRF from transient vibration signals

As before, we are able to prescribe a formula for obtaining a structure's FRF properties from measurements made during a transient vibration test. What is required is the calculation of the Fourier Transforms of both the excitation and the response signals. The ratio of these two functions (both of frequency) can be computed in order to obtain an expression for the corresponding frequency response function:

$$H(\omega) = \frac{X(\omega)}{F(\omega)} \qquad (2.143)$$

In practice, it is much more common to compute a Discrete Fourier Transform (DFT) or Series and thus to perform the same set of calculations as described in the previous section for periodic vibration. Indeed, such an approach using a DFT assumes that the complete transient event is quasi periodic. This is a realistic approach which can be applied to many transient-type signals — as opposed to continuous signals — provided that in the time period of the measurement both excitation and response signals are effectively zero at the start and the end of the sample.

Alternatively, the spectrum analyser may be used to compute the FRF in the same way as it would for random vibration (see below), namely by taking the ratio of the spectra. This is only useful if a succession of repeated transients are applied — nominally, the same as each other — in which case any noise on individual measurements will be averaged out.

2.11.3   Random Vibration
2.11.3.1 Random signals in time and frequency domains

We come now to the most complex type of vibration — where both excitation and response are described by random processes. Although it might be thought that this case could be treated in much the same way as the previous one — by considering the random signals to be periodic with infinite period — this is not possible because the inherent properties of random signals cause them to violate the Dirichlet condition. As a result, neither excitation nor response signals can be subjected to a valid Fourier Transform calculation and another approach must be found.

It will be necessary to introduce and to define two sets of parameters which are used to describe random signals: one based in the time domain — the Correlation Functions — and the other in the frequency domain — the Spectral Densities. We shall attempt here to provide some insight into these quantities without necessarily detailing all the background theory.

Consider a typical random vibration parameter, $f(t)$, illustrated in Fig. 2.40(a), which will be assumed to be ergodic*. We shall introduce and define the Autocorrelation Function, $R_{ff}(\tau)$, as the 'expected' (or average) value of the product $(f(t).f(t+\tau))$, computed along the time axis. This will always be a real and even function of time, and is written:

$$R_{ff}(\tau) = E\big[f(t).f(t+\tau)\big] \tag{2.144}$$

and will generally take the form illustrated in the sketch of Fig. 2.40(b). This correlation function, unlike the original quantity $f(t)$, does satisfy the requirements for Fourier transformation and thus we can obtain its Fourier Transform by the usual equation. The resulting parameter we shall call a Spectral Density, in this case the Auto- or Power Spectral Density (PSD), $S_{ff}(\omega)$, which is defined as:

---

*   'Ergodic' is a type of random process which requires only a single sample, albeit of considerable length, to portray all the statistical properties required for its definition.

$$S_{ff}(\omega) = (1/2\pi) \int_{-\infty}^{\infty} R_{ff}(\tau)e^{-i\omega\tau}d\tau \qquad (2.145)$$

The Auto-Spectral Density is a real and even function of frequency, and does in fact provide a description of the frequency composition of the original function, $f(t)$. It has units of (f^2/ω) and would generally appear as in the plot of Fig. 2.40(c).

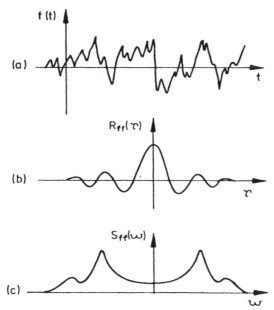

Fig. 2.40 Random signals.
(a) Time history; (b) Autocorrelation function; (c) Power spectral density

A similar concept can be applied to a pair of functions such as $f(t)$ and $x(t)$ to produce cross correlation and cross spectral density functions. The cross correlation function, $R_{xf}(\tau)$, is defined as:

$$R_{xf}(\tau) = E[x(t).f(t+\tau)] \qquad (2.146)$$

and the cross spectral density (CSD) is defined as its Fourier Transform:

$$S_{xf}(\omega) = (1/2\pi) \int_{-\infty}^{\infty} R_{xf}(\tau)e^{-i\omega\tau}d\tau \qquad (2.147)$$

Cross correlation functions are real, but not always even, functions of time, and cross spectral densities, unlike auto spectral densities, are generally complex functions of frequency with the particular conjugate property that:

$$S_{xf}(\omega) = S_{fx}^{*}(\omega) \qquad (2.148)$$

Now that we have established the necessary parameters to describe random processes in general, we are in a position to define the input/output relationships for systems undergoing random vibration. In deriving the final equations which permit the calculation of response from known excitations, we shall not present a full analysis; rather, we shall indicate the main path followed without detailing all the algebra. In this way, we hope to demonstrate the basis of the analysis and the origins of the final expressions, which are the only ones required in normal modal testing practice.

The analysis is based on the general excitation/response relationship in the time domain, quoted above in equation (2.141) and repeated here:

$$x(t) = \int_{-\infty}^{\infty} h(t - t') f(t') \, dt' \qquad (2.149)$$

Using this property, it is possible to derive an expression for $x(t)$ and another for $x(t+\tau)$ and thus to calculate the response autocorrelation, $R_{xx}(\tau)$:

$$R_{xx}(\tau) = E\left[x(t).x(t + \tau)\right] \qquad (2.150)$$

This equation can be manipulated to describe the response autocorrelation in terms of the corresponding property of the excitation, $R_{ff}(\tau)$, but the result is a complicated and unusable triple integral. However, this same equation can be transformed to the frequency domain to emerge with the very simple form:

$$S_{xx}(\omega) = \left| H(\omega) \right|^{2} S_{ff}(\omega) \qquad (2.151)$$

Although apparently very convenient, equation (2.151) does not provide a complete description of the random vibration conditions. Further, it is clear that it could not be used to determine the FRF from measurements of excitation and response because it contains only the modulus of $H(\omega)$, the phase information being omitted from this formula. A second equation is required and this may be obtained by a similar analysis based on the cross correlation between the excitation and the response, the frequency domain form of which is:

$$S_{fx}(\omega) = H(\omega)S_{ff}(\omega)$$

or, alternatively: (2.152)

$$S_{xx}(\omega) = H(\omega)S_{xf}(\omega)$$

So far, all the analysis in this section has been confined to the case of a single excitation parameter, although it is clear that several responses can be considered by repeated application of the equations (2.151) and (2.152). In fact, the analysis can be extended to situations where several excitations are applied simultaneously, whether or not these are correlated with each other. This analysis involves not only the autospectra of all the individual excitations, but also the cross spectra which link one with the others. The general input/output equation for this case is:

$$\left[S_{fx}(\omega)\right] = \left[S_{ff}(\omega)\right]\left[H(\omega)\right]$$ (2.153)

2.11.3.2 To derive FRF from random vibration signals
The pair of equations (2.152) provides us with the basis for a method of determining a system's FRF properties from the measurement and analysis of a random vibration test. Using either of them, we have a simple formula for determining the FRF from estimates of the relevant spectral densities:

$$H(\omega) = \frac{S_{fx}(\omega)}{S_{ff}(\omega)} \text{ , usually identified as } H_1(\omega)$$

or (2.154)

$$H(\omega) = \frac{S_{xx}(\omega)}{S_{xf}(\omega)} \text{ , usually identified as } H_2(\omega)$$

In fact, the existence of two equations (and a third, if we include (2.151)) presents an opportunity to check the quality of calculations made using measured (and therefore imperfect) data, as will be discussed in more detail in Chapter 3.

2.11.3.3 Instrumental variable model for FRF

In Chapter 3, we shall see that there are difficulties encountered in implementing some of the above formulac in practice because of noise and other limitations concerned with the data acquisition and processing. A number of techniques will be developed in order to circumvent or to minimise some of these problems and the first of these is presented below, under the title of its control systems origin, the 'instrumental variable' formulation for the FRF of a system. That formulation differs from those above in that three, rather than two, quantities are involved in the definition of the output/input ratio. The system considered can best be described with reference to Fig. 2.41 which shows first, in (a), the traditional two-channel, single-input, single-output (SISO), model upon which the formulae such as (2.154) are based. Then, in (b), is given a more detailed and representative model of the system which is used in a modal test to measure the input and output quantities in order to determine the FRF of the system, $H(\omega)$. In this diagram, we show separately the various items which make up the measurement system:

- exciter coil (voltage, $r(t)$)
- exciter armature (force, $p(t)$)
- input (i.e. force) transducer (force, $f(t)$)
- test system
- response transducer (response, $x'(t)$)

However, in this configuration it can be seen that there are two feedback mechanisms which apply, the first between the shaker and the structure and the second between the response of the system and the input transducer. Further, it is shown that both transducers' output signals are prone to noise, so that the quantities actually measured will be $f'(t)$ and $x'(t)$. We shall leave until Chapter 3 further discussion about noise, and signal processing, but shall use this introduction to define the new formula which is available for the determination of the system FRF from measurements of the input and output quantities in a system such as that shown in Fig. 2.41. The alternative formula is:

$$H(\omega) = \frac{S_{x'v}(\omega)}{S_{fv}(\omega)}, \text{ which is usually identified as } H_3(\omega) \qquad (2.155)$$

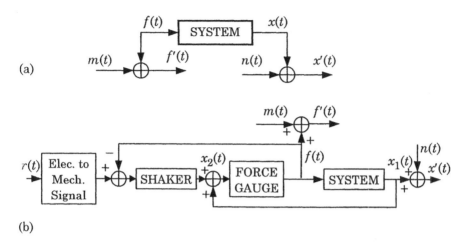

Fig. 2.41 System for FRF determination.
(a) Basic SISO model; (b) More complex SISO model with feedback

2.11.3.4 Derivation of FRF from MIMO data

A natural extension of the preceding section is to the more general case, already mentioned in passing in equation (2.153), where there are several simultaneous inputs (excitations) and where the responses at several points are obtained simultaneously. This is the general multi-input, multi-output (MIMO) case and is widely used as the basis for FRF measurements in modal tests of large systems.

A diagram for the general n-input case is shown in Fig. 2.42, taken from reference [25]. The algebra required to derive the required formulae is somewhat tedious and so the inquisitive reader is directed to the reference for full details which lead to the following:

$$\left[H_{xf}(\omega)\right]_{n\times n} = \left[S_{x'v}(\omega)\right]_{n\times n}\left[S_{f'v}(\omega)\right]_{n\times n}^{-1} \tag{2.156}$$

or, if the original, simpler, version is to be used, to:

$$\left[H_{xf}(\omega)\right]_{n\times n} = \left[S_{f'f'}(\omega)\right]_{n\times n}^{-1}\left[S_{x'f'}(\omega)\right]_{n\times n} \tag{2.157}$$

In practical application of both of these formulae, care must be taken to ensure the non-singularity of the spectral density matrix which is to be inverted and it is in this respect that the former version may be found to be more reliable. Also, the expressions need to be generalised to the

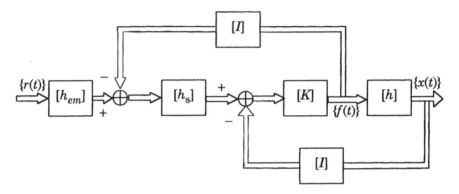

Fig. 2.42 System for FRF determination via MIMO model

case where there are a different number of response parameters to the number of inputs, or excitations, here defined as n. In this latter case, the matrices concerned will be rectangular and will require a generalised inversion, rather than the simple inverse shown above.

2.12   COMPLETE AND INCOMPLETE MODELS
2.12.1   Some Definitions
Most of the preceding theory has been concerned with complete models; that is, the analysis has been presented for an N degree-of-freedom system with the implicit assumption that all the mass, stiffness and damping properties are known and that all the elements in the eigenmatrices and the FRF matrix are available. While this is a valid approach for a theoretical study, it is less generally applicable for experimentally-based investigations where it is not usually possible to measure all the DOFs, or to examine all the modes possessed by a structure. Because of this limitation, it is necessary to extend our analysis to examine the implications of having access to something less than a complete set of data, or model, and this leads us to the concept of a 'reduced' or 'incomplete' type of model.

It is appropriate here to introduce a few additional definitions which will be used throughout this book when dealing with the various types of incomplete model. A complete model is one which is fully defined by its description. This can be achieved in any of the three types of model if all the individual mass stiffness and damping elements are included (spatial model), or if all the modes (natural frequencies and mode shapes) are included or if all the FRF data are known over a frequency range which includes all the modes (response model). This means that the full $N \times N$ matrices are available for the different mathematical descriptions. Models become incomplete when less than the above information is available, or displayed.

There are different types of incomplete model. There is the model which is reduced in size (from N to n) by simply deleting information about certain degrees-of-freedom. This process leads to a reduced model which retains full accuracy for the DOFs which are retained, but which loses access to those which have been deleted. The process can be applied only to the modal and response models and results in a modal model described by an $N \times N$ eigenvalue matrix but by an eigenvector matrix which is only $n \times N$. The corresponding response model is an incomplete FRF matrix of size $n \times n$, although all the elements of that reduced matrix are themselves fully accurate. Another type of reduced model is one in which the number of modes (i.e. natural frequencies) is reduced as well (from N to m), so that the eigenvalue matrix is only $m \times m$ in size. A consequence of this is that the elements in the reduced $n \times n$ FRF matrix in this case are only approximate (see further discussion below).

Another type of model reduction can be achieved by **condensation** from N to n DOFs. This is a process in which a number of DOFs are again eliminated from the complete description but an attempt is made to include the effects of the masses and stiffnesses which are thereby eliminated in the retained DOFs. This is the condensation process which is applied in the Guyan and other reduction techniques used to contain the size of otherwise very large finite element models. In such a condensed model, the spatial, modal and response models are all reduced to $n \times n$ matrices, and it must be noted that the properties of each are approximate in every respect.

Lastly, it should be mentioned that it is sometimes required to seek to recover a full-sized model of a structure's dynamics from the basis of an incomplete model. This can be attempted by one of various processes of interpolation and leads to an expanded model, usually of full size $(N \times N)$, but great care should be exercised in making any use of such a model.

2.12.2 Incomplete Response Models

As intimated, there are two ways in which a model can be incomplete — by the omission of some modes, and/or by the omission of some degrees-of-freedom — and we shall examine these individually, paying particular attention to the implications for the response model (in the form of the FRF matrix). Consider first the complete FRF matrix, which is $N \times N$:

$$\left[H(\omega)\right]_{N \times N}$$

and then suppose that we decide to limit our description of the system to include certain DOFs only (and thus to ignore what happens at the others, which is not the same as supposing they do not exist). Our reduced response model is now of order $n \times n$, and is written as:

$$\left[H^R(\omega)\right]_{n \times n}$$

Now, it is clear that as we have not altered the basic system, and it still has the same number of degrees-of-freedom even though we have foregone our ability to describe the system's behaviour at all of them. In this case, the elements which remain in the reduced FRF matrix are identical to the corresponding elements in the full $N \times N$ matrix. In other words, the reduced matrix is formed simply by retaining the elements of interest and removing or deleting those to be ignored.

At this point, it is appropriate to mention the consequences of this type of reduction on the impedance type of FRF data. The impedance matrix which corresponds to the reduced model defined by $[H^R]$ will be denoted as $[Z^R]$ and it is clear that

$$\left[Z^R(\omega)\right] = \left[H^R(\omega)\right]^{-1} \tag{2.158}$$

It is also clear that the elements in the reduced impedance matrix such as Z_{jk}^R are **not** the same quantities as the corresponding elements in the full impedance matrix and, indeed, a completely different impedance matrix applies to each specific reduction. Thus:

$$H_{ij}^R(\omega) = H_{ij}(\omega) \quad \text{but} \quad Z_{ij}^R(\omega) \neq Z_{ij}(\omega)$$

We can also consider the implications of this form of reduction on the other types of model, namely the modal model and the spatial model. For the modal model, elimination of the data pertaining to some of the DOFs results in a smaller eigenvector matrix, which then becomes rectangular or order $n \times N$. This matrix still retains N columns, and the corresponding eigenvalue matrix is still $N \times N$ because we still have all N modes included.

For the spatial model it is more difficult to effect a reduction of this type. It is clearly not realistic simply to remove the rows and columns corresponding to the eliminated DOFs from the mass and stiffness matrices as this would represent a drastic change to the system. It is possible, however, to reduce these spatial matrices by a number of methods which have the effect of redistributing the mass and stiffness (and damping) properties which relate to the redundant DOFs amongst those which are retained. In this way, the total mass of the structure, and its correct stiffness properties can be largely retained. The Guyan reduction procedure is perhaps the best known of this type although there are several modelling techniques (see later, Chapter 5). Such reduced spatial properties will be denoted as:

$$\left[M^R\right], \left[K^R\right]$$

Next, we shall consider the other form of reduction in which *only* m of the N modes of the system are included. Frequently, this is a necessary approach in that many of the high-frequency modes will be of little interest and almost certainly very difficult to measure. Consider first the FRF matrix and include initially all the DOFs but suppose that each element in the matrix is computed using not only m of the N terms in the summation, i.e.

$$\tilde{H}_{jk}(\omega) = \sum_{r=1}^{m \leq N} \frac{{}_r A_{jk}}{\omega_r^2 - \omega^2 + i\eta_r \omega_r^2} \tag{2.159}$$

In full, we can write the FRF matrix as:

$$\left[\tilde{H}(\omega)\right]_{N \times N} = \left[\Phi\right]_{N \times m} \left[\left(\lambda_r^2 - \omega^2\right)\right]_{m \times m}^{-1} \left[\Phi\right]_{m \times N}^T \tag{2.160}$$

Of course, both types of reduction can be combined when the resulting matrix would be denoted:

$$\left[\tilde{H}^R(\omega)\right]_{n \times n}$$

It can be seen from (2.160) that the FRF matrix thus formed, $[\tilde{H}^R(\omega)]$, will, in general, be rank deficient, and thus it will not be possible to obtain the impedance matrix by numerical inversion. This remains the case as long as $n > m$ (and can even be found in cases where $n < m$) and so the numerical condition of matrices of incomplete models is frequently found to be a cause for concern. In order to overcome these problems, it is often convenient to attempt to provide an approximate correction to the FRF data to compensate for the errors introduced by leaving out some of the terms. This is usually effected by adding a constant or 'residual' term to each FRF, as shown in the following equation:

$$[H(\omega)] = \left[\tilde{H}(\omega)\right] + [R] \tag{2.161}$$

148

The consequence of neglecting some of the modes on the modal model is evident in that the eigenvalue matrix becomes of order $m \times m$ and the eigenvector matrix is again rectangular, although in the other sense, and we have:

$$\left[\lambda_r^2\right]_{m \times m} \quad ; \quad \left[\Phi\right]_{N \times m}$$

2.12.3 Incomplete Modal and Spatial Models

It has been shown earlier that the orthogonality properties of the modal model provide a direct link between the modal model and the spatial model:

$$[\Phi]^T[M][\Phi] = [I] \quad ; \quad [\Phi]^T[K][\Phi] = \left[\omega_r^2\right] \tag{2.24}$$

which can be inverted to yield:

$$[M] = [\Phi]^{-T}[\Phi]^{-1}; \quad [K] = [\Phi]^{-T}\left[\omega_r^2\right][\Phi]^{-1} \tag{2.162}$$

If the modal model is incomplete, then we can note the implications for the orthogonality properties. First, if we have a modal incompleteness ($m < N$ modes included), then we can write:

$$\{\Phi\}_{m \times N}^T[M][\Phi]_{N \times m} = [I]_{m \times m} \quad ; \quad [\Phi]_{m \times N}^T[K][\Phi]_{N \times m} = \left[\omega_r^2\right]_{m \times m}$$

However, if we have a spatial incompleteness (only $n < N$ DOFs included), then we cannot express any orthogonality properties at all because the eigenvector matrix is not commutable with the system mass and stiffness matrices. In both reduced-model cases, it becomes impossible to use (2.162) to re-construct the system mass and stiffness matrices from an incomplete modal model. Even in the special case where $m = n$, in which case we have square reduced eigenvector and eigenvalue matrices and can — in theory — compute the inverses required by the equation (2.162), these matrices are generally singular and thus not invertible. Even if they are numerically non-singular, there is no theoretical basis for applying (2.162) in this case and any mass and stiffness matrices produced by such application have no physical significance and should not be used.

Fig. 2.43 shows the relationship between different forms of complete and incomplete models.

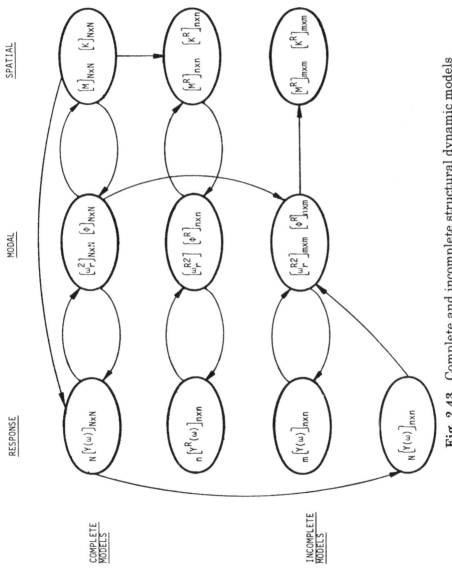

Fig. 2.43 Complete and incomplete structural dynamic models

2.13   SENSITIVITY OF MODELS
2.13.1   Introduction
There are an increasing number of applications of the structural dynamics models which we are in the course of developing that make use of the sensitivity properties of the models. These sensitivities describe the rates of change of some of the key properties — such as natural frequencies or mode shapes — with small changes in some of the model parameters, such as individual masses or stiffnesses. The model sensitivities are required for various purposes:

- they help to locate errors in models in updating applications;
- they are useful in guiding design optimisation procedures, and
- they are used in the course of curve-fitting for the purposes of testing the reliability of the modal analysis processes.

It will be helpful to include here a short summary of the main sensitivity parameters and to show how they may be deduced from both theoretically- and experimentally-derived models.

2.13.2   Modal Sensitivities
The most commonly-used sensitivities are those which describe the rates of change of the modal parameters with the individual mass and stiffness elements in the spatial model. These quantities are defined in general as follows:

$$\frac{\partial \omega_r}{\partial p} \quad \text{and} \quad \frac{\partial \{\phi\}_r}{\partial p}$$

where p represents any variable of interest.

(a)   SDOF system
It is useful to approach the general expressions for these parameters via a very simple example based on an undamped SDOF system. We can introduce the concept of sensitivity through the basic SDOF system comprising mass, m, and spring, k, and we can define the basic sensitivities of the system's natural frequency, ω_0, with respect to these two design parameters as:

$$\frac{\partial \omega_0}{\partial m} \quad \text{and} \quad \frac{\partial \omega_0}{\partial k}$$

respectively. We can readily show that these two sensitivities can be expressed as follows:

$$\frac{\partial \omega_0}{\partial m} = 2\omega_0 \frac{\partial \omega_0}{\partial m} = \frac{\partial}{\partial m}\left(k m^{-1}\right) = -k m^{-2}$$

$$\frac{\partial \omega_0}{\partial m} = -\frac{\sqrt{k}}{2\sqrt{m^3}}$$

and:

$$\frac{\partial \omega_0}{\partial k} = \frac{1}{2\sqrt{km}} \tag{2.163}$$

(b) MDOF systems — eigenvalue sensitivity

We may now approach the more general case of the undamped MDOF system which we have analysed in detail earlier in this chapter. From Section 2.4 we can recall the following equation:

$$\left([K] - \omega_r^2 [M]\right)\{\phi\}_r = \{0\}$$

which we can now differentiate with respect to an arbitrary variable, p, that might be an individual mass or stiffness in the original model, m_i or k_j, or which might be an individual element in the spatial model matrices, m_{ij} or k_{ij}.

$$\left([K] - \omega_r^2 [M]\right)\frac{\partial\{\phi\}_r}{\partial p} + \left(\frac{\partial[K]}{\partial p} - \omega_r^2 \frac{\partial[M]}{\partial p} - \frac{\partial\omega_r^2}{\partial p}[M]\right)\{\phi\}_r = \{0\} \tag{2.164}$$

Next, we can premultiply this expression by $\{\phi\}_r^T$:

$$\{\phi\}_r^T \left([K] - \omega_r^2 [M]\right)\frac{\partial\{\phi\}_r}{\partial p} + \{\phi\}_r^T \left(\frac{\partial[K]}{\partial p} - \omega_r^2 \frac{\partial[M]}{\partial p} - \frac{\partial\omega_r^2}{\partial p}[M]\right)\{\phi\}_r = 0$$

or

$$\{0\} + \{\phi\}_r^T \left(\frac{\partial[K]}{\partial p} - \omega_r^2 \frac{\partial[M]}{\partial p}\right)\{\phi\}_r - \frac{\partial\omega_r^2}{\partial p} = 0$$

and so

$$\frac{\partial \omega_r^2}{\partial p} = \{\phi\}_r^T \left(\frac{\partial [K]}{\partial p} - \omega_r^2 \frac{\partial [M]}{\partial p} \right) \{\phi\}_r \qquad (2.165)$$

This result can be checked against the earlier simple case for an SDOF system because in that case, ϕ is simply $(1/\sqrt{m})$ so that we have:

$$\frac{\partial \omega_r^2}{\partial m} = \{\phi\}_r^T \left(\frac{\partial [K]}{\partial m} - \omega_r^2 \frac{\partial [M]}{\partial m} \right) \{\phi\}_r = \frac{1}{\sqrt{m}} \left(0 - \omega_0^2 (1) \right) \frac{1}{\sqrt{m}}$$

$$= -\frac{k}{m^2}$$

as before.

(c) MDOF systems — eigenvector sensitivity
A similar, although somewhat lengthier, analysis can be made for the eigenvector sensitivity terms, as follows. We can use the previous equation (2.164) together with the fact that the eigenvector sensitivity term can always be expressed as a linear combination of the original eigenvectors:

$$\left([K] - \omega_r^2 [M] \right) \sum_{j=1}^N \gamma_{rj} \{\phi\}_j + \left(\frac{\partial [K]}{\partial p} - \omega_r^2 \frac{\partial [M]}{\partial p} - \frac{\partial \omega_r^2}{\partial p} [M] \right) \{\phi\}_r = \{0\} \quad (2.166)$$

This time, we shall premultiply the equation by $\{\phi\}_s^T$ and exploit the various orthogonality properties which apply to form:

$$\{\phi\}_s^T \left([K] - \omega_r^2 [M] \right) \sum_{j=1}^N \gamma_{rj} \{\phi\}_j + \{\phi\}_s^T \left(\frac{\partial [K]}{\partial p} - \omega_r^2 \frac{\partial [M]}{\partial p} - \frac{\partial \omega_r^2}{\partial p} [M] \right) \{\phi\}_r = 0$$

so

$$\left(\omega_s^2 - \omega_r^2 \right) \gamma_{rs} + \{\phi\}_s^T \left(\frac{\partial [K]}{\partial p} - \omega_r^2 \frac{\partial [M]}{\partial p} \right) \{\phi\}_r = 0$$

hence γ_{rs} and thence

$$\frac{\partial\{\phi\}_r}{\partial p} = \sum_{s=1}^{N} - \left(\frac{\{\phi\}_s^T \left(\frac{\partial[K]}{\partial p} - \omega_r^2 \frac{\partial[M]}{\partial p} \right) \{\phi\}_r}{\left(\omega_s^2 - \omega_r^2 \right)} \right) \{\phi\}_s \qquad (2.167)$$

2.13.3 FRF Sensitivities

It may be seen that it is also possible to derive FRF sensitivities, as well as the classical modal terms. If we consider first the simple SDOF system for which the receptance FRF, $\alpha(\omega)$, is given by:

$$\alpha(\omega) = \frac{1}{k + i\omega c - \omega^2 m}$$

Differentiate with respect to m, k to yield:

$$\frac{\partial\alpha(\omega)}{\partial m} = \frac{-\omega^2}{\left(k + i\omega c - \omega^2 m\right)^2} \quad \text{and} \quad \frac{\partial\alpha(\omega)}{\partial k} = \frac{1}{\left(k + i\omega c - \omega^2 m\right)^2} \qquad (2.168)$$

It can be shown that this simple expression can be extended to the more general case for MDOF systems as follows:

$$\frac{\partial[\alpha(\omega)]}{\partial p} = [\alpha(\omega)] \left(\left(\frac{\partial[K]}{\partial p} \right) + i\omega \left(\frac{\partial[C]}{\partial p} \right) - \omega^2 \left(\frac{\partial[M]}{\partial p} \right) \right) [\alpha(\omega)] \qquad (2.169)$$

from which equation it is possible to derive the simple case shown in equation (2.168) above as well as many other cases.

2.13.4 Modal Sensitivities from FRF Data

As mentioned above, most model sensitivity studies are restricted to use with the theoretical models which are required at various stages of structural dynamic analysis. However, there exists the possibility of deriving certain of the above-mentioned sensitivity parameters directly from FRF data such as can be measured in a modal test. Essentially, it is possible to derive expressions for the eigenvalue sensitivities to selected individual mass and stiffness parameters by analysing the point FRF properties at the selected DOFs.

Specifically, it can be shown [26] that an expression can be derived which describes the eigenvalue sensitivities for a particular DOF (*j*) in terms of the frequencies of resonance and antiresonance on the point FRF at that DOF, $H_{jj}(\omega)$:

$$\frac{d\omega_r}{dm_j} = \frac{C_{jj}\,\omega_r \prod\limits_{s=1}^{m}\left(\omega_s^2\right)\prod\limits_{i=1}^{m-1}\left(_{jj}\Omega_i^2 - \omega_r^2\right)}{\prod\limits_{i=1}^{m-1}\left(_{jj}\Omega_i^2\right)\prod\limits_{s=1;\neq r}^{m}\left(\omega_s^2 - \omega_r^2\right)} = -\frac{1}{\omega_r^2}\frac{d\omega_r}{dk_j} \qquad (2.170)$$

where

ω_r = natural frequency of mode r;

$_{jj}\Omega_i$ = ith antiresonance frequency on point FRF at DOF j: $H_{jj}(\omega)$

Clearly, the accuracy of this expression depends upon the number of resonance and antiresonance frequencies which are included in the frequency range of the measurement, and so the application of the formula will always result in an approximate answer. However, it must be remembered that the sensitivity expressions which are being discussed here are themselves only valid for infinitesimal changes of mass, or stiffness, and may not be applied at will for arbitrary mass and stiffness errors (in model updating) or modifications (in structural optimisation tasks). See Chapter 6 for a more detailed discussion of these applications.

The last point can be depicted graphically in the plots of Fig. 2.44 which show how each of the natural frequencies of a test structure will change as a mass of increasing magnitude is added to the specific test DOF to which the displayed FRF refers. It is clear that these natural frequencies reduce linearly at first, in accordance with the relevant sensitivity coefficient, but that they become progressively less and less effective in lowering these natural frequencies until, as these approach the antiresonance frequencies of the original structural configuration, a limiting value is reached.

2.14   ANALYSIS OF WEAKLY NON-LINEAR STRUCTURES
2.14.1   General — Approximate Analysis of Non-linear Structures

Before concluding our review of the theoretical basis of the subject, it is appropriate to include a consideration of the possibility that not all the systems or structures encountered in practice will be linear. All the preceding analysis and, indeed, the whole basis of the subject, assumes linearity, an assumption which has two main implications in the present context:

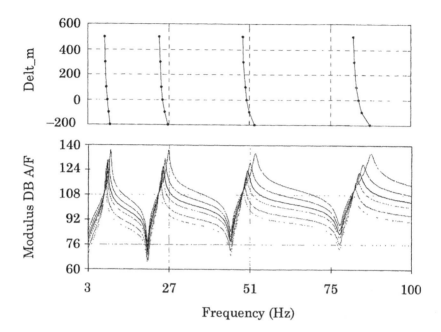

Fig. 2.44 Natural frequency dependence on added mass

(i)   that doubling the magnitude of the excitation force would simply result in a doubling of the response, and so on (response linearly related to excitation), and

(ii) that if two or more excitation patterns are applied simultaneously then the response thus produced will be equal to the sum of the responses caused by each excitation applied individually (i.e. the principle of linear superposition applies).

We shall now introduce some of the characteristics exhibited by weakly (i.e. slightly) non-linear systems, not in order to provide detailed analysis but so that such structural behaviour can be recognised and identified if encountered during a modal test. Thus we shall seek to derive and illustrate the frequency response characteristics of such systems.

The equation of motion for a single degree of freedom system with displacement- and/or velocity-dependent non-linearity and undergoing steady-state harmonic excitation, can be written as:

$$m\left(\ddot{x} + 2\zeta\,\omega_0\,\dot{x} + \omega_0^2\,x + \mu(x,\dot{x})\right) = f(t) = F_0\cos\omega t \qquad (2.171)$$

This equation can be expressed approximately as

$$m\left(\ddot{x} + \tilde{\lambda}\dot{x} + \tilde{\omega}_0^2\, x\right) = F_0\cos\omega t \tag{2.172}$$

where $\tilde{\lambda}$ and $\tilde{\omega}_0^2$ depend on the amplitude (X_0) and frequency (ω) of vibration, according to:

$$\tilde{\lambda} = -\frac{1}{\pi a\omega}\mu\left(X_0\cos\phi, -X_0\omega\sin\phi\right)\sin\phi\, d\phi$$

$$\tilde{\omega}_0^2 = \omega_0^2 + \frac{1}{\pi a}\mu\left(X_0\cos\phi, -X_0\omega\sin\phi\right)\cos\phi\, d\phi \tag{2.173}$$

This is sometimes referred to as the **Harmonic Balance Method** and it provides an approximate solution to the complicated situation in which the simple harmonic excitation produces a response which contains several harmonic components (including, but not only, one at the excitation frequency, ω). The approximate solution yields the amplitude of the fundamental component of the response (i.e. the component at the excitation frequency, ω) and this is of interest because it is a quantity that we shall be able to measure in practical tests.

We shall confine our attention here to two specific cases of practical interest, namely

(a) cubic stiffness, where the spring force is given by $k(x + \beta x^3)$, and
(b) coulomb friction, where (some of) the damping is provided by dry friction.

2.14.2 Cubic Stiffness Non-linearity
In this case, we have a basic equation of motion of the form

$$m\left(\ddot{x} + 2\zeta\,\omega_0\,\dot{x} + \omega_0^2\left(x + \beta x^3\right)\right) = f(t)$$

$$m\left(\ddot{x} + \tilde{\lambda}\,\dot{x} + \tilde{\omega}_0^2\, x\right) = f(t)$$

where

$$\tilde{\lambda} = 2\zeta\,\omega_0$$

and $\hfill$ (2.174)

$$\tilde{\omega}_0^2 = \omega_0^2\left(1 + 3\beta\, X_0^2/4\right)$$

In the case of harmonic excitation, $f(t) = Fe^{i\omega t}$, we find that in fact the response $x(t)$ is not simply harmonic — the non-linearity causes the generation of some response (generally small) at multiples of the excitation frequency — but it is convenient to examine the component of the response at the excitation frequency. (Note that this is just what would be measured in a sinusoidal test, though not with a random, periodic or transient test.) Then we shall find:

$$x(t) \approx X_0 \, e^{i\omega t} \quad ; \quad X_0 = \frac{F/m}{\tilde{\omega}_0^2 - \omega^2 + i\tilde{\lambda}\omega} \tag{2.175}$$

where $\tilde{\omega}_0^2$ and $\tilde{\lambda}$ are themselves functions of X_0. Equation (2.175) needs further processing to find an explicit expression for X_0. This is obtained from:

$$\left(3\beta \, \omega_0^2/4\right)^2 X_0^6 - \left(3\left(\omega_0^2 - \omega^2\right)\beta \, \omega_0^2/2\right)X_0^4$$

$$+\left(\left(\omega_0^2 - \omega^2\right)+ (2\zeta \, \omega_0 \, \omega)^2\right)X_0^2 - (F/m)^2 = 0 \tag{2.176}$$

which shows that there can either be one or three real roots. All cubic stiffness systems exhibit this characteristic although the existence of the two types of solution (one possible value for X_0, or three) depends upon the frequency and the magnitude of the excitation and/or the non-linearity. Some typical plots are shown in Fig. 2.45, computed using the above expressions.

Another way of interpreting the approximation involved in this approach is by examining the force/displacement characteristic for this system, shown in Fig. 2.46. It can be seen that when the system is undergoing steady-state vibration with amplitude $\pm X_0$, the system has an effective stiffness given by the slope of the straight line joining the two extremities on the load-displacement curve. While the true characteristics of the structure are clearly more complicated than this simple representation, it does provide a practical means of representing the non-linear system by an equivalent linear one for the purpose of applying linear modal analysis theory to an essentially non-linear system.

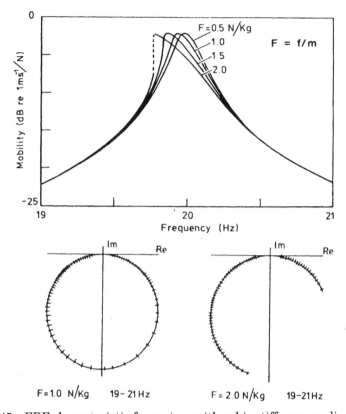

Fig. 2.45 FRF characteristic for system with cubic stiffness non-linearity

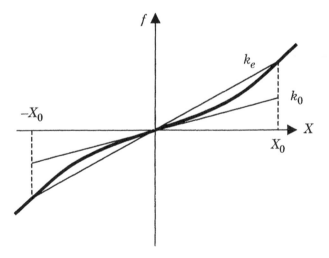

Fig. 2.46 Force deflection characteristic of typical non-linear system

2.14.3   Coulomb Friction Non-linearity

In this case, we have a basic equation of motion of the type:

$$m\left(\ddot{x} + 2\zeta\,\omega_0\,\dot{x} + \omega_0^2\,x\right) + \frac{Rx}{|\dot{x}|} = f(t)$$

or   (2.177)

$$m\left(\ddot{x} + \tilde{\lambda}\dot{x} + \tilde{\omega}_0^2\,x\right) = f(t) = m\,F(t)$$

where

$$\tilde{\lambda} = 2\zeta\,\omega_0 + \frac{4R}{(\pi\,X_0\,\omega_0\,m)}$$

$$\tilde{\omega}_0^2 = \omega_0^2$$ (2.178)

Once again, a harmonic excitation will produce a more complex, though periodic, response whose fundamental component is given by

$$x(t) \approx X_0\,e^{i\omega t}$$

where X_0 is obtained from

$$\left(\left(\omega_0^2 - \omega^2\right)^2 + \left(2\zeta\,\omega_0\,\omega\right)^2\right)X_0^2 + \left(8\zeta\,\omega^2\,R/\pi\right)X_0$$

$$+ \left(\left(4R\,\omega/\pi\,\omega_0\right)^2 - \left(F/m\right)^2\right) = 0$$ (2.179)

Further analysis of this expression shows that only a single value of X_0 applies. Plots of (X_0/F) for various values of F and/or R are illustrated in Fig. 2.47.

2.14.4 Other Non-linearities and Other Descriptions (Higher-order FRFs)

Other nonlinearities which are likely to be encountered in modal testing of real structures include:

- Backlash
- Bilinear stiffness
- Microslip friction damping
- Quadratic (and other power law) damping

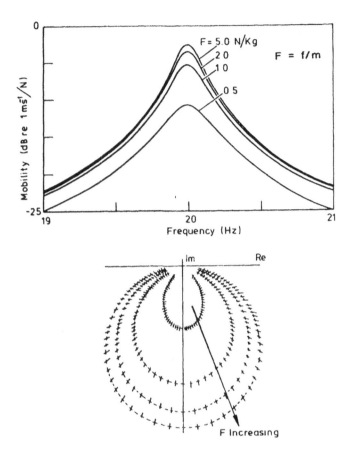

Fig. 2.47 FRF characteristic for system with dry friction damping non-linearity

Details of the essential characteristics of these elements can be found in specialist texts on non-linear behaviour (e.g. [27]).

We have here limited our interest to the approximate approach in which just the first-order FRF functions have been derived, and that is a necessary restriction if we are to limit our activities to the methods of conventional modal testing. It is, however, appropriate to mention that the essential feature of non-linear systems — that a single frequency excitation produces a multi-frequency response — gives rise to a family of higher-order response functions than the type we have been dealing with hitherto. Although the treatment of these higher-order FRFs is well outside the scope of this work, it is appropriate to indicate the direction in which a more detailed and accurate analysis of this class of system will take. The processing and interpretation of these complicated functions constitutes a specialist study in its own right and, for the time being, is beyond the grasp of the conventional modal analysis and testing tools. The interested reader — who has encountered non-linear effects on a scale that cannot be treated by the approximate procedures advocated in this book — is directed towards references such as [25] for a more thorough explanation of these advanced concepts.

CHAPTER 3

FRF Measurement Techniques

3.1 INTRODUCTION AND TEST PLANNING
3.1.1 Introduction

In this chapter we shall be concerned with the measurement techniques which are used for modal testing. First, it is appropriate to consider vibration measurement methods in general in order to view the context of those used for our particular interest here. Basically, there are two types of vibration measurement:

(i) those in which just one type of parameter is measured (usually the response levels), and

(ii) those in which both input and response output parameters are measured.

Recalling the basic relationship:

$$\boxed{\text{RESPONSE}} \; = \; \boxed{\text{PROPERTIES}} \; \times \; \boxed{\text{INPUT}}$$

we can see that only when two of the three terms in this equation have been measured can we define completely what is going on in the vibration of the test object. If we measure only the response, then we are unable to say whether a particularly large response level is due to a strong excitation or to a resonance of the structure. Nevertheless, both types of measurement have their applications and much of the equipment and instrumentation used is the same in both cases.

We shall be concerned here with the second type of measurement, where both excitation and response are measured simultaneously so that the basic equation can be used to deduce the system properties directly from the measured data. Within this category there are a number of different approaches which can be adopted but we shall concentrate first on one which we refer to as the 'single-point excitation' method. Later we shall progress to discussing the more general approach which involves simultaneous excitation at several points on the structure, although we shall see that this approach carries with it both advantages and disadvantages. Our interest will first be focused on the more straightforward approach (at least from the viewpoint of the experimenter) where the excitation is applied at a single point (although in the course of a test, this point may be varied around the structure). Measurements of this type are often referred to as 'mobility measurements' or 'FRF measurements', and these are the names we shall use throughout this work. Indeed, our goal in the measurement phase of a modal test is to measure the FRF data which are necessary for the subsequent modal analysis and modelling phases. Responses obtained using a single excitation yield FRF data directly, simply by 'dividing' the measured responses by the measured excitation force. These are referred to as SISO (single-input, single-output) or SIMO (single-input, multiple-output) tests. Responses obtained using several simultaneous excitations yield operating deflection shapes (ODSs) from which it is necessary to extract the required FRF data by sometimes-complicated analysis procedures. These are referred to as MIMO (multi-input, multi-output) tests.

3.1.2   Test Planning

We shall see that the choice of method we use to perform the test, and which data we measure, will be of paramount importance to the success of the venture, taking due account of the reasons why each test is being performed in the first place. It is appropriate to recall from Chapter 1 that a series of five Levels of Test have been identified [12], and that each modal test should be classified according to its application objectives before any measurements are even planned. These Test Levels are:

Level 0: estimation of natural frequencies and damping factors; response levels measured at few points; very short test times.

Level 1: estimation of natural frequencies and damping factors; mode shapes defined qualitatively rather than quantitatively.

Level 2: measurements of all modal parameters suitable for tabulation and mode shape display, albeit un-normalised.

Level 3: measurements of all modal parameters, including normalised mode shapes; full quality checks performed and model usable for model validation.

Level 4: measurements of all modal parameters and residual effects for out-of-range modes; full quality checks performed and model usable for all response-based applications, including modification, coupling and response predictions.

It is clear that there will need to be an extensive test planning phase before full-scale measurements are made and decisions taken concerning the methods of excitation, signal processing and data analysis, as well as the proper selection of which data to measure, where to excite the structure and how to prepare and support it for those measurements. In this Chapter we shall be concerned with the details of how each type of procedure or device works and less with the making of these choices. To some extent, such decisions can only be made on the basis of experience of using the various alternatives but some guidance in their formulation may be found in such references as the DTA Handbook [12] as well as in the various User Guides for the software and hardware systems which are used for this type of test. Accordingly, our aim here is to explain the basis and operation of the various devices and techniques in question.

Having said that, it is appropriate to note that recent developments in modal testing have led to a series of procedures for the determination of the optimum choice of various parameters, including exciter and response transducer locations. These methods are based on the pre-test existence of some form of theoretical model of the test structure which they use to identify the most (and least) important or propitious areas of the structure from consideration of the various demands of:

* support points (to minimise external influences);
" excitation points (to ensure effective excitation of all modes); and — most important of all —
" response points (to ensure adequate coverage of all mode shapes so as to permit clear identification and discrimination of those in the test range).

These methods are discussed in more detail in Chapter 6, Section 6.6.

3.1.3   Checking the Quality of Measured Data

Nowadays, concern with quality assurance has heightened awareness of the needs to ensure the highest reliability and accuracy of the data we measure and of the results we obtain by its subsequent processing. To this end, it is appropriate at the start of this Chapter to itemise some of the features which need constant checking throughout a modal test, in order to enable results to be obtained of the highest quality possible by the equipment and methods used.

3.1.3.1 Signal quality

Usually, the first concern is to the acquisition of signals of sufficient strength and clarity, and that are free of excessive noise. Apart from the obvious issues of proper selection of transducer and conditioning electronics, it is sometimes found that the dynamic range of the measured quantities is extreme, especially when the frequency range being covered is wide — more than a decade in frequency. What often happens is that there is a very large component of signal in one frequency range and this dictates the gain settings on amplifiers and analysers such that lower-level components of the signal are difficult to measure accurately. In these circumstances, it is preferable to restrict the frequency range of the measurements so that all components of interest have sufficient signal strength to permit their accurate measurement. It should be noted that in following such advice, it is important to ensure that the signal offered to the amplifier/analyser has been suitably amended, and not just the range of the measurement (i.e. if there is a large component in a signal at 500 Hz which is 'drowning-out' much smaller components at, say, 100 Hz, then it is not sufficient just to change the measurement range to 0-200 Hz: the component at 500 Hz must be removed from the signal altogether).

3.1.3.2 Signal fidelity

Another, similar, problem can arise whereby the signals obtained do not truly represent the quantity which is to be measured. An example of this problem is provided by the transverse sensitivity property exhibited by most accelerometers and other response transducers used in modal tests. Sometimes, motion of the transducer is much greater in a direction perpendicular to its measurement axis (than in the measurement direction) with the result that the output is heavily contaminated by the transverse sensitivity component, sometimes giving quite misleading indications as to the motion of the structure.

There are other circumstances which can lead to erroneous measurements, including the incorrect labelling or connection of transducers, cables and channels to the data acquisition system. These can only be eradicated by careful housekeeping during testing but there is one other check which can be applied, at least on selected measurements. That is the check of the basic pattern of resonances and antiresonances visible on the FRF curves. It has already been noted that for any point FRF (excitation and response at the same DOF), resonances and antiresonances must alternate. Also, the incidence of antiresonances in an FRF curve is related, loosely, to the degree of separation of the two DOFs to which that FRF refers: excitation and response points which are well-separated on the test structure will tend to possess fewer antiresonances than will those which have these two points much closer together. This is not a rigorous check, nor a fail-safe one, but it is useful in assuring that gross errors have not been introduced.

3.1.3.3 Measurement repeatability

One obvious and essential check for any modal test is that which tests the repeatability of certain measurements. Certain FRFs should be re-measured from time to time, just to check that neither the structure nor the measurement system have experienced any significant changes. There are several reasons why the test structure might change its properties perceptibly throughout a test programme, and if these changes are non-trivial, their consequences can have serious effects on the subsequent analysis processes to which the measured data are subjected.

3.1.3.4 Measurement reliability

This is a slightly different check to that of repeatability, in which nominally-identical measurements are made, and it seeks to establish that the measured data are independent of the measuring system. In such a reliability check, a given quantity (usually, an FRF) would be measured using a slightly different setup, or procedure such as a different excitation signal, to that used for the original measurement. Such checks are more expensive than simple repeatability ones, but are very important to demonstrate the underlying validity of the measurement method being used.

3.1.3.5 Measured data consistency, including reciprocity

Yet another variation on the repeatability/reliability theme is that of consistency. The various FRF data measured on a given structure should (in some cases, must) exhibit consistency, by which is meant that the underlying natural frequencies, damping factors and mode shapes visible in the FRF data must all derive from a common modal model. It is unacceptable if the natural frequency of mode 1 appears to be slightly different in FRF H_{ij} to the value which is apparent in a second FRF, H_{pq}. Equally, the reciprocity expected to exist between FRFs such as H_{jk} and H_{kj} should be checked and found to be at an acceptable level. Of course, it may be difficult to prescribe what is 'acceptable' in such circumstances, but the checks should be made anyway, so that the level of consistency is known, even if it is difficult to make a judgement on its acceptability.

There are several reasons why the measured data may exhibit a degree of inconsistency, and these will depend heavily on the measurement method(s) employed in the test. If the FRFs which make up the complete set are measured individually, one at a time and

168

sequentially, then there are two common causes for the resulting data to be inconsistent: (i) because during the finite time it takes for the data to be gathered, the structure has actually changed its properties, perhaps due to variations in temperature, or to the consequences of continued vibration; and/or (ii) because the test setup conditions are changed in order to excite or to measure at different points of interest on the structure. Changes in the local added mass or stiffness which result from attaching exciters or transducers will cause the structure which is tested to differ slightly from FRF curve to FRF curve. These effects can result in a degree of inconsistency in the data which needs at best to be known about and at worst to be removed.

3.2   BASIC MEASUREMENT SYSTEM

The experimental setup used for mobility measurement is basically quite simple although there exist a great many different variants on it, in terms of the specific items used. There are three major items:

(i)   an excitation mechanism;
(ii)   a transduction system (to measure the various parameters of interest); and
(iii)   an analyser, to extract the desired information (in the presence of the inevitable imperfections which will accumulate on the measured signals).

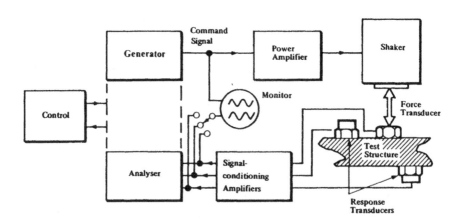

Fig. 3.1   General layout of FRF measurement system

Fig. 3.1 shows a typical layout for the measurement system, detailing some of the 'standard' items which are usually found. The component in this illustration labelled 'Controller' is nowadays usually provided by a complex software package but should be read so as to include the user him- or herself since the effective use of such test equipment retains the need for intelligent input and choices by the test engineer, notwithstanding the power and speed of today's computers. Many of the detailed procedures in FRF measurements are repetitive and tedious, and so some form of automation is essential but, if provided by a computer, this can also serve to process the measured data as required for the modal analysis stage, later in the overall process.

The main elements in the measurement chain are, then:

(a)   A source for the excitation signal. This will depend on the type of test being undertaken and can be any of the following:

   • sinusoidal (from an oscillator),
   • periodic (from a special signal generator capable of producing a specific frequency content).
   • random (from a noise generator),
   • transient (from a special pulse or burst signal generating device, or by applying an impact with a hammer),

(b)   Power amplifier. This component will be necessary in order to drive the actual device used to vibrate the structure which, in turn, will take one of a number of different forms, as discussed below. The power amplifier will necessarily be selected to match the excitation device.

(c)   Exciter. The structure can be excited into vibration in several ways, although the two most commonly (and successfully) used are by an attached shaker or by a hammer blow. Other possibilities exist by step relaxation (releasing from a deflected position) and by ambient excitation (such as wave, wind or roadway excitations), but these are relatively special cases which are only used when the more conventional methods are not possible.

(d)   Transducers. Here again, there are a great many different possibilities for the devices available to measure the excitation forces and the various responses of interest. For the most part, piezoelectric transducers are widely used for both types of parameter although strain gauges are often found to be convenient because of their minimal interference with the test object. More recently, lasers have become commonplace in the modal testing laboratory, both to provide recordable and analysable holographic images of vibrating structures and, most recently, to provide a non-contact response transducer in the form of the (sometimes, scanning) Laser Doppler Velocimeter (LDV).

(e)   Conditioning Amplifiers. The choice of amplifier depends heavily on the type of transducer used and should, in effect, be regarded as part of it. In all cases, its role is to strengthen the (usually) small signals generated by the transducers so that they can be fed to the analyser for measurement.

(f)   Analyser. The function of this item is simply to measure the various signals developed by the transducers in order to ascertain the magnitudes of the excitation force(s) and responses. In essence, it is a voltmeter but in practice it is a very sophisticated one. There are different types of analyser available and the choice will depend on the type of excitation which has been used: sinusoidal, random, transient, periodic. The two most common devices are Spectrum (Fourier) Analysers and Frequency Response Analysers although the same functions as provided by these can be performed by a tuneable narrow-band filter, a voltmeter and a phase meter plus a great deal of time and patience!

3.3   STRUCTURE PREPARATION
3.3.1   Free and Grounded Supports
One important preliminary to the whole process of FRF measurement is the preparation of the test structure itself. This is often not given the attention it deserves and the consequences which accrue can cause an unnecessary degradation of the whole test.

The first decision which has to be taken is whether the structure is to be tested in a 'free' condition or 'grounded'.

3.3.1.1   Free supports
By 'free' is meant that the test object is not attached to ground at any of its coordinates and is, in effect, freely suspended in space. In this condition, the structure will exhibit rigid body modes which are determined solely by its mass and inertia properties and in which there is no bending or flexing at all. Theoretically, any structure will possess six rigid body modes and each of these has a natural frequency of 0 Hz. By testing a structure in this free condition, we are able to determine

the rigid body modes and thus the mass and inertia properties which can themselves be very useful data.

In practice, of course, it is not feasible to provide a truly free support — the structure must be held in some way — but it is generally feasible to provide a suspension system which closely approximates to the free condition. This can be achieved by supporting the testpiece on very soft 'springs', such as might be provided by light elastic bands, so that the rigid body modes, while no longer having zero natural frequencies, have values which are very low in relation to those of the bending modes ('very low' in this context means that the highest rigid body mode frequency is less than 10 to 20 per cent of that for the lowest bending mode). If we achieve a suspension system of this type, then we can still derive the rigid body (inertia) properties from the very low frequency behaviour of the structure without having any significant influence on the flexural modes that are the object of the test. (In fact, there are several instances where a test of this type may be carried out only to examine the rigid body modes as this is an effective way of determining the full inertia properties of a complex structure.) One added precaution which can be taken to ensure minimum interference by the suspension on the lowest bending mode of the structure — the one most vulnerable — is to attach the suspension as close as possible to nodal points of the mode in question. At the same time, particular attention should be paid to the possibility of the suspension adding significant damping to otherwise lightly-damped testpieces.

As a parting comment on this type of suspension, it is necessary to note that any rigid body will possess no less than six modes and it is necessary to check that the natural frequencies of all of these are sufficiently low before being satisfied that the suspension system used is sufficiently soft. To this end, suspension wires, etc. should generally be normal to the primary direction of vibration, as in Fig. 3.2(b) rather than in the same direction as these supports.

Selection of the optimum suspension points is one of the features offered by the test planning procedures referred to above in 3.1.2, and described in detail later, in Chapter 6.

3.3.1.2 Grounded supports

The other type of support is referred to as 'grounded' because it attempts to fix selected points on the structure to ground. While this condition is extremely easy to apply in a theoretical analysis, simply by deleting the appropriate coordinates, it is much more difficult to implement in the practical case. The reason for this is that it is very difficult to provide a base or foundation on which to attach the test structure which is sufficiently rigid to provide the necessary grounding. All structures have a finite impedance (or a non-zero mobility) and thus cannot be regarded as truly rigid but whereas we are able to approximate the free condition by a soft suspension, it is less easy to approximate the grounded condition without taking extraordinary precautions when designing the support structure. Perhaps the safest procedure to follow is to measure the mobility FRF of the base structure itself over the frequency range for the test and to establish that this is a much lower mobility than the corresponding levels for the test structure at the point of attachment. If this condition can be satisfied for all the coordinates to be grounded then the base structure can reasonably be assumed to be grounded. However, as a word of caution, it should be noted that the coordinates involved will often include rotations and these are notoriously difficult to measure.

From the above comments, it might be concluded that we should always test structures in a freely supported condition. Ideally, this is so but there are numerous practical situations where this approach is simply not feasible and again others where it is not the most appropriate (note the comments in Section 3.3.1.3, below). For example, very large testpieces, such as parts of power generating stations or civil engineering structures, could not be tested in a freely-supported state. Further, in just the same way that low frequency properties of a freely supported structure can provide information on its mass and inertia characteristics, so also can the corresponding parts of the mobility curves for a grounded structure yield information on its static stiffness. Another consideration to be made when deciding on the format of the test is the environment in which the structure is to operate. If we consider a turbine blade, for example, it is clear that in its operating condition the vibration modes of interest will be much closer to those of a cantilevered root fixing than to those of a completely free blade. Whereas it is possible to test and to analyse a single blade as a free structure, the modes and frequencies which will then form the basis of the test/analysis comparison will be quite different from those which obtain under running conditions. Of course, theoretically, we can validate or obtain a model of the blade using its free properties and expect this to be equally applicable when the root is grounded, but in the real world, where we are dealing with approximations and less-than-perfect data, there is additional comfort to be gained from a comparison made using modes which are close to those of the functioning structure, i.e. with a grounded root.

3.3.1.3 Loaded boundaries

A compromise procedure can be applied in some cases in which the test object (such as the blade in the preceding paragraph) is connected at certain coordinates to another simple component of known mobility, such as a specific mass. This modified or 'loaded' testpiece is then studied experimentally and the effects of the added component 'removed' analytically. This is a device which is used increasingly in place of simple free supports. When a structure is tested in a free-free condition, it is often found that there are many fewer modes observed in a given frequency range than there will be for that same component when it is installed in its final habitat (usually, as one part of a structural assembly). Moreover, the modes of the structure as an isolated free component are often quite different in form to those of the installed condition — the free ends often contain virtually no strain energy, for example — and so there may be questions as to the suitability of a test in the free-free condition. It is possible to resolve some of these questions by the device of loading the free boundaries of the structure by adding simple masses there and taking account of these masses in any theoretical model with which the test data may be compared.

3.3.1.4 Perturbed boundary conditions

There is an extension of the above idea of loading the boundary surfaces of a component known as the 'perturbed boundary condition' approach. In a number of applications, including validation and updating of theoretical models, there is often a dirth of experimental data available from the modal test and so additional tests are sought wherever possible. The data base for a given structure can be extended, or 'enriched', by the repetition of the modal test for different boundary conditions. This, in effect, means testing several different structures, but each is simply related to the others by the differences in the boundary loads, and these are known exactly, and so a multiplicity of test data can be derived from just the one test structure.

3.3.1.5 Summary

In the above paragraphs, we have presented a number of considerations which must be made in deciding what is the best way to support the test structure for FRF measurements. There is no universal method: each test must be considered individually and the above points taken into account. Perhaps, as a final comment for those cases in which a decision is difficult, we should observe that, at least from a theoretical standpoint, it is always possible to determine the grounded structure's properties from those in a free or loaded condition while it is not possible to go in the opposite direction. (This characteristic comes from the fact that the free support involves more degrees of freedom, some of which can later be deleted, while it is not possible — without the addition of new data — to convert the more limited model of a grounded structure to one with greater freedom as would be necessary to describe a freely-supported structure.)

Examples of both types of test configuration are shown in Fig. 3.2.

3.3.2   Local Stiffening
If it is decided to ground the structure, care must be taken to ensure that no local stiffening or other distortion is introduced by the attachment, other than that which is an integral part of the structure itself. In fact, great care must be paid to the area of the attachment if a realistic and reliable test configuration is to be obtained and it is advisable to perform some simple checks to ensure that the whole assembly gives repeatable results when dismantled and reassembled again. Such attention to detail will be repaid by confidence in the eventual results.

3.4   EXCITATION OF THE STRUCTURE
3.4.1   General
Various devices are available for exciting the structure and several of these are in widespread use. Basically, they can be divided into two types: contacting and non-contacting. The first of these involves the connection of an exciter of some form which remains attached to the structure throughout the test, whether the excitation type is continuous

Fig. 3.2   Examples of (a) grounded, and (b) freely-supported structures

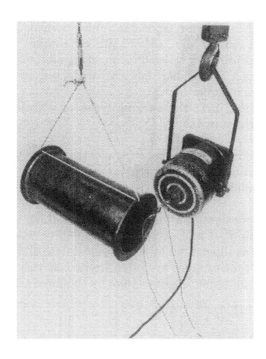

Fig. 3.2   Examples of (a) grounded, and (b) freely-supported structures

(sinusoidal, random etc.) or transient (pulse, chirp). The second type includes devices which are either out of contact throughout the vibration (such as provided by a non-contacting electromagnet) or which are only in contact for a short period, while the excitation is being applied (such as a hammer blow).

We shall discuss first the various types of vibrator, or shaker, of which there are three in use:

- mechanical (out-of-balance rotating masses);
- electromagnetic (moving coil in magnetic field);
- electrohydraulic.

Each has its advantages and disadvantages — which we shall attempt to summarise below — and each is most effective within a particular operating range, as illustrated by some typical data shown in Fig. 3.3. It should be noted that exciters are often limited at very low frequencies by the stroke (displacement) rather than by the force generated.

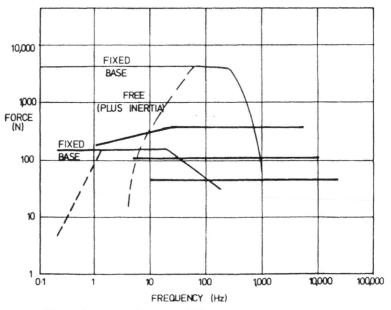

Fig. 3.3   Typical exciter characteristics

3.4.2   Mechanical Exciters

The mechanical exciter, which uses rotating out of balance masses, is capable of generating a prescribed force at a variable frequency although there is relatively little flexibility or control in its use. The magnitude of the force is restricted by the out-of-balance and is only variable by making adjustments to this quantity — not something which can be done while the vibration is continuing. Also, this type of excitation mechanism is relatively ineffective at low frequencies because of the speed-squared dependence. However, unless the amplitude of vibration caused by the exciter becomes large relative to the orbit of the out-of-balance masses, the magnitude and phase of the excitation force is known quite accurately and does not need further measurement, as is the case for the other types of exciter.

3.4.3   Electromagnetic Exciters

The electromagnetic principle has long been used to generate vibrations in structures for the purpose of what we today call 'modal testing'. The simplest of these applications is through the direct application of a magnetic force on the structure to be excited, without any direct physical contact. This method has its attractions but also presents severe problems of control and is not widely used, except in special cases. Some of these are of interest in the testing of rotating machines and are discussed separately below, in Section 3.4.8.

However, perhaps the most common type of exciter is the electromagnetic (or 'electrodynamic') shaker in which the supplied input signal is converted to an alternating magnetic field in which is placed a coil which is attached to the drive part of the device, and to the structure. In this case, the frequency and amplitude of excitation are controlled independently of each other, giving more operational flexibility — especially useful as it is generally found that it is better to vary the level of the excitation as resonances are passed through. However, it must be noted that the electrical impedance of these devices varies with the amplitude of motion of the moving coil and so it is not possible to deduce the excitation force from a measurement of the voltage applied to the shaker. Nor, in fact, is it usually appropriate to deduce the excitation force by measuring the current passing through the shaker because this measures the force applied not to the structure itself, but to the assembly of structure and shaker drive. Although it may appear that the difference between this force (generated within the shaker) and that applied to the structure is likely to be small, it must be noted that just near resonance very little force is required to produce a large response and what usually happens is that without altering the settings on the power amplifier or signal generator, there is a marked reduction in the force level at frequencies adjacent to the structure's natural frequencies. As a result, the true force applied to the structure becomes the (small) difference between the force generated in the exciter and the inertia force required to move the drive rod and shaker table and is, in fact, much smaller than either. See Fig. 3.4(a).

178

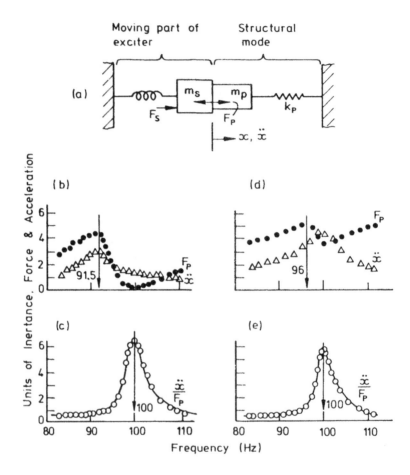

Fig. 3.4 Variations in measured parameters around resonance for shaker attached to structure.
(a) Shaker-structure model; (b) Measured data at point 1;
(c) Measured FRF at point 1; (d) Measured data at point 2;
(e) Measured FRF at point 2

As this is an important feature of most attached-shaker tests using continuous sinusoidal, random or periodic excitation, it is worth describing the point in some detail using the following example.

Suppose we are testing a plate and are trying to determine the properties of one of its modes. In one measurement, where the excitation and response are measured at the same point (a point FRF) and in the immediate vicinity of a natural frequency, the plate behaves very much like a single-degree-of-freedom oscillator with an apparent mass of m_{p_1} and an apparent stiffness of k_{p_1}. (Note that the natural frequency of this mode is given by $(k_{p_1}/m_{p_1})^{1/2}$.) Suppose also that the mass of the moving part of the shaker and its connection to the structure (which is not part of the structure proper) is m_s. Now, let the force generated in the shaker be $f_s(t)$ and the force actually applied to the structure (the one we want to measure) be $f_p(t)$. If the acceleration of the structure is denoted by $\ddot{x}(t)$, and we consider the vibration test to be conducted at various sinusoidal frequencies, ω, then we may write the simple relationship:

$$f_p(t) = f_s(t) - m_s \ddot{x}(t)$$

Taking some typical data, we show in Fig. 3.4(b) the magnitudes of the various quantities which are, or which could be, measured. Also shown in Fig. 3.4(c) is the curve for the m_{FRF} quantity of interest, in this case the accelerance (or inertance), $\ddot{X}/F_p$, and it is particularly interesting to see how the true natural frequency (indicated when the accelerance reaches a maximum) is considerably displaced from that suggested by the apparent resonance when the response alone reaches a maximum.

We now move to a different point on the structure which, for the same mode, will have different values for the apparent mass and stiffness, m_{p_2} and k_{p_2}, although these two quantities will necessarily stay in the same ratio (i.e. $k_{p_1}/m_{p_1} = k_{p_2}/m_{p_2} = \omega_0^2$). Another plot of the various quantities in this case is shown in Figs. 3.4(d) and 3.4(e) from which it is clear that although the FRF shows the natural frequency to be at the same value as before, the system resonance is now at a different frequency to that encountered in the first measurement, and this occurs simply because of the different balance between the structure's apparent properties (which vary from point to point) and those of the shaker (which remain the same throughout).

This example serves to illustrate the need for a direct measurement of the force applied to the structure as close to the surface as possible in order to obtain a reliable and accurate indication of the excitation level, and hence the mobility properties. It also illustrates a characteristic which gives rise to some difficulties in making such measurements: namely, that the (true) applied excitation force becomes very small in the vicinity of the resonant frequency with the consequence that it is particularly vulnerable to noise or distortion, see Fig. 3.4(b). It is clear from comparison of Figs. 3.4(b) and (d) that the extent of the force 'drop-out' (as it is sometimes called) depends on the relative magnitudes of the shaker mass, m_s, and the apparent mass of the structural mode being excited, m_{p_i}, at the point of excitation, i. Whereas the first of these quantities is fixed (by the choice of shaker), the second is variable and depends directly upon how close the excitation point is to a node of the mode of vibration in question. At an antinode, the apparent mass and stiffness are both at their minimum values, while near a node they increase to very large values (approaching infinity actually on the node), all the time maintaining the same relative values, dictated by the natural frequency, $\omega_0 = (k_{p_i}/m_{p_i})^{1/2}$.

Generally, the larger the shaker, the greater the force which may be generated for exciting the structure. However, besides the obvious penalty of expense incurred by using too large an exciter, there is a limitation imposed on the working frequency range. The above discussion, which shows how the force generated in the exciter itself finds its way out to the structure, applies only as long as the moving parts of the shaker remain a rigid mass. Once the frequency of vibration approaches and passes the first natural frequency of the shaker coil and drive platform then there is a severe attenuation of the force which is available for driving the test object and although some excitation is possible above this critical frequency, it does impose a natural limit on the useful working range of the device. Not surprisingly, this frequency is lower for the larger shakers. Fig. 3.3 shows, approximately, the relationship between maximum force level and upper frequency limit for a typical range of shakers of this type.

3.4.4 Electrohydraulic Exciters

The next type of exciter to be considered is the hydraulic (electrohydraulic, to be precise). In this device, the power amplification to generate substantial forces is achieved through the use of hydraulics and although more costly and complex than their electromagnetic counterparts, these exciters do have one potentially significant advantage. That is their ability to apply simultaneously a static load as well as the dynamic vibratory load and this can be extremely useful when testing structures or materials whose normal vibration environment is combined with a major static load which may well change its dynamic properties or even its geometry. Without the facility of applying both static and dynamic loads simultaneously, it is necessary to make elaborate arrangements to provide the necessary static forces and so in these cases hydraulic shakers have a distinct advantage.

Another advantage which they may afford is the possibility of providing a relatively long stroke, thereby permitting the excitation of structures at large amplitudes — a facility not available on the comparably-sized electromagnetic shakers. On the other hand, hydraulic exciters tend to be limited in operational frequency range and only very specialised ones permit measurements in the range above 1 kHz, whereas electromagnetic exciters can operate well into the 30-50 kHz region, depending on their size. Also mentioned, as earlier, hydraulic shakers are more complex and expensive, although they are generally compact and lightweight compared with electromagnetic devices.

The comments made above concerning the need to measure force at the point of application to the structure also apply to this type of exciter, although the relative magnitudes of the various parameters involved will probably be quite different.

3.4.5 Attachment to the Structure
3.4.5.1 Push rods or stingers

For the above excitation devices, it is necessary to connect the driving platform of the shaker to the structure, usually incorporating a force transducer. There are one or two precautions which must be taken at this stage in order to avoid the introduction of unwanted excitations or the inadvertent modification of the structure. The first of these is perhaps the most important because it is the least visible. If we return to our definition of a single mobility or frequency response parameter, $H_{jk}(\omega)$, we note that this is the ratio between the harmonic response at point or DOF j caused by a single harmonic force applied in DOF k. There is also a stipulation in the definition that this single force must be the **only** excitation of the structure and it is this condition that we must be at pains to satisfy in our test. Although it may seem that the exciter is capable of applying a force in one direction only — it is essentially a unidirectional device — there exists a problem on most practical structures whose motion is generally complex and multidirectional. The problem is that when pushed in one direction — say, along the x axis — the structure responds not only in that same direction but also in others, such as along the y and z axes and also in the three rotation directions. Such motion is perfectly in order and expected but it is possible that it can give rise to a secondary form of excitation if the shaker is incorrectly attached to the structure. It is usual for the moving part of the shaker to be very mobile along the axis of its drive but for it to be quite the opposite (i.e. very stiff) in the other directions. Thus, if the structure wishes to respond in, say, a lateral direction as well as in the line of action of the exciter, then the stiffness of the exciter will cause resisting forces or moments to be generated which are, in effect, exerted on the structure in the form of a secondary excitation. The response transducers know nothing of this and they pick up the total response, which is that caused not only by the intended driving force (which is known) but also by the secondary and unknown forces.

The solution is to attach the shaker to the structure through a drive rod or similar connector which has the characteristic of being stiff in one direction (that of the intended excitation) while at the same time being relatively flexible in the other five directions. One such device is illustrated in Fig. 3.5(a). Care must be taken not to over-compensate: if

Fig. 3.5 Exciter attachment and drive rod assembly.
(a) Practical example

the drive rod or 'stinger' is made too long, or too flexible, then it begins to introduce the effects of its own resonances into the measurements and these can be very difficult to extricate from the genuine data. For most general structures, an exposed length of some 5-10 mm of 1 mm diameter wire is found to be satisfactory, although by experience rather than by formal analysis. Various alternative arrangements are sometimes found, as illustrated in Figs. 3.5(b), (c), (d) and (e). Of these, (b) is unsatisfactory while (c) and (d) are acceptable, if not ideal, configurations. It is always necessary to check for the existence of an internal resonance of the drive rod — either axially or in flexure — as this can introduce spurious effects on the measured FRF properties. Furthermore, in the case of an axial resonance, it will be found that

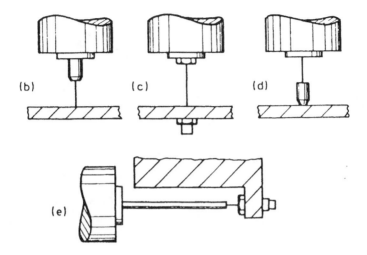

Fig. 3.5   Exciter attachment and drive rod assembly.
(b) Unsatisfactory assembly with impedance head; (c) Acceptable
assembly; (d) Acceptable assembly; (e) Use of extension rod

very little excitation force will be delivered to the test structure at
frequencies above the first axial mode. (This should be noted as it also
applies to cases where a non-flexible extension rod is used to overcome
problems of access, Fig 3.5(e).)

3.4.5.2   Excitation of rotating shafts

Increasingly, the techniques of modal testing are applied to machines
which contain rotating components. Some of these tests involve
excitation of the stationary, or non-rotating components, and for these
all the above comments apply directly (except for the expectation of
reciprocity between two points). However, it is sometimes necessary to
excite the rotor itself and this can present extra difficulties. The
possibilities of using non-contacting excitation are discussed in more
detail in Section 3.4.8, but a simple application of the conventional
shaker-pushrod-transducer configuration is worth mentioning here.
Fig. 3.6 shows a modal test in progress on a rotating shaft, with the
excitation applied through a free-spinning bearing which also houses
the force and response transducers. Correction for the mass of the non-
rotating components is, of course, possible and the setup shown is
entirely feasible, always provided that access to the shaft is possible.

3.4.5.3   Support of shaker

Another consideration which concerns the shaker is the question of how
it should be supported, or mounted, in relation to the test structure. Of

Fig. 3.6   Exciters applied to rotating shaft via auxiliary bearing

the many possibilities, some of which are illustrated in Fig. 3.7, two are
generally acceptable while others range from 'possible-with-care' to
unsatisfactory. The setup shown in Fig. 3.7(a) presents the most
satisfactory arrangement in which the shaker is fixed to ground while
the test structure is supported by a soft suspension. Fig. 3.7(b) shows an
alternative configuration in which the shaker itself is resiliently
supported. In this arrangement, the structure can be grounded or
ungrounded, but it may be necessary to add an additional inertia mass
to the shaker in order to generate sufficient excitation forces at low
frequencies. The particular problem which arises here is that the
reaction force causes a movement of the shaker body which, at low
frequencies, can be of large displacement. This, in turn, causes a
reduction in the force generation by the shaker so that its effectiveness
at driving the test structure is diminished.

In the cases shown in Figs. 3.7(a) and 3.7(b), we have sought to
ensure that the reaction force imposed on the shaker (equal and
opposite to that applied to the drive rod) is not transmitted to the test
structure. Fig. 3.7(c) shows a setup which does not meet that
requirement with the result that an invalid FRF measurement would be
obtained because the response measured at A would not be due solely to
the force applied at B (which has been measured), but would, in part, be
caused by the (unmeasured) force applied at C.

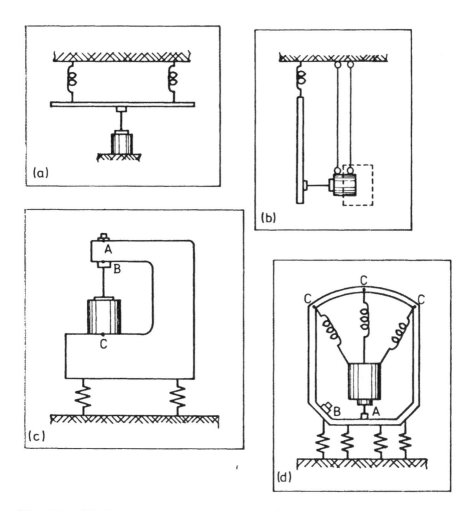

Fig. 3.7 Various mounting arrangements for exciter.
(a) Ideal configuration; (b) Suspended shaker plus inertia mass;
(c) Unsatisfactory configuration; (d) Compromise configuration

The final example, Fig. 3.7(d), shows a compromise which is sometimes necessary for practical reasons. In this case, it is essential to check that the measured response at A is caused primarily by the directly applied force at B and that it is not significantly influenced by the transmission of the reaction on the shaker through its suspension at C. This is achieved by ensuring that the frequency range for the measurements is well above the suspension resonance of the shaker: then, the reaction forces will be effectively attenuated by normal vibration isolation principles.

186

3.4.6 Hammer or Impactor Excitation

Another popular method of excitation is through use of an impactor or hammer. Although this type of test places greater demands on the analysis phase of the measurement process, it is a relatively simple means of exciting the structure into vibration. The equipment consists of no more than an impactor, usually with a set of different tips and heads which serve to extend the frequency and force level ranges for testing a variety of different structures. The useful range may also be extended by using different sizes of impactor. Integral with the impactor there is usually a load cell, or force transducer, which detects the magnitude of the force felt by the impactor, and which is assumed to be equal and opposite to that experienced by the structure. When applied by hand, the impactor incorporates a handle — to form a hammer (Fig. 3.8(a)). Otherwise, it can be applied with a suspension arrangement, such as is shown in Fig. 3.8(b), or even as a spring-loaded pistol device.

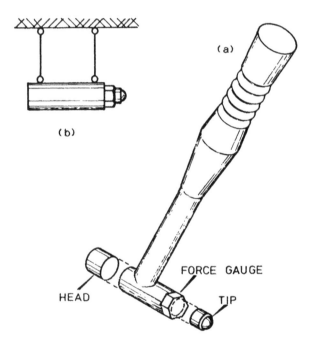

Fig. 3.8   Impactor and hammer details

Basically, the magnitude of the impact is determined by the mass of the hammer head and the velocity with which it is moving when it hits the structure. Often, the operator will control the velocity rather than the force level itself, and so an appropriate way of adjusting the order of the force level is by varying the mass of the hammer head.

The frequency range which is effectively excited by this type of device is controlled by the stiffness of the contacting surfaces and the mass of the impactor head: there is a system resonance at a frequency given by (contact stiffness / impactor mass)$^{1/2}$ above which it is difficult to deliver energy into the test structure. When the hammer tip impacts the test structure, this will experience a force pulse which is substantially that of a half-sine shape, as shown in Fig. 3.9(a). A pulse of this type can be shown to have a frequency content of the form illustrated in Fig. 3.9(b) which is essentially flat up to a certain frequency (ω_c) and then of diminished and uncertain strength thereafter. Clearly, a pulse of this type would be relatively ineffective at

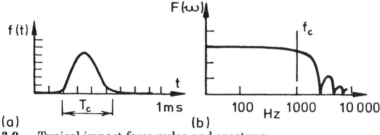

Fig. 3.9   Typical impact force pulse and spectrum.
(a) Time history; (b) Frequency spectrum

exciting vibrations in the frequency range above ω_c and so we need to have some control over this parameter. It can be shown that there is a direct relationship between the first cut-off frequency, ω_c, and the duration of the pulse, T_c, and that in order to raise the frequency range it is necessary to induce a shorter pulse length. This, in turn, can be seen to be related to the stiffness (not the hardness) of the contacting surfaces and the mass of the impactor head. The stiffer the materials, the shorter will be the duration of the pulse and the higher will be the frequency range covered by the impact. Similarly, the lighter the impactor mass the higher the effective frequency range. It is for this purpose that a set of different hammer tips and heads are used to permit the regulation of the frequency range to be encompassed. Generally, as soft a tip as possible will be used in order to inject all the input energy into the frequency range of interest: using a stiffer tip than necessary will result in energy being input to vibrations outside the range of interest at the expense of those inside that range.

On a different aspect, one of the difficulties of applying excitation using a hammer is ensuring that each impact is essentially the same as the previous ones, not so much in magnitude (as that is accommodated in the force and response measurement process) as in position and orientation relative to the normal to the surface. At the same time, multiple impacts or 'hammer bounce' must be avoided as these create difficulties in the signal processing stage.

Yet another problem to be considered when using the hammer type of excitation derives from the essentially transient nature of the vibrations under which the measurements are being made. We shall return to this characteristic later but here it is appropriate to mention the possibility of inflicting an overload during the excitation pulse, forcing the structure outside its elastic or linear range.

3.4.7 Step Relaxation

One variant on the hammer type of transient excitation is that referred to as 'step relaxation' or 'step release'. In this procedure, which is often used for large civil engineering types of structure, a large steady load is gradually applied to the structure under test, usually by means of a steel cable or a rope, until the required level of initial deflection is achieved — see Fig. 3.10. Alternatively, a steady load can be applied by

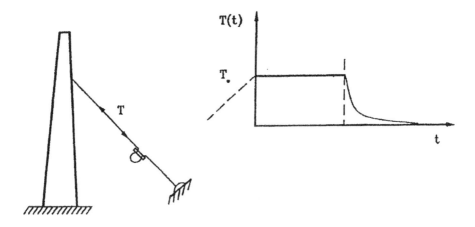

Fig. 3.10 Setup for excitation by step relaxation

other means, such as a dead weight applied to the structure. In order to initiate the vibration which will be used to measure the structure's properties, the static load is suddenly released, by cutting the cable or removing the steady load. The actual load should ideally be measured (by a load cell in the cable) as must the ensuing free vibration response so that the FRF can be derived in the usual way.

It will be realised that the relevant signal processing must be done to convert time histories to frequency spectra so that the FRF can be computed. It will be shown later that there is a requirement in this process that the signals from both excitation and response must be effectively zero at both ends of the time record. This would appear to be difficult in this case because both the initial load and the displacement (although not the velocity or acceleration) are finite at $t=0$ and zero at the end of the measurement, thereby introducing a discontinuity in the pseudo periodic signals assumed by the FRF calculation. This problem can be circumnavigated in one of two ways: either (a) the whole record of the (slow) loading phase is included in the time histories of both input force and resulting deflection or (b) the signals are differentiated so that it is the rate of change of force that is used as input and velocity rather than displacement (or acceleration rather than velocity) that is used to record the response. In this way, the step change in the actual force and deflection is converted to an impulsive signal which can be more readily processed, as is done in hammer testing.

3.4.8   Non-Contact Magnetic Excitation

As mentioned previously, there are occasions where a non-contacting excitation is desirable and this can sometimes be supplied by a simple electromagnet positioned so as to apply a controllable force on a chosen site on the test structure. Of course, the test object must be magnetic, or carry a magnetic strip at the excitation point, and both it and the magnet be grounded so that the d.c. or steady component of the magnetic field does not simply draw the magnet and structure together, otherwise the method is impossible. However, given the essential features, it is possible to design an electromagnet with a small, localised pole which can be placed close to the test structure and driven at the desired frequency to generate an alternating force, see Fig. 3.11(a). The force applied to the structure cannot be measured directly, but its reaction on the body of the magnet can be measured with a conventional force transducer, as shown in Fig. 3.11(b). Because of the current-squared nature of the force generated by the magnet, it is found that a sinusoidal input signal to the magnet results in a more complex periodic force signal, as illustrated in Fig. 3.11(c). The harmonics thus generated can be eliminated, if required, by the suitable generation of a multi-harmonic input signal. However, this is quite a complex task and not to be undertaken unless absolutely essential.

The one application where non-contact electromagnetic excitation has been developed to an advanced state is in the use of Active Magnetic Bearings (AMBs) in some types of high-speed rotating machinery. These devices have the dual role of (i) supporting the rotor and (ii) injecting any desired dynamic loading to the rotor, perhaps to compensate for an out-of-balance or to excite the rotor for vibration measurements such as are of interest to us here. Such units are sufficiently special-purpose that they will be not be considered further

190

Electro-magnet Designs.

Force end

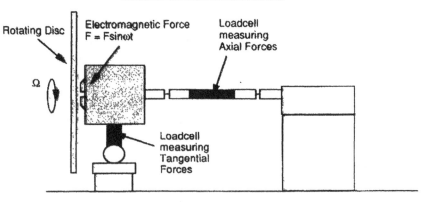

Target

Iron core

Force Gauge

Reaction Mass

Copper windings

Large force but acts over area.
Very heavy unit.

Iron strip cover

Target

Lower force but
compact & light weight

Iron wires

Electro Magnetic Exciter

Rotating Disc

Electromagnetic Force
F = Fsinωt

Loadcell measuring Axial Forces

Ω

Loadcell measuring Tangential Forces

(a)

Fig. 3.11 Excitation via non-contacting electromagnetic exciter.
(a) Exciter

Force Measurement with Electro-magnets

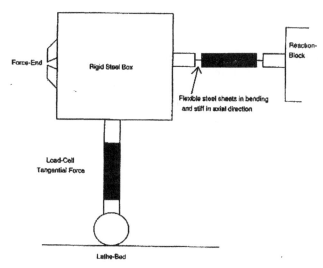

(b)

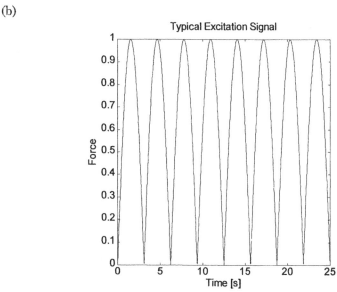

(c)

Fig. 3.11 Excitation via non-contacting electromagnetic exciter.
(b) Reaction force measurement; (c) Typical excitation signal

here but the interested reader is directed to [28] for further study and consideration of these devices.

3.4.9 Base Excitation

There exist a class of vibration tests in which a structure is excited in order to assess its vulnerability to some form of dynamic loading by driving it at its 'base'. Perhaps the simplest example of this is provided by the case of a model building which is placed on a slip table or bedplate to represent the ground which is then vibrated, laterally, to simulate the excitation this structure might experience in practice, see Fig. 3.12. This is the so-called 'base-excitation' type of test. Although,

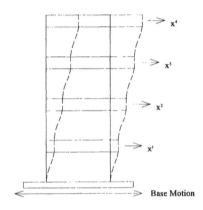

Fig. 3.12 Basic excitation configuration

strictly, the structure is treated as a free-free structure with a loaded boundary (the load provided by the mass of the slip table), the model which is under investigation is, in effect, that of the structure with its base grounded. In this type of test, only responses are measured — and **not** the input force — and the question which arises is whether or not it is possible to perform a modal test in this manner.

In fact, it **is** possible to carry out a limited modal test using the procedure and data measured in such a base excitation test, although it requires some additional information in the form of the mass distribution of the model in order to extract normalised mode shapes as well as the natural frequencies and damping factors. The analysis procedure necessary to achieve this goal will be described in Chapter 5.

3.4.10 Excitation of Rotating Structures

We have already mentioned (in Section 3.4.5.3) the mechanics of attaching a conventional exciter (shaker) to a rotating shaft on order to provide the necessary excitation to this class of structure but it is appropriate here to extend those comments to consider this requirement in more detail. The application of modal testing techniques to structures and machines that contain rotating components constitutes a growing area of interest, but it is one that is beset with several theoretical and practical obstacles. Providing a suitable excitation is one of these and one solution — using conventional exciters applied through an additional bearing — has already been presented. A second alternative has also been mentioned in passing, and that is to use non-contacting magnetic exciters. There are two possibilities in this category: (i) to use AMBs (active magnetic bearings) — where these are available — not only to support the rotor but also to inject an excitation signal for the purpose of stimulating vibrations that are then measured and used to derive the desired response functions (although it is important to note the complications that can arise when mixing quantities measured in the rotating frame of reference (e.g. responses) with others measured in the fixed frame of reference (e.g. excitation forces) — see Section 2.8), or (ii) to use separate magnetic exciters of the type described above in Section 3.4.8. In both cases, there is an inherent difficulty in accurately determining the force which is applied to the moving structure. While most AMBs have integral force-measuring devices, other magnetic exciters do not and it is often necessary to infer the applied excitation forces by measuring the reaction forces experienced by the support structure, after making any appropriate compensation for the motion of the magnet itself.

A third alternative, although one which does not readily lend itself to measuring the FRF data required for modal analysis, is to use an auxiliary rotating out-of-balance excitation device which is mounted co-centrically with the rotating component of interest. Such a device can be rotated synchronously or non-synchronously with the spinning test structure and can deliver an excitation which comprises simultaneous 'horizontal' and 'vertical' harmonic forces that are in quadrature with each other. As was shown in Section 2.8, the response to such out-of-balance excitations can be of interest in attempting to describe the dynamic properties of such a structure, although it is difficult to extract the FRFs necessary for a full modal analysis and modelling application.

A last possibility is to excite the rotating structure by hitting it with a hammer or equivalent device. While this will undoubtedly have the required effect of exciting vibration in the rotor, once again there will be a difficulty in establishing exactly what the excitation force(s) is(are). In particular, it will be clear that there will be a tangential force generated and applied in addition to the radial one and, further, that it will be very difficult to determine the magnitude of that component of excitation. If it is significant (i.e. if it produces response levels which are of similar magnitude to those from the primary (radial) excitation component (which is — presumably — measured by an internal force transducer)), then it will present a major obstacle to the task of determining FRF data because it will be impossible to extract the response components due to each of the excitation components individually, as is required for FRFs. Nevertheless, in view of the practical difficulties of applying any of the other excitations to a rotating structure, this simple option may be the most realistic option.

3.5   TRANSDUCERS AND AMPLIFIERS
3.5.1   General
The piezoelectric type of transducer is by far the most popular and widely-used means of measuring the parameters of interest in modal tests. Only in special circumstances are alternative types used and thus we shall confine our discussion of transducers to these piezoelectric devices.

Three types of piezoelectric transducer are available for mobility measurements — force gauges, accelerometers and impedance heads (impedance heads being simply a combination of force- and acceleration-sensitive elements in a single unit). The basic principle of operation makes use of the fact that an element of piezoelectric material (either a natural or synthetic crystal) generates an electrical charge across its end faces when subjected to a mechanical stress. By suitable design, such a crystal may be incorporated into a device which induces in it a stress proportional to the physical quantity to be measured (i.e. force or acceleration).

3.5.2   Force Transducers
The force transducer is the simplest type of piezoelectric transducer. The transmitted force F (see Fig. 3.13(a)), or a known fraction of it, is applied directly across the crystal, which thus generates a corresponding charge, q, proportional to F. It is usual for the sensitive crystals to be used in pairs, arranged so that the negative sides of both are attached to the case, and the positive sides are in mutual contact at their interface. This arrangement obviates the need to insulate one end of the case from the other electrically. One important feature in the design of force gauges is the relative stiffness (in the axial direction) of the crystals and of the case. The fraction of F which is transmitted through the crystals depends directly upon this ratio. In addition, there exists the undesirable possibility of a cross sensitivity — i.e. an electrical output when there is zero force F but, say, a transverse or shear loading — and this is also influenced by the casing.

The force indicated by the charge output of the crystals will always be slightly different from the force applied by the shaker, and also from that transmitted to the structure. This is because a fraction of the force

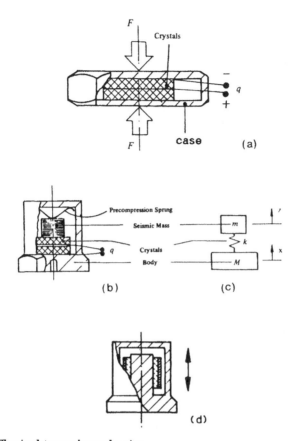

Fig. 3.13 Typical transducer basics.
(a) Force transducer; (b) Compression-type of piezoelectric accelerometer; (c) Shear-type of accelerometer; (d) Simple dynamic model for accelerometer

detected by the crystals will be used to move the small amount of material between the crystals and the structure. The implications of this effect are discussed later in a section on mass cancellation (Section 3.10), but suffice it to say here that for each force gauge, one end will have a smaller mass than the other, and it is this (lighter) end which should be connected to the structure under test.

3.5.3 Accelerometers

In an accelerometer, transduction is indirect and is achieved using an auxiliary, or seismic, mass (see Fig. 3.13(b) and (c)). In this configuration, the force exerted on the crystals is the inertia force of the seismic mass (i.e. $m\ddot{z}$). Thus, so long as the body and the seismic mass move together (i.e. $\ddot{z}$ and $\ddot{x}$ are identical), the output of the transducer will be proportional to the acceleration of its body ($\ddot{x}$), and thus of the structure to which it is attached. Analysis of a simple dynamical model for this device (see Fig. 3.13(d)) shows that the ratio ($\ddot{x}/\ddot{z}$) is effectively unity over a wide range of frequency from zero upwards until the first resonant frequency of the transducer is approached. At 20 per cent of this resonant frequency, the difference (or error in the ratio) $\ddot{x}/\ddot{z}$ is 0.04, and at 33 per cent it has grown to 0.12. Thus, in order to define the working range of an accelerometer, it is necessary to know its lowest resonant frequency. However, this property will depend to some extent upon the characteristics of the structure to which it is fixed, and indeed, upon the fixture itself. Manufacturers' data usually include a value for the 'mounted resonant frequency' and this is the lowest natural frequency of the seismic mass on the stiffness of the crystal when the body is fixed to a rigid base. In the simple model above, this frequency is given by $\sqrt{k/m}$. This value must be regarded as an upper limit (and thus not a conservative one) since in most applications the accelerometer body is attached to something which is less than rigid and so the transducer may possess a lower resonant frequency than that quoted. In any event, the actual attachment to the test structure must always be as rigid as possible and the manufacturers' advice to this end should be followed.

As with force transducers, there is a problem of cross- or transverse-sensitivity of accelerometers which can result from imperfections in the crystal geometry and from interaction through the casing. Modern designs aim to minimise these effects and one configuration which tends to be better in this respect is the shear type, a very simple arrangement of which is illustrated in the sketch in Fig. 3.13(c).

3.5.4   Selection of Accelerometers

Accelerometer sensitivities vary between 1 and 10,000 pC/g. How is one to choose the most suitable for any given application? In general, we require as high a sensitivity as possible, but it must be noted that the higher the sensitivity, the heavier and larger the transducer (thus interfering more with the structure) and furthermore, the lower is the transducer's resonant frequency (and thus the maximum working frequency). These considerations, together with any particular environmental requirements, will usually narrow the choice to one within a small range. For accurate measurements, especially on complex structures (which are liable to vibrate simultaneously in several directions), transducers with low transverse sensitivity (less than 1-2 per cent) should be selected.

It must be realised that the addition of even a small transducer to the structure imposes additional and unwanted forces on that structure. The loads are basically the inertia forces and moments associated with the transducer's motion along with the structure and although it may be possible to compensate for some of these (see the section on Mass Cancellation, below), it is not possible to account for them all and so, especially at high frequencies and/or for small structures, care should be taken to use the smallest transducer which will provide the necessary signals. To this end, some of the newer transducers with built-in amplifiers offer a considerable improvement.

3.5.5   Impedance Heads
It has been found convenient for some applications to combine both force- and acceleration-measuring elements in a single housing, thereby forming an impedance head. The main reason for using such a device is to facilitate the measurement of both parameters at a single point. We shall discuss the implications of this particular detail of experimental technique later, and confine our attention here to the performance characteristics of impedance heads. A typical, although not unique, construction for an impedance head is shown in Fig. 3.14. It is desirable

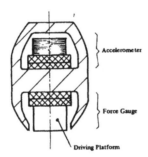

Fig. 3.14   Impedance head

to have both elements as close as possible to the structure — the force gauge in order to minimise the mass 'below the crystal', and the accelerometer to ensure as high a base stiffness as possible. Clearly, a design of the form shown must be a compromise and, accordingly, the specifications of these heads should be carefully scrutinised. In particular, the extent of any cross-coupling between the force and acceleration elements should be established, since this can introduce errors in certain frequency and/or mobility ranges.

3.5.6   Conditioning Amplifiers
One of the advantages of the piezoelectric transducer is that it is an active device, and does not require a power supply in order to function. However, this means that it cannot measure truly static quantities and

so there is a low frequency limit below which measurements are not practical. This limit is usually determined not simply by the properties of the transducer itself, but also by those of the amplifiers which are necessary to boost the very small electrical charge that is generated by the crystals into a signal strong enough to be measured by the analyser.

Two types of amplifier are available for this role — voltage amplifiers and charge amplifiers — and the essential characteristic for either type is that it must have a very high input impedance. Detailed comparisons of voltage and charge amplifiers are provided by manufacturers' literature. To summarise the main points: voltage amplifiers tend to be simpler (electronically) and to have a better signal/noise characteristic than do charge amplifiers, but they cannot be used at such low frequencies as the latter and the overall gain, or sensitivity, is affected by the length and properties of the transducer cable whereas that for a charge amplifier is effectively independent of the cable.

Increasingly, however, transducers are being constructed with integral amplifiers which make use of microelectronics. These devices require a low voltage power supply to be fed to the transducer but, in return, offer marked advantages in terms of a lower sensitivity to cable noise and fragility.

3.5.7   Attachment and location of Transducers
3.5.7.1   Attachment

The correct location and installation of transducers, especially accelerometers, is most important. There are various means of fixing the transducers to the surface of the test structure, some more convenient than others. Some of these methods are illustrated in Fig. 3.15 and range from a threaded stud, which requires the appropriate modification of the test structure (not always possible), through various adhesives in conjunction with a stud, to the use of a wax, which is the simplest and easiest to use. These forms of attachment become less reliable as the convenience improves, although it is generally possible to define the limits of the usefulness of each and thus to select the correct one in any particular application. Also shown on Fig. 3.15 are typical frequency limits for each type of attachment. The particularly high frequency capability of the screwed stud attachment can only be attained if the transducer is affixed exactly normal to the structure surface so that there is a high stiffness contact between the two components. If, for example, the axis of the tapped hole is not normal to the surface, then the misalignment which results will cause a poor contact region with a corresponding loss of stiffness and of high-frequency range.

Another consideration when attaching the transducer is the extent

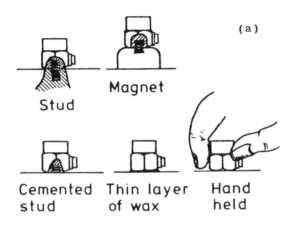

(a)

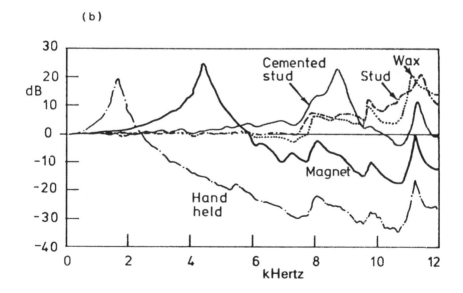

Fig. 3.15   Accelerometer attachment characteristics.
(a) Alternative attachment methods; (b) Frequency response characteristics of different attachments

of local stiffening which is introduced by its addition to the structure. If this is being fixed to a relatively flexible plate-like surface, then there is a distinct possibility that the local stiffness will be increased considerably. The only real solution to this difficulty is to move the transducer to another, more substantial, part of the structure.

3.5.7.2 Location

Another problem which may require the removal of the transducer to another location is the possibility (or even probability) that it is positioned at or very close to a node of one or more of the structure's modes. In this event, it will be very difficult to make an effective measurement of that particular mode. However, if the measurement points are already determined (for example, by matching those on a finite element grid), then it is necessary to make what measurements are possible, even if some of these are less than ideal.

Most modal tests require a point mobility measurement as one of the measured frequency response functions, and this can present a special problem which should be anticipated and avoided. Clearly, in order to measure a true point FRF, both force and response transducers should be at the same point on the structure and, equally clearly, this may well be hard to achieve. Three possibilities exist, viz:

(i)   use an impedance head (see Section 3.5.5);
(ii) place the force and acceleration transducers in line but on opposite sides of the structure as shown in Fig. 3.16(a) (this is only possible if the structure is locally thin);
(iii) place the accelerometer alongside, as close as possible to the force gauge, as shown in Fig. 3.16(b).

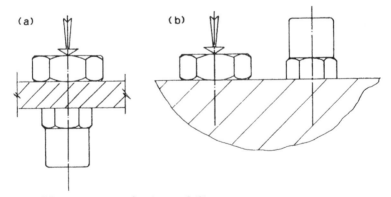

Fig. 3.16 Measurement of point mobility.
(a) Ideal configuration; (b) Compromise configuration

(Note that there will be a phase difference of 180° between acceleration measurements made by (ii) and (iii) due to the inversion of the accelerometer.)

It is the third alternative which presents the problem, since particular care is required to ensure that the resulting measurements really are representative of a point mobility. A practical example illustrates the problem very clearly. Fig. 3.17 shows the result of measuring the 'point' mobility on a model ship's hull structure where the third technique (iii) above was used. Fig. 3.17(a) illustrates a detail of the measurement area, showing four possible sites for the accelerometer. Fig. 3.17(b) presents four curves obtained by placing the accelerometer at each of these four points in turn. It is immediately

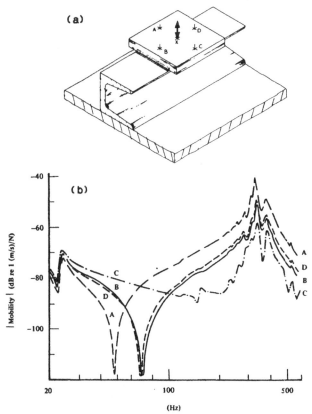

Fig. 3.17 Examples of measurements of 'point' mobility.
(a) Test structure layout; (b) Measured FRF curves

apparent that two of the positions (B and D) result in very similar curves, and it can be shown that both give a good indication of the motion of the point of interest, X. The other two curves, corresponding to the accelerometer being placed at either A or C, show markedly different response characteristics and indeed introduce enormous errors. In this particular case, the trends of these results could be explained by detailed examination of the local stiffness of the structure near the measurement point, but it is possible that similar errors could be incurred in other examples where such an explanation might be less readily available.

One possible approach to reduce errors of this type is to average the outputs of two accelerometers, such as A and C or B and D, but some caution would be advisable when interpreting any results thus obtained.

3.5.8   Laser Transducers
3.5.8.1   General

There is increasing use being made of optical and, in particular, of laser transducers which have a major advantage in the non-intrusive nature of their operation. While lasers have been used for holographic measurements of vibration for some time, these are not ideally suited for incorporation in a modal test because they require a major image processing task to convert the detailed full-field measurement of the structure's vibration response to useful numerical data, and because they generally do not readily provide adequate phase information as is required for the post-measurement analysis phase. A popular version of laser holography for mode shape (strictly, operating deflection shape) visualisation is provided by the ESPI systems which give a real-time on-screen display of fringe patterns from which the vibration pattern can be deduced. A second type of laser measurement system which has been proposed and used in connection with modal testing is the double-pulse laser holograph technique. This also requires post-measurement image processing but has the advantage that phase information is fully captured in the measurement.

However, the most readily usable laser transducers are those based on the laser Doppler velocimeter (LDV) concept, and we shall describe briefly the two types of LDV in current use: the standard single-point version and the scanning version (or SLDV)

3.5.8.2 Laser Doppler velocimeter (LDV)

The basic LDV transducer is a device which is capable of detecting the instantaneous velocity of the surface of a structure. The velocity measurement is made by directing a beam of laser light at the target point and measuring the Doppler-shifted wavelength of the reflected light which is returned from the moving surface, using an interferometer. The measurement made is of the velocity of the target point along the line of the laser beam. The sketch in Fig. 3.18 shows the

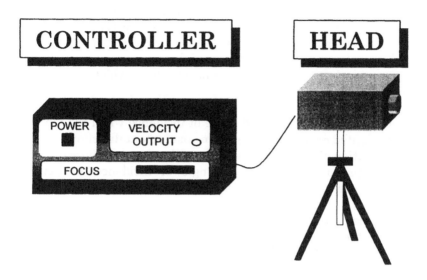

Fig. 3.18   Sketch of LDV measurement setup

basic layout of an LDV measurement system and it should be noted that
the main requirement is for a line of sight to the target measurement
point and a surface which is capable of reflecting the laser beam
adequately (this does not require a highly polished finish: and reflecting
materials are only necessary for very distant targets). There are
relatively few of these devices on the market at the present time and so
there is not a large range of specifications or performance to describe.
Characteristics for a typical device are:

*   frequency range: 0-250kHz
*   vibration velocity range: 0.01-20000 mm/s
*   target distance: 0.2-30 m
*   signal/noise: depends on target, type of scan and measurement
    range
*   sensitivity: 1-1000 mm/s/V

The main limitations of the LDV as a general-purpose response
transducer are (i) the line of sight requirement, and (ii) the problems
associated with speckle noise, a phenomenon which results in occasional
drop-outs, or null measurements, that have to be rejected from the data
set acquired in a measurement using this device. However, against
these disadvantages are distinct advantages for measurements which
have to be made in hostile environments, especially on surfaces which
have such high temperatures that conventional transducers cannot be

employed. Successful measurements have been made using an LDV on components with surface temperatures in excess of 1000C.

The practical details of using an LDV for accurate vibration measurements as required for modal testing can be studied in references such as [29,30]. The considerations which must be made include:

- correction for parallax and angle of incidence when the vibration is not oriented exactly in the beam measurement direction;
- registration of the structure's axes with respect to those of the LDV;
- determination of the actual angle of incidence, particularly if mirrors are used for indirect line of sight measurement points.

A typical measurement made side by side with a conventional accelerometer, using the same force transducer measurement in both cases, is shown in Fig. 3.19, from which it can be seen that the two transducers have similar performance in this case.

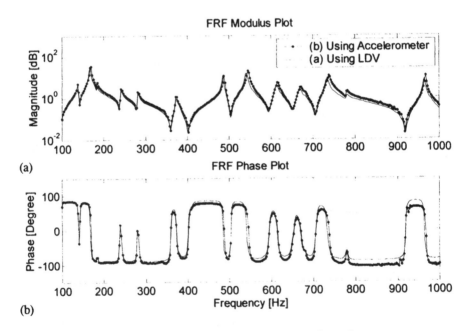

Fig. 3.19 Plot of FRFs using LDV and conventional accelerometer.

The conventional LDV can be directed towards its target site either (i) manually, by positioning the device so that the measurement beam falls on the desired measurement point, or (ii) by remote control of positioning mirrors, a pair of which can be integrated into the laser housing as a means of directing the laser beam at a series of different points without having to move the instrument. With this type of device, it is possible to specify the target site(s) via a computer which then directs the laser beam by suitable adjustment of the two mirrors, one controlling the x-deflection of the beam and the other controlling the y-direction movement. Typical applications of this type of device include (i) measuring at a large number of points comprising a mesh covering the surface of a structure of interest and (ii) rapidly visiting and measuring the response at a smaller number of carefully-selected sites, determined for their efficacy in some subsequent application (see Section 6.6 for a specific application of just this capability).

3.5.8.3   Scanning Laser Doppler Velocimeter (SLDV)

A natural extension of the capability described above for directing the measurement beam in the LDV is to incorporate a dynamic feature in the location mechanism. This means that we can exploit the ability to locate the laser beam direction on demand by devising a scanning process which moves the beam from one measurement site to the next in a controlled way. In its simplest form, the scanning velocimeter simply moves the beam to the first measurement point, makes a measurement, and then moves to the next measurement point and repeats the process. The faster this 'stepping' can be done, the shorter will be the total measurement time. However, the speed of such a procedure is limited by a number of factors: (i) those concerning the time required to dwell at a measurement point in order to have sufficient information to characterise its behaviour (that is a property of the motion to be measured and not of the measurement system) and (ii) those determined by the physical limitations inside the transducer, such as the inertia of the mirrors which must be moved in order to bring about the desired change of direction of the beam. In effect, the latter group constitute a major barrier to faster measurements of this type, especially at high frequencies of vibration, and the former at lower frequencies.

However, there is a definite advantage to be gained by using the scanning capability as an additional controllable variable in the measurement process. Within the abovementioned constraints, it is possible to control the mirror drives so that the measurement point moves across the surface of the structure in a controlled and prescribed *continuous* manner: a scan in the correct sense of the word and referred to as a CSLDV device. In general, a line scan can be acquired much faster this way than a series of discrete points. It can be seen that if the structure is undergoing steady-state harmonic vibration, then as the laser beam is scanned across its surface, the output signal from the transducer will exhibit an amplitude-modulated harmonic signal, such

206

as that shown in Fig. 3.20(a). If such a signal is subjected to a frequency analysis, it will be found to possess several frequency components, not simply the frequency of the steady vibration of the structure, which reflect that vibration signal, the scan rate or speed, and the amplitude variation along the scanned line (i.e. the operating deflection shape): see Fig. 3.20(b). As the frequencies of vibration and scanning are known, the measured signal can be used to extract valuable information about the amplitude variation across the structure. This is a feature that we shall be able to exploit in various applications, some of which are discussed in Sections 3.11 and 3.14.

LDV Output Time Signal   **FFT of LDV Output**

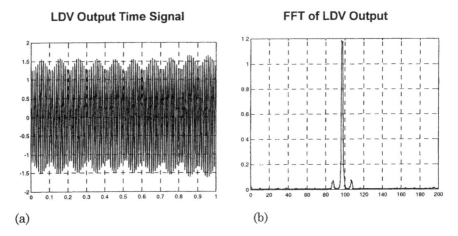

(a)   (b)

Fig. 3.20 Signals form scanned LDV measurement.
(a) Time record; (b) Frequency spectrum of (a)

3.5.8.4 Tracking LDV

The latest prospect which is offered by the scanning LDV is that of tracking a specific point, or points, on a moving structure and performing continuous or scanned measurements on such a structure while it is moving as well as vibrating. The two most immediate applications of such a device are (i) rotating machines, where it may be desirable to measure at selected points on a rotating disc, or other component, and (ii) vehicles, where it may be desired to make measurements continuously during the motion of the vehicle. Other applications, such as windscreen wipers, belts, etc. clearly show the demand for such a measurement device, if it can be developed. It is the first of these applications that is the most advanced at the time of writing. Further details of these possibilities can be found in reference [30], which describes vibration measurements made on the surface of a rotating disc using an SLDV whose point of measurement was continuously controlled by the signal from an encoder on the rotating shaft carrying the disc. Measurements were made during synchronous scanning (with the measurement point 'locked' on to a prescribed point on the disc), and during non-synchronous scanning, where the measurement point was scanned continuously around the spinning disc so as to address all points at a given radius (radii) during the continuous motion of the rotor. One problem is that, unless the measurement beam is exactly perpendicular to the direction of (continuous) motion, it detects a proportion of this, which may be very large, in addition to the required vibration velocity. While these are advanced applications at the time of writing, it is likely that they will become much more widely employed in the near future and so the interested reader is encouraged to scour the relevant literature for developments of both hardware, software and applications of this interesting branch of the technology.

3.6 ANALYSERS
3.6.1 Role of the Analyser
Each FRF measurement system incorporates an analyser in order to measure the specific parameters of interest — force and response levels. In principle, each analyser is a form of voltmeter although the signal processing required to extract the necessary information concerning magnitude and phase of each parameter leads to very complex and sophisticated devices. Different measurement systems employ different types of analyser. There are two of these in widespread use:

* frequency response analysers;
* spectrum analysers.

Nowadays, these are digital devices, although some analogue units may still be in use. In all cases, the data are supplied to the analyser in analogue form but with a digital instrument the first stage of the signal processing is analogue-to-digital (A-D) conversion so that the quantities to be processed are then in the form of a string of discrete values, as opposed to a continuous function. The subsequent processing stages are then performed digitally, as in a computer, using a variety of special-purpose routines which are usually hard-wired into the analyser in the form of a microprocessor for maximum speed of execution.

It is appropriate to describe briefly each of these types of device.

3.6.2 Frequency Response Analyser (FRA)
The frequency response analyser (FRA) is a development of the tracking filter concept in that it also is used with sinusoidal excitation. However, in these devices, the heart of the processing is performed digitally.

The source or 'command' signal — the sine-wave at the desired frequency — is first generated digitally within the analyser and then output as an analogue signal via a D-A converter. Within the same device, the two (or more) input signals (from the force and response transducers) are digitised via an A-D converter and then, one at a time, correlated numerically with the outgoing signal in such a way that all the components of each incoming signal other than that at exactly the frequency of the command signal are eliminated. This is, in effect, a digital filtering process and, when completed, permits the accurate measurement of the component of the transducer signals at the current frequency of interest. As with all such instruments, the accuracy of the measurements can be controlled to a large extent by the time spent in analysis. In the FRA, it is possible to improve the non-synchronous component rejection simply by performing the correlation (or filtering) over a longer period of time. Sometimes, this is quantified by the number of cycles (of the command signal) during which the computation takes place. Fig. 3.21 illustrates the type of dependence of rejection effectiveness versus integration time.

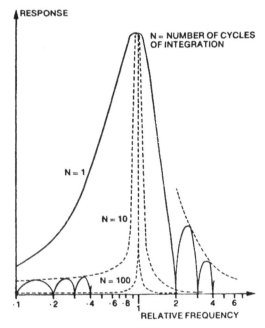

Fig. 3.21 Typical FRA filtering capability

3.6.3 Spectrum Analysers

The spectrum analyser (or frequency analyser, as it is sometimes called) is a quite different type of instrument to the FRA device described above. Whereas the FRA was concerned with extracting just one frequency component at a time, the spectrum analyser seeks to measure simultaneously **all** the frequency components present in a complex time-varying signal. Its output consists of a spectrum, usually a discrete one containing a finite number of components, describing the relative magnitudes of a whole range of frequencies present in the signal. Perhaps the simplest way to visualise the spectrum analyser is as a set of FRA units, each tuned to a different frequency and working simultaneously.

The current generation of digital spectrum analysers available have proven to be the workhorse of signal processing used in modal testing for the past two to three decades. The digital spectrum analyser (or 'Digital Fourier Analyser' or 'Digital Fourier Transform Analyser') is capable of computing a wide range of properties of incoming signals, including those required for FRF measurements, all of which are based on the Discrete Fourier Transform. In order to appreciate how best to use analysers of this type, it is necessary to understand the basics of their operation and we shall devote the next section to a study of some of the main features. However, it should be remembered that our objective in so doing is specifically to facilitate our use of the analyser to measure the quantities required for mobility measurements. These are listed in Chapter 2; in Section 2.11, equations (2.138), (2.143) and (2.154).

As a postscript here, it should be noted that, increasingly, the software for digital signal processing is becoming widely available in PCs and other computers and so special hard-wired analysers are less essential than a few years ago. While the do-it-yourself software option may appear attractive, and much cheaper than the purchase of a special-purpose instrument, care is required to ensure that all the functions provided by the latter devices are included in the simpler package. Tasks such as the anti-aliasing filtering which is so important to reduce aliasing errors must still be undertaken and this can only be effectively done prior to the A-D conversion, which means before the data enter the computer.

3.7 DIGITAL SIGNAL PROCESSING

3.7.1 Objective

The tasks of the spectrum analyser which concern us here are those of estimating the Fourier Transforms or Spectral Densities of signals which are supplied as inputs.

The basic theory of Fourier analysis is presented in Appendix 5 but it is appropriate here to relate the two most relevant versions of the fundamental Fourier transformation between the time and frequency domains. In its simplest form, this states that a function $x(t)$, periodic in time T, can be written as an infinite series:

$$x(t) = \frac{a_0}{2} + \sum_{n=1}^{\infty}\left(a_n\cos\left(\frac{2\pi n t}{T}\right)+b_n\sin\left(\frac{2\pi n t}{T}\right)\right) \tag{3.1a}$$

where a_n and b_n can be computed from knowledge of $x(t)$ via the relationships (3.1b):

$$a_n = \left(\frac{2}{T}\right)\int_0^T x(t)\cos\left(\frac{2\pi n t}{T}\right)dt$$

$$b_n = \left(\frac{2}{T}\right)\int_0^T x(t)\sin\left(\frac{2\pi n t}{T}\right)dt$$

(3.1b)

In the situation where $x(t)$ is discretised and of finite duration, so that it is defined only at a set of N particular values of time (t_k; $k=1,N$), we can write a finite Fourier series:

$$x_k\left(=x(t_k)\right) = \frac{a_0}{2} + \sum_{n=1}^{N/2}\left(a_n\cos\left(\frac{2\pi n t_k}{T}\right)+b_n\sin\left(\frac{2\pi n t_k}{T}\right)\right); k = 1, N \tag{3.1c}$$

The coefficients a_n, b_n are the Fourier or Spectral coefficients for the function $x(t)$ and they are often displayed in modulus (and phase) form, $c_n(= X_n) = (a_n^2 + b_n^2)^{1/2}$ (and $\phi_n = tg^{-1}(-b_n/a_n)$). This is the form of the Fourier transform with which we are concerned throughout the practical application of the theory used in this subject.

The signals (accelerometer or force transducer outputs) originate in the time domain and the desired spectral properties are in the frequency domain. Fig. 3.22 shows the various types of time history encountered, their Fourier Series or Transforms or Spectral Density, and the approximate digitised (or discrete) approximations used and produced by a Discrete Fourier Transform (DFT) analysis.

3.7.2 Basics of the DFT
In each case, the input signal is digitised (by an A-D converter) and recorded as a set of N discrete values, evenly spaced in the period T during which the measurement is made. Then, assuming that the

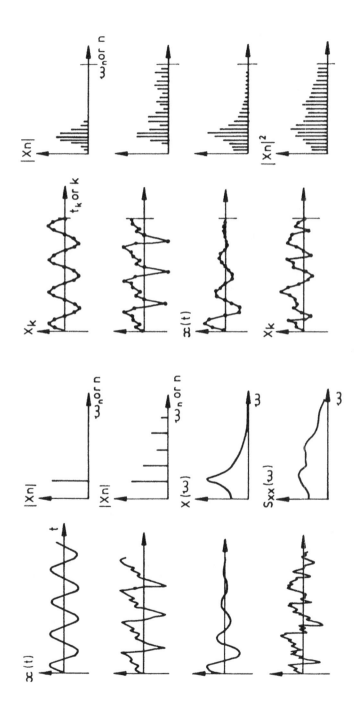

Fig. 3.22   Time histories and their spectral analyses: continuous and discrete signals

sample in time T is periodic, a Finite Fourier Series (or Transform) is computed according to (3.1c) above as an estimate to the required Fourier Transform. There is a basic relationship between the sample length T, the number of discrete values N, the sampling (or digitising) rate ω_s and the range and resolution of the frequency spectrum (ω_{max}, $\Delta\omega$). The range of the spectrum is 0-ω_{max} (ω_{max} is the Nyquist frequency) and the resolution of lines in the spectrum is $\Delta\omega$, where

$$\omega_{max} = \frac{\omega_s}{2} = \frac{1}{2}\left(\frac{2\pi N}{T}\right) \tag{3.2}$$

$$\Delta\omega = \frac{\omega_s}{N} = \frac{2\pi}{T} \tag{3.3}$$

As the size of the transform (N) is generally fixed for a given analyser (and is usually, though not always, a power of 2: 512, 1024 etc.), the frequency range covered and the resolution of the spectral lines is determined solely by the time length of each sample. This fact introduces constraints on the use of these analysers, as will be seen later.

The basic equation which is solved in order to determine the spectral composition derives from that given in Appendix 5:

$$\begin{Bmatrix} x_1 \\ x_2 \\ x_3 \\ \vdots \\ x_N \end{Bmatrix} = \begin{bmatrix} 0.5 & \cos(2\pi/T).. \\ 0.5 & \cos(4\pi/T).. \\ 0.5 & \cos(6\pi/T).. \\ \vdots & \vdots & \vdots \\ 0.5 & \cos(2N\pi/T) \end{bmatrix} \begin{Bmatrix} a_0 \\ a_1 \\ b_1 \\ \vdots \\ : \end{Bmatrix} \quad \text{or} \quad \{x_k\} = [C]\{a_n\} \tag{3.4}$$

Thus, we use $\{a_n\} = [C]^{-1}\{x_k\}$ to determine the unknown spectral or Fourier coefficients contained in $\{a_n\}$. Much of the effort in optimising the calculation of spectral analysis is effectively devoted to equation (3.4) and the most widely used algorithm is the 'Fast Fourier Transform' developed by Cooley and Tukey in the 1960s (Reference [29]). That method requires N to be an integral power of 2 and the value usually taken is between 256 and 4096.

There are a number of features of digital Fourier analysis which, if not properly treated, can give rise to erroneous results. These are generally the result of the discretisation approximation and of the need to limit the length of the time history. In the following sections we shall discuss the specific features of **aliasing, leakage, windowing,**

filtering, zooming and **averaging**.

3.7.3   Aliasing

There is a problem associated with digital spectral analysis known as 'aliasing' and this results from the discretisation of the originally continuous time history (Fig. 3.22). With this discretisation process, the existence of very high frequencies in the original signal may well be misinterpreted if the sampling rate is too slow. In fact, such high frequencies will appear as low frequencies or, rather, will be indistinguishable from genuine low frequency components. In Fig. 3.23, it can be seen that digitising a 'low' frequency signal (Fig. 3.23(a)) produces exactly the same set of discrete values as result from the same process applied to a higher frequency signal, (Fig. 3.23(b)).

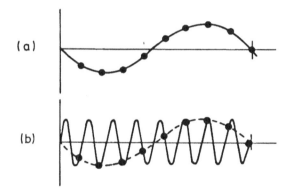

Fig. 3.23   The phenomenon of aliasing.
(a) Low-frequency signal; (b) High-frequency signal

Thus, a signal of frequency ω and one of $(\omega_s - \omega)$ are indistinguishable when represented as a discretised time history and this fact causes a distortion of the spectrum measured via the DFT, even when that is computed exactly.

A signal which has a true frequency content shown in Fig. 3.24(a) will appear in the DFT as the distorted form shown in Fig. 3.24(b). Note from Section 3.7.2 that the highest frequency which can be included in the spectrum (or transform) is $(\omega_s/2)$ and so the indicated spectrum should stop at that frequency, irrespective of the number of discrete values. (It should be noted that sometimes the spectrum is drawn over a wider frequency range than $0-(\omega_s/2)$. However, when this is done, the spectrum beyond $(\omega_s/2)$ is simply a reflection of that between $0-(\omega_s/2)$ and contains exactly the same information, but falsely labelled.) The distortion evident in Fig. 3.24(b) towards the upper end of the valid

214

frequency range can be explained by the fact that the part of the signal which has frequency components above $(\omega_s/2)$ will appear reflected or 'aliased' in the range $0\text{-}(\omega_s/2)$. Thus we see a Fourier Transform composed as illustrated in Fig. 3.24(b).

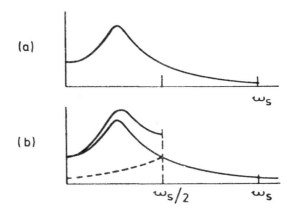

Fig. 3.24 Alias distortion of spectrum by discrete Fourier transform (DFT).
(a) True spectrum of signal; (b) Indicated spectrum from DFT

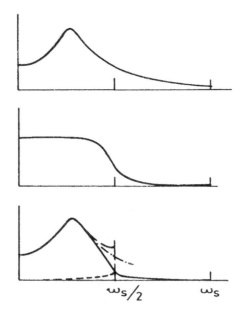

Fig. 3.25 Anti-aliasing filter process

The solution to the problem is to use an **anti-aliasing** filter which subjects the original time signal to a low-pass, sharp cut-off filter with a characteristic of the form shown in Fig. 3.25. This has the result of submitting a modified time history to the analyser. Because the filters used are inevitably less than perfect, and have a finite cut-off rate, it remains necessary to reject the spectral measurements in a frequency range approaching the Nyquist frequency, $(\omega_s/2)$. Typical values for that rejected range vary from 0.5-1.0 $(\omega_s/2)$ for a simple filter to 0.8-1.0 $(\omega_s/2)$ for a more advanced filter design. It is for this reason that a 2048-point transform does not result in the complete 1024-line spectrum being given on the analyser display: typically, only the first 800 lines will be shown as the higher ones are liable to be contaminated by the imperfect anti-aliasing.

It is essential that the correct anti-aliasing precautions are taken and so they are usually provided as a non-optional feature of the analyser.

3.7.4   Leakage

Leakage is a problem which is a direct consequence of the need to take only a finite length of time history coupled with the assumption of periodicity. The problem is best illustrated by the two examples shown in Fig. 3.26 in which two sinusoidal signals of slightly different

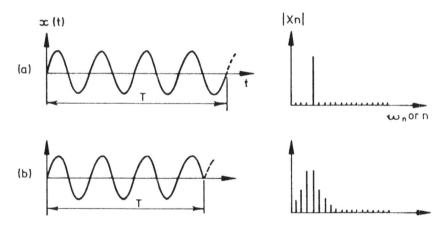

Fig. 3.26   Sample length and leakage of spectrum.
(a) 'Ideal' signal; (b) 'Awkward' signal

frequencies are subjected to the same analysis process. In the first case, (a), the signal is perfectly periodic in the time window, T, and the resulting spectrum is quite simply a single line — at the frequency of the sine wave. In the second case, (b), the periodicity assumption is not strictly valid and there is a discontinuity implied at each end of the sample. As a result, the spectrum produced for this case does **not** indicate the single frequency which the original time signal possessed — indeed, that frequency is not actually represented in the specific lines of the spectrum. Energy has 'leaked' into a number of the spectral lines close to the true frequency and the spectrum is spread over several lines or windows. The two examples above represent a best case and a worst case although the problems become most acute when the signal frequencies are lower.

Leakage is a serious problem in many applications of digital signal processing, including FRF measurement, and ways of avoiding or minimising its effects are important refinements to our techniques. There are various possibilities, which include:

- changing the duration of the measurement sample length to match any underlying periodicity in the signal (i.e. by changing T in the example in Fig. 3.26(b) so as to 'capture' an exact number of cycles of the (obviously sinusoidal) signal. Although such a solution can remove the leakage effect altogether, it can only do so if the signal being analysed **is** periodic — which is not always the case — **and** if the period of that signal can be determined — which is often difficult and, anyway, may well have been the objective of the analysis in the first place;
- increasing the duration of the measurement period, T, so that the separation between the spectral lines — the frequency 'resolution' — is finer (this does not remove but does reduce the severity of the leakage effect);
- adding zeroes to the end of the measured sample ('zero padding'), thereby partially achieving the preceding result but without requiring more data; or
- by modifying the signal sample obtained in such a way as to reduce the severity of the leakage effect. This process is referred to as 'windowing' and is widely employed in signal processing.

3.7.5   Windowing

In many situations, the most practical solution to the leakage problem involves the use of windowing and there are a range of different windows for different classes of problem.

Windowing involves the imposition of a prescribed profile on the time signal prior to performing the Fourier transform and the profiles or 'windows' are generally depicted as a time function, $w(t)$, as shown in Fig. 3.27. The analysed signal is $x'(t) = x(t).w(t)$. The result of using a window is seen in the third column of Fig. 3.27 and, for the case previously shown in Fig. 3.26(b), this produces the improved spectrum shown in Fig. 3.28. The Hanning (b) or Cosine Taper (c) windows are typically used for continuous signals, such as are produced by steady periodic or random vibration, while the Exponential window (d) is used for transient vibration applications where much of the important information is concentrated in the initial part of the time record and

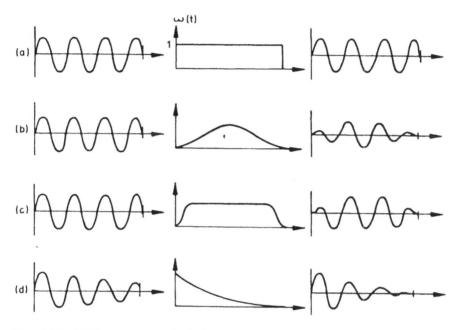

Fig. 3.27 Different types of window.
(a) Boxcar; (b) Hanning; (c) Cosine-taper; (d) Exponential

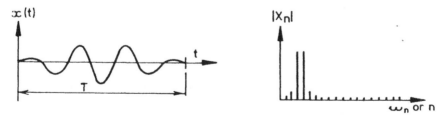

Fig. 3.28 Effect of Hanning window on discrete Fourier transform

would thus be suppressed by either of the above choices.

In all cases, a rescaling is required to compensate for the attenuation of the signals by the application of the window. However, if both response and excitation signals are subjected to the **same** window, and the results are used only to compute an FRF ratio, then the rescaling is not necessary. Care should be taken in some cases, especially with transient tests, as the two signals may be treated differently, using different windows on the two channels.

We have shown the effect of applying a window to the time domain signal which can benefit from such modification prior to undergoing its Fourier transformation calculation. It is possible, also, to witness the effect of applying a window by examining the same process in the frequency domain and, although this is more complex than the direct multiplication we have just made in the time domain, it serves a useful role to make such a parallel study.

It is a simple matter to make a Fourier transform of the time function which defines the window, $w(t)$, and to define the corresponding frequency-domain function, $W(\omega)$. Of course, because $w(t)$ is continuous, its spectrum, $W(\omega)$, will also be continuous. For the specific case of the Hanning window (in which $w(t) = 0.5(1 - \cos(2\pi t / T))$), we obtain the spectrum shown in Fig. 3.29. Similar spectra can be obtained for the other common windows used, and these are also shown in Fig. 3.29.

In seeking to define the spectrum of a signal **after** windowing, it must be noted that this cannot be obtained simply by multiplying the original signal spectrum by the spectrum of the window. Instead, it is necessary to perform a convolution of these two frequency-domain quantities so that the required output spectrum, $X'(\omega)$, is expressed in terms of the input spectrum, $X(\omega)$, and that of the window, $W(\omega)$, by the relationship:

$$X'(\omega) = X(\omega) * W(\omega) \qquad (3.5)$$

where * denotes the convolution process. However, it is appropriate to note again, at the end of this section, that a signal conditioning process which involves multiplication in the time domain (such as windowing) requires convolution to carry out the same process in the frequency domain. We shall see in the next section how the reverse is also true: a process which requires simple multiplication of spectra demands convolution in the time domain in order to define the modified signal's time-history.

3.7.6 Filtering

There is another signal conditioning process which has a direct parallel with windowing, and that is the process of filtering. In fact, we have already described one type of filter in our discussion of the aliasing problem and seen there that a filter is rather like a window, except that it is applied in the frequency domain rather than the time domain. We saw in Fig. 3.25 how the spectrum of a modified signal was obtained by

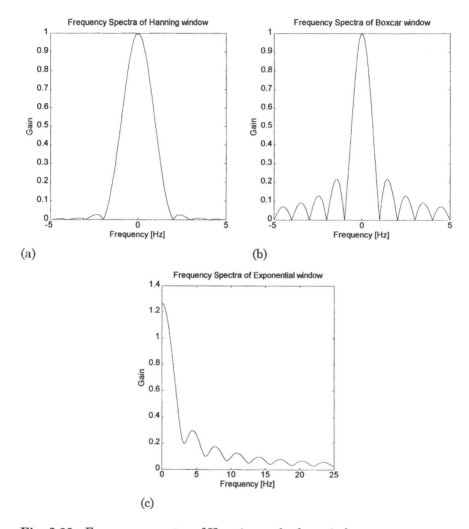

Fig. 3.29 Frequency spectra of Hanning and other windows.
(a) Hanning; (b) Boxcar; (c) Exponential

the simple process of multiplying the original signal spectrum by the frequency characteristic of the filter. This filter is of the low-pass type and other common filters are:

- high-pass
- band-limited
- narrow-band
- notch

220

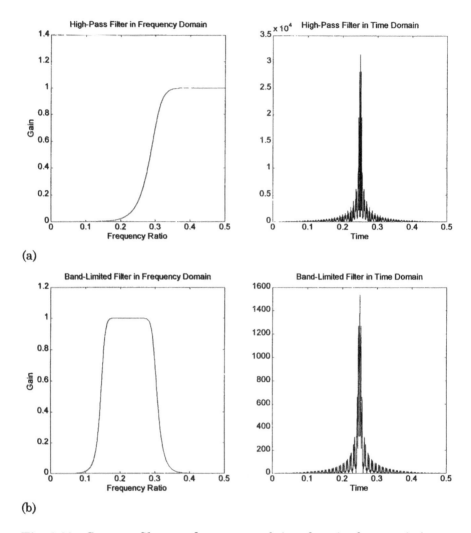

Fig. 3.30 Common filters — frequency and time domain characteristics.
(a) High-pass; (b) Band-limited

all of which are illustrated in Fig. 3.30. In practice, all filters will have a finite frequency range over which they function as designed and, although shown as having clear cut-off frequencies, will exhibit roll-off features near these critical frequency regions, the clarity of which determines their quality.

In the same way that the time-domain characteristic of a window could be transformed to the frequency domain, so also can a filter's characteristic be represented in the time domain, and this is done in the

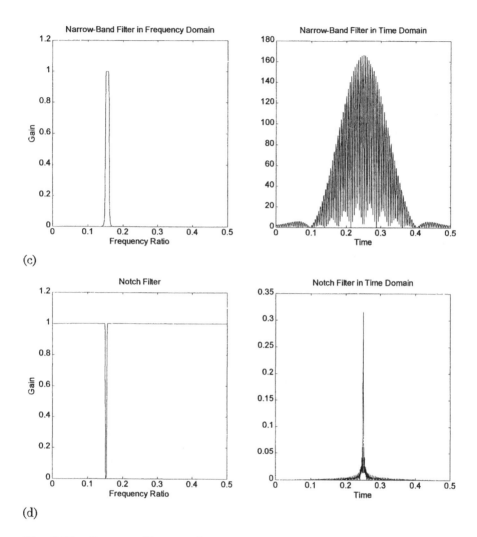

Fig. 3.30 Common filters — frequency and time domain characteristics.
(c) Narrow-band; (d) Notch

second part of Fig. 3.30. In order to derive expressions for the time domain descriptions of signals which have been filtered, it is necessary to use the convolution procedure (see Appendix 5) so that, in this case of filtering, we can write:

$$X'(\omega) = X(\omega).W(\omega)$$

and

$$x'(t) = x(t) * w(t)$$

However, it is not usual to want to perform calculations in this way as the direct frequency formula is quite convenient.

3.7.7   Improving Resolution

So far, we have been concerned with the basic DFT which is often found to have limitations of inadequate frequency resolution, especially at the lower end of the frequency range and especially for lightly-damped systems. This problem arises because of the constraints imposed by the limited number of discrete points available (N), the maximum frequency range to be covered and/or the length of time sample available or necessary to provide good data. We shall now address possible actions to improve the resolution available to us and to the precision with which we can make the required spectral analysis of our data.

3.7.7.1   Increasing transform size

An immediate solution to this problem would be to use a larger transform but, although giving finer frequency resolution around the regions of interest, this carries the penalty of providing more information than is required and, anyway, the size of transform is not always selectable. Until quite recently, the time and storage requirements to perform the DFT were a limiting factor in deciding the size of the transform but, nowadays, computer power is such that this is less and less of a consideration. There comes a point, however, where increasing the fineness of the spectrum **overall** is counterproductive (there are too many data points to handle) and transform sizes of the order of 2000 to 8000 are standard (that means 1000 to 4000 frequency lines, not all of which may be presented).

3.7.7.2   Zero padding

While a simple expansion of the transform size is indeed a way of increasing the spectral resolution, it carries with it the requirement that a correspondingly longer sample of signal must be available (to maintain the same overall frequency range, but to increase resolution by n times, demands a signal sample of n times the duration). Sometimes, this is simply not feasible, for a variety of reasons, and in these cases it may be possible to achieve the same resolution increase by adding a series of zeroes to the short sample of actual signal so as to create a new sample which is longer than the original measurement and which thus provides the desired finer resolution. Care must be taken in such a procedure because, in keeping with the fact that no additional data have been provided, the apparently greater detail in the spectrum is achieved at a price. It is not a genuinely finer spectrum; rather, it is the coarser spectrum which is available from the actual measured data interpolated and smoothed by the extension of the analysed record.

An example of the effects, and potential dangers, of using zero padding is shown in Fig. 3.31. In the first case, (a), a standard DFT is

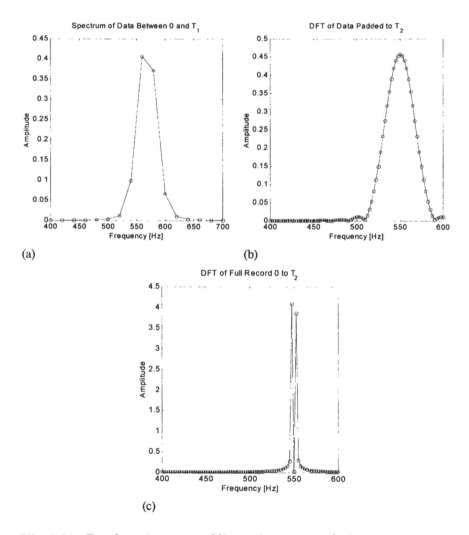

(a)   (b)

(c)

Fig. 3.31 Results using zero padding to improve resolution.
(a) DFT of data between 0 and T_1; (b) DFT of data padded to T_2; (c) DFT of full record 0 to T_2

shown for the sample of measured data between 0 and T_1. In the second plot, (b), a more detailed transform is performed by sampling at the same rate but on the longer modified sample, of duration T_2 produced by adding a string of zeros to the original measured sample. The finer resolution around the frequency of interest is clearly seen. In the third plot (c), we show the result of performing a larger transform on a complete record of length T_2, and this reveals the presence of two frequency components in the signal; a fact which was not evident from either of the first two analyses.

3.7.7.3 Zoom

The common solution to the need for finer frequency resolution is to 'zoom in' on the frequency range of interest and to concentrate all the spectral lines (400 or 800 etc.) into a narrow band between ω_{min} and ω_{max} (instead of between 0 and ω_{max}, as hitherto). There are various ways of achieving this result but perhaps the one which is easiest to understand physically is that which uses a frequency shifting process coupled with a controlled aliasing device.

Suppose the signal to be analysed, $x(t)$, has a spectrum, $X(\omega)$, of the type shown in Fig. 3.32(a) and that we are interested in a detailed (zoom) analysis around the second peak — between ω_1 and ω_2. If we apply a band-pass filter to the signal, as shown in Fig. 3.32(b), and

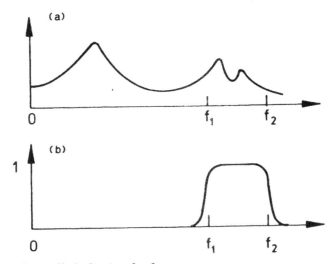

Fig. 3.32 Controlled aliasing for frequency zoom.
(a) Spectrum of signal; (b) Band-pass filter

perform a DFT between 0 and $(\omega_2 - \omega_1)$, then because of the aliasing phenomenon described earlier, the frequency components between ω_1

and ω_2 will appear aliased in the analysis range 0 to $(\omega_2-\omega_1)$ (see Fig. 3.33) with the advantage of a finer resolution. In the example shown here, the resolution is four times finer than in the original baseband analysis.

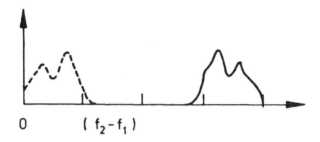

Fig. 3.33   Effective frequency translation for zoom

When using zoom to measure FRF in a narrow frequency range, it is important to ensure that there is as little vibration energy as possible **outside** the frequency range of interest. This means that wherever possible the excitation supplied to drive the structure should be band-limited to the analysis range; a feature not provided automatically on some analysers.

This is not the only way of achieving a zoom measurement, but it serves to illustrate the concept. Other methods are based on effectively shifting the frequency origin of the spectrum by multiplying the original time history by a $\cos(\omega_1 t)$ function and then filtering out the higher of the two components thus produced. For example, suppose the signal to be analysed is:

$$x(t) = A\sin(\omega t)$$

Multiplying this by $\cos(\omega_1 t)$ yields:

$$x'(t) = A\sin\omega t.\cos\omega_1 t = \frac{A}{2}\left(\sin\left(\omega-\omega_1\right)t + \sin\left(\omega+\omega_1\right)t\right) \tag{3.6}$$

and if we then filter out the second component we are left with the original signal translated down the frequency range by ω_1. The modified signal is then analysed in the range 0 to $(\omega_2-\omega_1)$ yielding a zoom measurement of the original signal between ω_1 and ω_2. In this method, it is clear that sample times are multiplied by the zoom magnification factor (of 4 times, or 10 times, etc.), but that sampling is carried out at the slower rate (also 4 times, or 10 times, etc.) dictated by the new effective frequency range.

There are, as mentioned earlier, other methods of zoom but this is the one most often used in standard modal testing applications. The other one worth mentioning is the so-called 'non-destructive zoom' method. This is a method which requires the capacity to store all the data obtained when the original signal is discretised at the original rate but over the longer time window to be used to gain the finer resolution. (If this were to be a zoom factor of 10 times, then this would represent an increase of 10 times the quantity of data acquired in a standard baseband measurement.) In this method, the full set of data points are assembled into smaller segments, each the length of a standard sample, by combining the 1st, 11th, 21st, etc. points (or the 3rd, 13th, 23rd, ...). Each of these constructed segments represents a small frequency band within the overall range (in fact, 10 per cent of it) and a direct DFT of each one reveals the spectrum in that limited frequency band but with a resolution which is 10 times finer than before. Details of this method, and other aspects of the DFT in general, can be found in [32].

3.7.8   Averaging
3.7.8.1 Need for averaging

We now turn our attention to another feature of digital spectral analysis that concerns the particular requirements for processing random signals. (So far, we have dealt with deterministic data.) When analysing random vibration signals, it is not sufficient to compute Fourier transforms — strictly, these do not exist for a random process — and we must instead obtain estimates for the spectral densities and correlation functions which are used to characterise this type of signal. Although these properties are computed **from** the Fourier transforms, there are additional considerations concerning their accuracy and statistical reliability which must be given due attention. Generally, it is necessary to perform an averaging process, involving several individual time records, or samples, before a result is obtained which can be used with confidence. The two major considerations which determine the number of averages required are:

*   the statistical reliability; and
*   the removal of spurious random noise from the signals.

Detailed guidance on the use of DFT analysers for valid random signal processing may be obtained from specialist texts, such as Newland (Reference [33]) and Bendat and Piersol (Reference [34]). However, an indication of the requirements from a statistical standpoint may be provided by the 'statistical degrees of freedom' (κ) which is provided by

$$\kappa = 2BT_t$$

where B = frequency bandwidth
$\quad T_t$ = total time encompassing all data
$\quad (= mT$ for m samples each of T duration)

As a guide, this quantity κ should be a minimum of 10 (in which case there is an 80 per cent probability that the estimated spectrum lies between 0.5 and 1.5 times the true value) and should approach 100 for reasonably reliable estimates (i.e. an 80 per cent probability that the measured value is within 18 per cent of the true value).

3.7.8.2 Types of average
There are several options which can be selected when setting an analyser into average mode: peak hold, exponential, linear, and so on. Each simply introduces a different weighting to the different samples in computing the relevant mean values.

3.7.8.3 Overlap averaging
The reference above to m samples, each of duration T, implies that these are mutually exclusive, as shown by Fig. 3.34(a). However, the computing capabilities of modern analysers mean that the DFT is calculated in an extremely short time and, consequently, that a new transform could be computed rather sooner than a new complete sample of data has been collected. In this case, it is sometimes convenient to perform a second transform as soon as possible, using the most recent N data points, even though some of these may have been used in the previous transform. This procedure is depicted in Fig. 3.34(b) and it is clear that 100 averages performed in this way cannot have the same statistical properties as would 100 completely independent samples.

Nevertheless, the procedure **is** more effective than if all the data points are used only once and it manifests this extra processing by producing smoother spectra than would be obtained if each data sample were used only once.

3.8　USE OF DIFFERENT EXCITATION SIGNALS
It is now appropriate to discuss the different types of excitation signals which can be used to drive the test structure so that measurements can be made of its response characteristics. There are, in fact, three different classes of signal used, and these are:

- periodic
- transient
- random

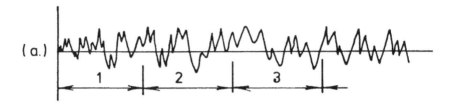

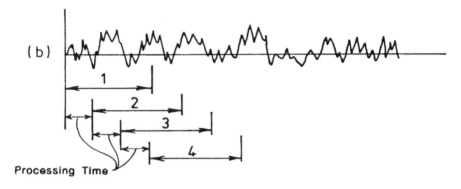

Processing Time

Fig. 3.34 Different interpretations of multi-sample averaging
(a) Sequential; (b) Overlap

These are widely used for modal testing excitation signals, specific
types of which include:

- periodic
  - *stepped sine*
  - *slow sine sweep*
  - *periodic*
  - *pseudo-random*
  - *periodic random*

- transient
  - *burst sine*
  - *burst random*
  - *chirp*
  - *impulse*

- random
  - *(true) random*
  - *white noise*

– *narrow-band random*

All of these are in widespread use, each having its own particular merits and drawbacks, and examples of each are shown in Fig. 3.35.

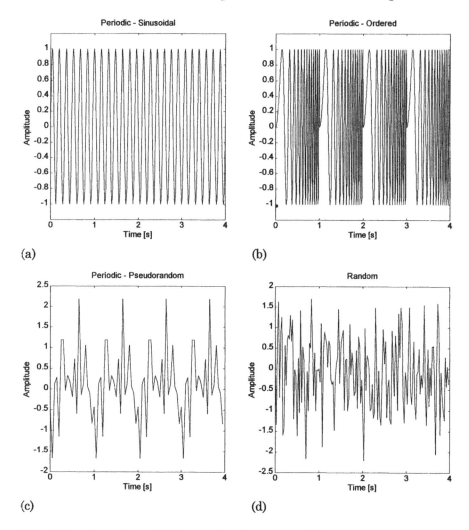

(a)   (b)

(c)   (d)

Fig. 3.35 Examples of different signal types used for excitation.
(a) Periodic — sinusoidal; (b) Periodic — ordered; (c) Periodic — pseudo-random; (d) Random

230

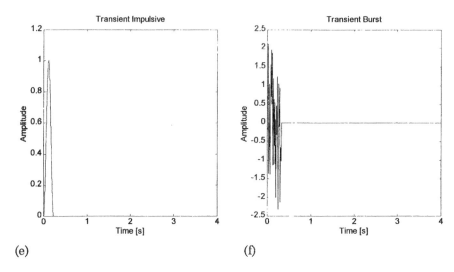

(e)   (f)

Fig. 3.35 Examples of different signal types used for excitation.
(e) Transient — impulsive; (f) Transient — burst

3.8.1   Stepped-Sine Testing

Stepped-sine (or step-sine) testing is the name given to the classical
method of measuring a frequency response function in which the
command signal supplied to the exciter is a discrete sinusoid with a
fixed amplitude and frequency. Strictly speaking, sine excitation is
simply a particular type of periodic signal, but it has several singular
features, and uses different analysis equipment to the other periodic
signals, so it is appropriate to treat it separately.

In order to encompass a frequency range of interest, the command
signal frequency is stepped from one discrete value to another in such a
way as to provide the necessary density of points on the frequency
response plot. Invariably driving through an attached shaker, the
excitation force and response(s) are measured, usually with a Frequency
Response Analyser. In this technique, it is necessary to ensure that
steady-state conditions have been attained before the measurements
are made and this entails delaying the start of the measurement
process for a short while after a new frequency has been selected as
there will be a transient response as well as the steady part. The extent
of the unwanted transient response will depend on:

(i)   the proximity of the excitation frequency to a natural frequency of
        the structure;
(ii)   the abruptness of the changeover from the previous command
        signal to the new one; and

(iii)   the lightness of the damping of the nearby structural modes.

The more pronounced each of these features is, the more serious is the transient effect and the longer must be the delay before measurements are made. In practice, it is only in the immediate vicinity of a lightly damped resonance that the necessary delay becomes significant when compared with the actual measurement time and at this condition, extra attention is usually required anyway because there is a tendency for the force signal to become very small and to require long measurement times in order to extract an accurate estimate of its true level.

One of the advantageous features of the discrete or stepped-sine test method is the facility of taking measurements just where and as they are required. For example, the typical FRF curve has large regions of relatively slow changes of level with frequency (away from resonances and antiresonances) and in these regions it is sufficient to take measurements at relatively widely spaced frequency points. By contrast, near the resonance and antiresonant frequencies, the curve exhibits much more rapid changes and it is more appropriate to take measurements at more closely spaced frequencies. It is also more efficient to use less delay and measurement time away from these critical regions, partly because there are less problems there but also because these data are less likely to be required with great accuracy for the modal analysis phases later on. Thus, we have the possibility of optimising the measurement process when using discrete sinusoidal excitation, especially if the whole measurement is under the control of a computer or processor, as is now generally the case. Fig. 3.36 shows a typical FRF curve measured using discrete sine excitation, in which two sweeps have been made through the range of interest: one a rapid coarse sweep with a large frequency increment, followed by a set of small fine sweeps localised around the resonances of interest using a much finer frequency increment and taking more care with each measured point. Of course, accurate detail of a resonance peak will only be possible if a sufficiently small increment is used for the fine sweeps. As a guide to the required increment, the following table shows the largest error that might be incurred by taking the maximum FRF value as the true peak value for that resonance.

3.8.2   Slow Sine Sweep Testing

This is the traditional method of mobility (or frequency response) measurement and involves the use of a sweep oscillator to provide a sinusoidal command signal, the frequency of which is varied slowly but continuously through the range of interest. As before, it is necessary to check that progress through the frequency range is sufficiently slow to

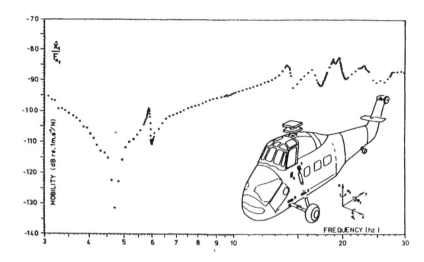

Fig. 3.36 Typical FRF data from stepped-sine test

Number of Frequency Intervals Between Half-Power Points*	Largest Error	
	%	dB
1	30	3
2	10	1
3	5	0.5
5	2	0.2
8	1	0.1

* For definition, see Section 4.2

check that steady-state response conditions are attained before measurements are made. If an excessive sweep rate is used, then distortions of the FRF plot are introduced, and these can be as severe as those illustrated in Fig. 3.37(a), which shows the apparent FRF curves produced by different sweep rates, both increasing and decreasing in frequency through a resonance region. One way of checking the suitability of a sweep rate is to make the measurement twice, once sweeping up and the second time sweeping down through the frequency range. If the same curve results in the two cases, then there is a probability (though not an assurance) that the sweep rate is not excessive.

It is possible to prescribe a 'correct' or optimum sweep rate for a given structure, taking due account of its prevailing damping levels. In theory, any sweep rate is too fast to guarantee that the full steady state

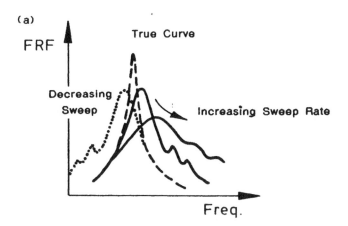

Fig. 3.37 FRF measurements by sine sweep test
(a) Distorting effect of sweep rate

response level will be attained, but in practice we can approach very close to this desired condition by using a logarithmic or similar type of sweep rate as indicated by the graphs in Figs. 3.37(b), (c) and (d). Alternatively, the ISO Standard (Reference [35]) prescribes maximum linear and logarithmic sweep rates through a resonance as follows (with the natural frequency, ω_r defined in Hz):

Linear sweep

or
$$S_{max} < 54(\omega_r)^2(\eta_r)^2$$
$$< 216(\omega_r)^2(\zeta_r)^2$$
Hz/min

Logarithmic sweep

or
$$S_{max} < 78(\omega_r)(\eta_r)^2$$
$$< 310(\omega_r)(\zeta_r)^2$$
octaves/min

234

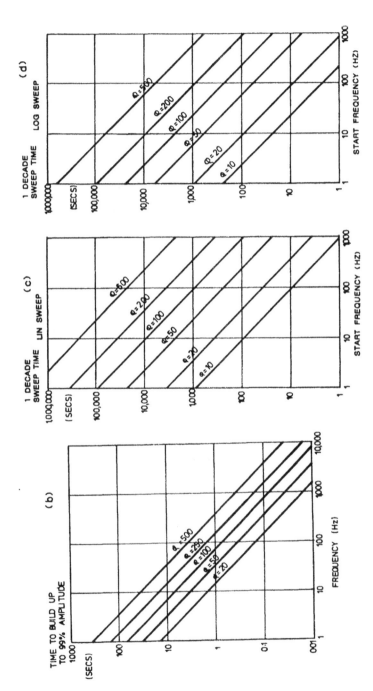

Fig. 3.37 FRF measurements by sine sweep test. (b) Time to build up to resonance; (c,d) Recommended sweep rates

3.8.3 Periodic Excitation

With the facility of the spectrum analyser to provide simultaneous information on all the frequency components in a given range, it is a natural extension of the sine wave test methods to use a complex periodic input signal which contains not one but all the frequencies of interest. This is nothing more complicated than a superposition of several sinusoids simultaneously, with the spectrum analyser capable of extracting the response to each of these components, again simultaneously.

The method of computing the FRF is quite simple: the discrete Fourier transform is computed of both the force and response signals and the ratio of these transforms gives the FRF, just as in equation (2.137). Since both signals are represented by discrete Fourier series, defined only at the set frequencies of that series, it follows that the FRF determined in this way is also defined only at those specific frequencies.

Two types of periodic signal are used, and both are usually generated within the analyser in order to ensure perfect synchronisation with the analysis part of the process. One is clearly deterministic and is an aperiodic signal in which all the components are mixed with ordered amplitude and phase relationships (e.g. a square wave), some of which will inevitably be relatively weak, while the other is a pseudo-random type of signal. This latter category involves the generation of a random mixture of amplitudes and phases for the various frequency components and may be adjusted to suit a particular requirement — such as equal energy at each frequency. This pseudo-random sequence is generated for a duration which equals the period of one sample in the analysis process, and is output repeatedly for several successive cycles and a satisfactory measurement is then made. A particular advantage of this type of excitation is its exact periodicity in the analyser bandwidth, resulting in zero leakage errors and therefore requiring no windows to be applied before its spectral analysis. (Note that one complete cycle must coincide exactly with a complete sample for the input to the analyser in order for the process to be truly periodic. It is possible for the original signal to be periodic in the literal sense but if this condition is not met then it is not seen as periodic by the analysis process and its spectral analysis will suffer from leakage as a result.) Having said that, however, it is appropriate here to quote a derivative of an old wisdom which, in the present context insists: 'when it is not necessary to use a window, then it is necessary not to use a window'. If, as is the case for truly periodic signals, there is no need to use a window of any form, then it is very important not to use one.

3.8.4   Random Excitation

3.8.4.1   FRF estimates using random excitation

For a truly random excitation, a different approach is required in order to determine the FRF although it is possible to undertake such measurements using the same spectrum analyser as before. Often, the source of a random command signal is found in an external device such as a noise generator and not in the analyser itself, although some types do contain independent noise sources for this purpose. In either event, it is important that the signal be different from the 'pseudo-random' type just mentioned as one of the family of periodic signals which are used for this purpose. It is usual for random excitation to be applied through an attached shaker.

The principle upon which the FRF is determined using random excitation has been explained in the theory chapter (Section 2.11) and relies on the following equations which relate the spectral density properties of the excitation and the response of a system undergoing random vibration:

$$S_{xx}(\omega) = |H(\omega)|^2 S_{ff}(\omega)$$
$$S_{fx}(\omega) = H(\omega).S_{ff}(\omega) \qquad (3.7)$$
$$S_{xx}(\omega) = H(\omega).S_{xf}(\omega)$$

where $S_{xx}(\omega)$, $S_{ff}(\omega)$, $S_{xf}(\omega)$ are the autospectra of the response and excitation signals and the cross spectrum between these two signals, respectively, and $H(\omega)$ is the frequency response function linking the quantities x and f. (The curious reader might wonder at this point why the FRF formulation is so much more complicated in the case of random vibration than for periodic vibration. There is a simple explanation for this: in steady-state periodic vibration, the response in any one sample period can be related identically to the excitation in that same sample time period. In random vibration, the response in any individual sample is **not** due entirely to the excitation in that same period. Thus, the FRF is not simply the ratio of the Fourier transforms of the matching samples of random vibration, as **is** the case in periodic vibration.)

The spectrum analyser has the facility to **estimate** these various spectral densities, although it must be appreciated that such parameters can never be measured exactly with only a finite length of data. However, in this case we have the possibility of providing a cross check on the results by using more than one of the equations (3.7). We can obtain an estimate to the required FRF using the second equation in (3.7), for example, and we shall denote this estimate as $H_1(\omega)$:

$$H_1(\omega) = \frac{S_{fx}(\omega)}{S_{ff}(\omega)} \qquad\qquad (3.8a)$$

We can also compute a second estimate for the FRF using the third equation in (3.7) and this we shall denote as $H_2(\omega)$:

$$H_2(\omega) = \frac{S_{xx}(\omega)}{S_{xf}(\omega)} \qquad\qquad (3.8b)$$

Now, because we have used different quantities for these two estimates, we must be prepared for the eventuality that they are not identical as, according to theory, they should be and to this end we shall introduce a quantity γ^2, which is usually called the 'coherence' and which is defined as:

$$\gamma^2 = \frac{H_1(\omega)}{H_2(\omega)} \qquad\qquad (3.9)$$

The coherence can be shown to be always less than or equal to 1.0.

Clearly, if all is well with the measurement, the coherence should be unity and we shall be looking for this condition in our test to reassure us that the measurements have been well made. In the event that the coherence is not unity, it will be necessary to establish why not, and then to determine what is the correct value of the FRF. It should be noted at this stage that many commercial analysers provide only one of these two FRF estimates as standard and, because it is fractionally easier to compute, this is generally $H_1(\omega)$. Of course, given the coherence as well, it is a simple matter to deduce the other version, $H_2(\omega)$, from equation (3.9) and it is an interesting exercise to overlay the two estimates on the analyser screen. One feature which will be noted is the fact that the phase is identical for both estimates, only the magnitude is different in the two versions, but that may be found to differ by very large amounts in some regions of the response function. When this occurs it should be the source of considerable concern because it means that the FRF estimate obtained is unreliable.

3.8.4.2 Noisy data

There are several situations in which an imperfect measurement might be made, and a low coherence recorded. There may well be noise on one or other of the two signals which could degrade the measured spectra: near resonance this is likely to influence the force signal so that $S_{ff}(\omega)$ becomes vulnerable while at antiresonance it is the response signal which will suffer, making $S_{xx}(\omega)$ liable to errors. In the first of these cases, $H_1(\omega)$ will suffer most and so $H_2(\omega)$ might be a better indicator near resonance while the reverse applies at antiresonance, as shown in the following equations:

$$H_1(\omega) = \frac{S_{fx}(\omega)}{S_{ff}(\omega) + S_{nn}(\omega)} \qquad (3.10a)$$

$$H_2(\omega) = \frac{S_{xx}(\omega) + S_{mm}(\omega)}{S_{xf}(\omega)} \qquad (3.10b)$$

where $S_{mm}(\omega)$ and $S_{nn}(\omega)$ are the autospectra of the noise on the output and input, $m(t)$ and $n(t)$, respectively. One suggestion which has been made for an alternative, and closer to optimum, formula for the FRF is defined as the geometric mean of the two standard estimates, identified as $H_V(\omega)$ and defined by:

$$H_V(\omega) = \sqrt{H_1(\omega)H_2(\omega)}$$

Here again the phase is identical to that in the two basic estimates.

A second possible problem area arises when more than one excitation is applied to the structure. In this case, the response measured cannot be directly attributed to the force which is measured and the cross checks afforded by the above procedure will not be satisfied. Such a situation can arise all too easily if the coupling between the shaker and the structure is too stiff and a lateral or rotational constraint is inadvertently applied to the testpiece, as discussed in Section 3.4.5.

Yet another possible source of low coherence arises when the structure is not completely linear. Here again, the measured response cannot be completely attributed to the measured excitation, and hence a less-than-unity coherence will result.

3.8.4.3   Noise-free FRF estimates

It was mentioned in Chapter 2 that a third estimator for the FRF could be defined in cases of random excitation which could have advantages in certain situations: this was the so-called 'instrumental variable' estimate, or $H_3(\omega)$. This formula for the FRF is only possible if more than the usual two channels are being measured simultaneously, and so it can only be applied on multi-channel systems. However, these are increasingly common and so the formula is of interest because it does provide an estimate for the FRF which is unbiased by noise on either the force or the response transducer signals. The formula is:

$$H_3(\omega) = \frac{S_{xv}(\omega)}{S_{fv}(\omega)}$$

where $v(t)$ is a third signal in the system, such as the voltage supplied to the exciter, and it exploits the fact that noise on either input (force) or output (response) channels does not contaminate cross-spectral density estimates in the way that auto spectra are affected.

3.8.4.4 Leakage

It is known that a low coherence can arise in a measurement where the frequency resolution of the analyser is not fine (small) enough to describe adequately the very rapidly changing functions such as are encountered near resonance and antiresonance on lightly-damped structures. This is known as a 'bias' error and it can be shown that when it occurs near resonance, the $H_2(\omega)$ estimate is always the more accurate of the two, although this can itself be seriously in error relative to the correct value. This is often the most likely source of low coherence on lightly-damped structures.

Fig. 3.38(a) shows a typical measurement made using random excitation, presenting the standard FRF, $H_1(\omega)$ and the coherence γ^2. This shows the trend for a great many practical cases, demonstrating a good coherence everywhere except near resonance and antiresonance. Fig. 3.38(b) shows a detail from the previous plot around one of the resonances, and included this time are both the FRF estimates, $H_1(\omega)$ and $H_2(\omega)$. Here it can be seen that the second alternative shows a larger and more distinct modal circle and is in fact a much more accurate representation of the true response function. When this situation is encountered (low coherence near resonance), the best solution is usually to make a zoom measurement. This is the procedure described in Section 3.7.7 whereby the standard number of lines available on the spectrum analyser may be applied to **any** frequency range, not just to a 'baseband' range of 0 to ω_{max} Hz. Thus, we can analyse in more detail between ω_{min} and ω_{max}, thereby improving the resolution and often removing one of the major sources of low coherence.

It is worth just illustrating this aspect of zoom measurements by means of a simple example. Suppose we have made a measurement, such as that in Fig. 3.38(a), over a frequency range from 0 to 200 Hz. This measurement gives us a frequency resolution of 0.78 Hz (for a 1024-point transform) which is clearly too coarse for the sharp resonance regions, especially around the lowest one where the frequency increment is a particularly large fraction of the natural

240

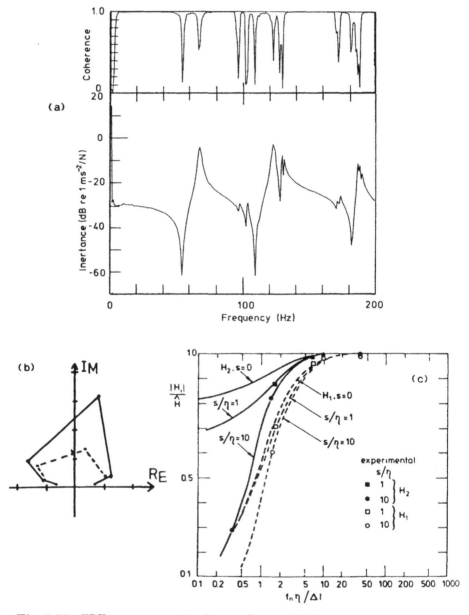

Fig. 3.38 FRF measurement using random excitation.
(a) FRF modulus and phase plots; (b) Nyquist plot detail of H_1
and H_2 FRF estimates; (c) Maximum errors in resonance peak
estimates [PC]

frequency concerned. We can improve the resolution by, for example, eight times by using a 25 Hz bandwidth zoom analysis with the analyser set to measure 25 Hz range between 113 and 138 Hz. By this device, we can greatly enhance the accuracy of the measurements around resonance as is illustrated by the sequence of plots in Fig. 3.39. It must be remembered, however, that this extra accuracy is gained at the cost of a longer measurement time: the analyser is now working approximately eight times slower than in the first measurement as the sample length (and thus data acquisition time) is linked to the overall frequency bandwidth.

Exactly how fine a frequency increment is required in order to reduce this source of error to a minimum depends on several factors, including the damping of the structure and the shaker/structure interaction discussed in Section 3.4. Fig. 3.38(c) indicates the maximum error which might be incurred in assuming that the true peak value of FRF is indicated by the maximum value on the measured spectrum.

3.8.4.5 Postscript

As a parting comment in this section on random excitation, we should mention the need to make several successive measurements and to accumulate a running average of the corresponding FRF estimates and coherence. It is sometimes thought that a poor coherence can be eliminated by taking a great many averages but this is only possible if the reason for the low coherence is random noise which can be averaged out over a period of time. If the reason is more systematic than that, such as the second and third possibilities mentioned above, then averaging will not help. A sequence of plots shown in Fig. 3.40 help to reinforce this point. However, for these cases or frequency ranges where the coherence genuinely reflects a statistical variation, then some guidance as to the required number of averages for a given level of confidence can be obtained from Fig. 3.41, based on the ISO Standard (Reference [35]).

Lastly, mention should be made here of a type of excitation referred to as 'periodic random' which is, in fact, a combination of pseudo-random (Section 3.8.3) and 'true' random. In this process, a pseudo-random (or periodic) excitation is generated and after a few cycles, a measurement of the input and the now steady-state response is made. Then, a different pseudo-random sequence is generated, the procedure repeated and the result treated as the second sample in what will develop to be an ensemble of random samples. The advantage over the simple random excitation is that due to the essential periodic nature of each of the pseudo-random (periodic) samples, there are no leakage or bias errors in any of the measurements. However, the cost is an increase in the measurement time, since $\frac{2}{3}$ or $\frac{3}{4}$ of the available data is

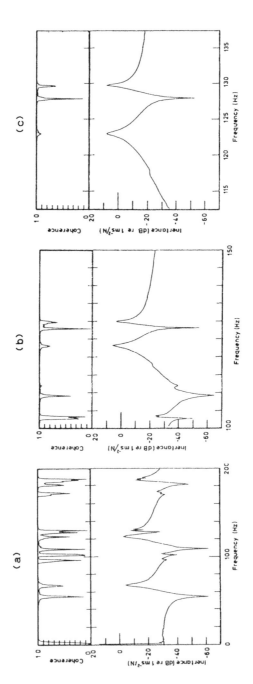

Fig 3.39   Use of zoom spectrum analysis
(a) 0(0.5)200 Hz; (b) 100(0.125)150 Hz; (c) 113(0.0625)138 Hz

243

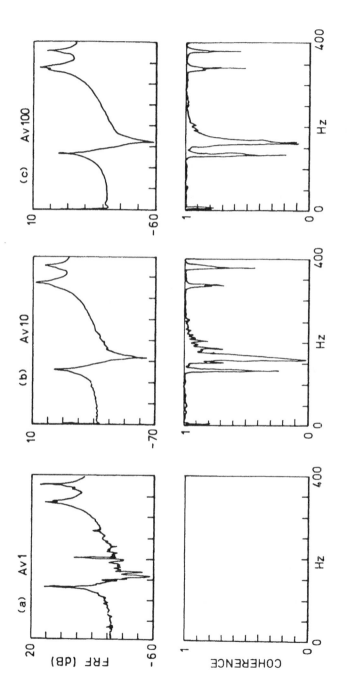

Fig. 3.40 Effect of averaging on FRF measurement using random excitation.
(a) 1 average; (b) 10 averages; (c) 100 averages

244

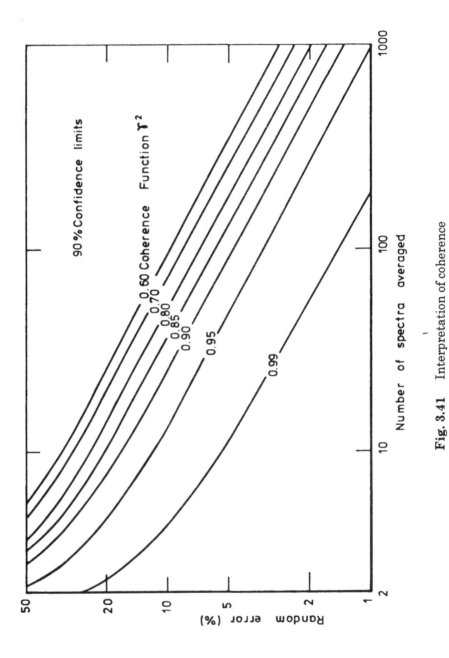

Fig. 3.41 Interpretation of coherence

unused while steady response conditions are awaited for each new sample.

3.8.5   Transient Excitation

There are three types of excitation to be included in this section because they all share the same principle for their signal processing, and they are (i) the 'burst' — a short section of signal (random or sine), (ii) the rapid sine sweep (or 'chirp', after the sound made by the input signal), and (iii) the impact from a hammer blow. The first and second of these generally require an attached shaker, just as do the previous periodic and random methods, but the last one can be implemented with a hammer or similar impactor device which is not permanently attached to the structure.

The principle which all these signals share is that the excitation and the consequent response are completely contained within the single sample of measurement which is made, a feature illustrated in Fig. 3.42(a). In practice, it is common to repeat the transient event more than once and to average the results to get the final result, but the measurement is based on the above principle, which means that the FRF can be derived from the simple ratio of the Fourier transforms of the response and excitation signals: $H(\omega) = X(\omega)/F(\omega)$. How they differ is in the exact form of the transient excitation signal and in the nature of the repeated application. In the burst type of signal, we have an excitation which is applied and analysed as if it were a continuous signal (such as periodic, or random), taking the successive samples for averaging one immediately after the other. For the chirp and impulse excitations, each individual sample is collected and processed before making the next one, and averaging, often deciding whether or not to accept the individual samples, and including them in the average, or rejecting them.

3.8.5.1   Burst excitation signals

Burst excitation signals consist of short sections of an underlying continuous signal — which may be a sine wave, a sine sweep or a random signal — followed by a period of zero output, resulting in a response which shows a transient build-up followed by a decay. Examples of burst random excitation and response are shown in Fig. 3.42(b). The duration of the burst is under the control of the operator and it is selected so as to provide the ideal signal processing conditions, which are essentially that the response signal has just died away by the end of the measurement period. If this condition has not been attained (burst too long), then leakage errors will result; if it has been reached well before the end of the period (burst too short), then the signal quality will be poor.

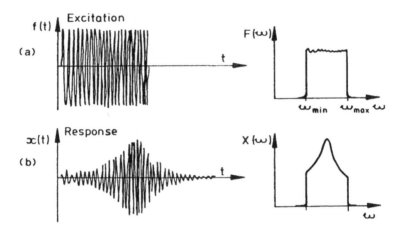

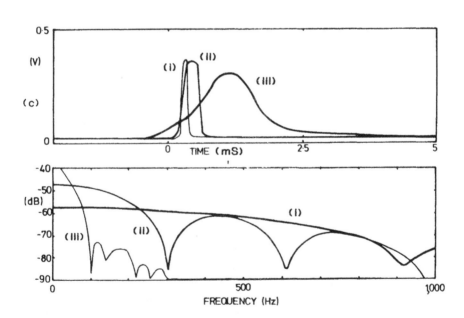

Fig. 3.42 Signals and spectra for excitations.
(a) Typical transient test data

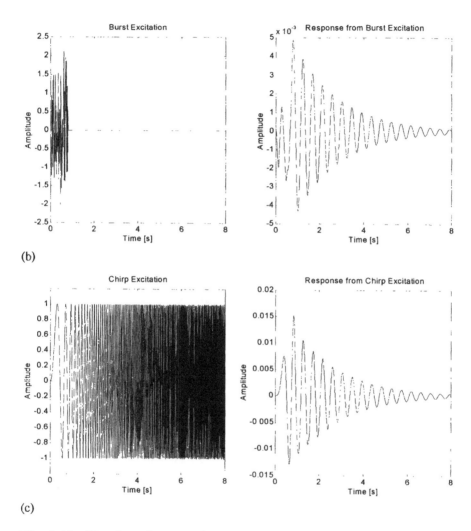

Fig. 3.42 Signals and spectra for excitations.

     (b) Burst excitation and response signals; (c) Chirp excitation and response signals

    The final measurement will be the result of averaging several samples, as with continuous periodic or random signals. In the case of the burst sine excitation, each sample would be expected to be identical so that the averaging serves only to remove noise on the signals. In the case of the burst random, however, each individual burst will be different to the others and so in this case there is an element of averaging randomly varying behaviour; a feature which is believed in

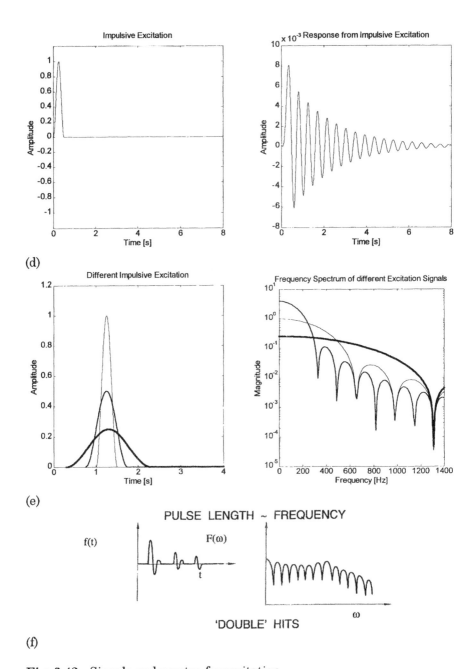

Fig. 3.42 Signals and spectra for excitation
(d) Impulsive excitation and response signals; (e) Different
impulsive excitation signals; (f) Signals and spectra for double-
hit case

some cases to enhance the measurement in the presence of weak non-linearities in the test structure.

3.8.5.2   Chirp excitation

The chirp consists of a short duration signal which has the form shown in Fig. 3.42(c) and which produces a response such as that shown in the accompanying graph. The frequency content of the chirp can be precisely chosen by the starting and finishing frequencies of the sweep, as illustrated in the figure and, as with the burst signal, the period of excitation can be tailored to optimise the conditions for the signal processing.

As in other cases, the final measurement will generally be the result of averaging several repeated samples.

3.8.5.3   Impulsive excitation

The hammer blow produces an input and response as shown in the companion plot of Fig. 3.42(d). From the analysis point of view, this and the previous case are very similar, the main difference being that the chirp offers the possibility of greater control of both amplitude and frequency content of the input and also permits the input of a greater amount of vibration energy. The spectrum of a chirp or a burst signal, such as those shown in Fig. 3.42(c) can be strictly controlled to be within the range between the starting and finishing frequencies of the rapid sinusoidal sweep or of the narrow band random signal. Although that of the hammer blow is dictated by the materials involved (as described in the earlier section) and is rather more difficult to control. However, it should be recorded that in the region below the first 'cut-off' frequency induced by the elasticity of the hammer tip/structure contact, the spectrum of the force signal tends to be very flat — see Fig. 3.42(e). Whereas that for a chirp, when applied through a shaker, suffers similar problems to those described earlier in Section 3. One practical difficulty which is sometimes encountered in using the hammer excitation is that of the 'double-hit'. On some structures, the movement of the structure in response to the hammer blow can be such that it returns and rebounds on the hammer tip before the user has had time to move that out of the way. In such cases, illustrated by the example in Fig. 3.42(f), the spectrum of the excitation (and hence the response) is seen to have 'holes' in it at certain frequencies, found on closer inspection to be derived from the time gap between the initial impact and the rebound. Although such holes in the spectrum are not intrinsically a problem, they result in erroneous results being obtained in a later stage of the measurement where the ratio is computed between a zero (or noisy) response level and a zero (or noisy) force level, again with an illustration shown in Fig. 3.42.

In order to perform the required Fourier analysis of all these cases of transient signals, an assumption is made that the data obtained from a single event (transient input and output) can be regarded as representing one period of a quasi-periodic process. This means that if exactly the same input was applied T seconds after the first one (where T is the duration of the transient event and the period of the measurement), then exactly the same response would be observed as had been captured in the T seconds of the actual measurement. This assumption is illustrated in Fig. 3.43 where it can be seen that:

(i)   the excitation pulse in (a) clearly satisfies the assumption;

(ii)   the response history in (b) also satisfies the assumption; but that

(iii) the response history in (c) does **not** satisfy the assumption, at least for the period T used to define the sample length.

This last example demonstrates a major difficulty encountered in processing the data from some transient tests. If the data as shown in Fig. 3.43(c) are used directly, then a form of leakage will result and the required spectral properties will be erroneously estimated. (A physical explanation for this error is readily provided: in the last case, if the excitation transient were *actually* to be applied a second time, then the response *actually* measured in the second sample period T would **not** be the same as that measured in the first. Clearly, in this case, the assumption of periodicity in the first measured samples would be invalid.) It would appear that the solution in cases such as that shown in Fig. 3.43(c) — which represents the case for most lightly-damped structures — is to lengthen the period, T, but often this is not easily changeable (it is determined by the frequency range required in the resulting spectrum) and so other solutions are required. Once again, the device of a window applied to the raw data provides a practical solution. This time, it is recommended to apply an exponential window (see Fig. 3.27) to the signals (it is preferable to apply the same window to both signals) with the result shown in Fig. 3.43(d). By choosing an appropriate exponential decay rate (equivalent, in effect, to adding numerical damping to the structure), the modified signal can be made to have effectively died away by the end of the prescribed measurement period, thereby satisfying the signal processing needs.

This method of signal conditioning is widely used in modal testing but it is not without its problems. Often, curiously complex modes are extracted from data treated in this way and so the method should be used sparingly in cases where accurate modal data are required for the eventual application.

An alternative to this problem is to use the zoom facility mentioned earlier. It is possible to exploit the fact that in a zoom measurement the

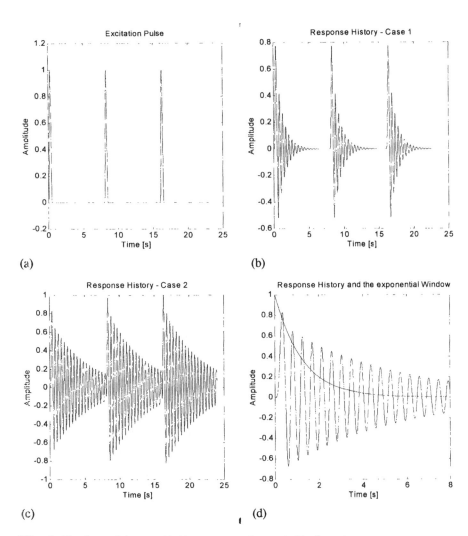

Fig. 3.43 Impulsive excitation as pseudo-periodic function.
(a) Excitation pulse; (b) Response history — Case 1; (c) Response history — Case 2; (d) Response history and the exponential window

measurement time is prolonged by the zoom factor and this very extension of the sample length can be used to advantage in cases where the baseband measurement is handicapped in the way described above. One of the consequences of using a zoom is that the frequency band is reduced by a proportionate amount and, at first sight, it might look as

252

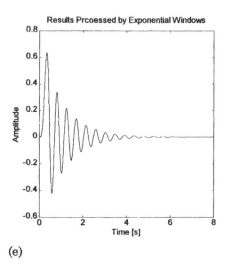

Results Prcoessed by Exponential Windows

(e)

Fig. 3.43 Impulsive excitation as pseudo-periodic function
(e) Results processed by exponential window

though we shall lose frequency range by extending the measurement period. However, by making a number (equal to the zoom factor) of separate measurements, each one for a different centre frequency for the zoomed band, it is possible to construct an FRF over the entire frequency range of interest with the twin advantages of it being a window-free measurement and having a much finer frequency resolution than the original base-band measurement would provide.

Once the necessary conditions of pseudo periodicity have been satisfactorily established, a discrete Fourier series description can be obtained of both the input force signal, $F(\omega_k)$, and of the response signal, $X(\omega_k)$, and the frequency response function can be computed from:

$$H(\omega_k) = \frac{X(\omega_k)}{F(\omega_k)} \tag{3.11}$$

Alternatively, the force signals can be treated in the same way as for random excitation, and the formulae in equation (3.8) are used. Usually, therefore, a spectrum analyser must be used for measurements made by transient excitation. Users of the chirp tend to favour the former approach while those more familiar with the impactor method advocate the latter method. However, considerable care must be exercised when interpreting the results from such an approach since the coherence function — widely used as an indicator of measurement quality — has a different significance here. One of the parameters indicated by coherence (although not the only one) is the statistical reliability of an estimate based on a number of averages of a random process. In the case of an FRF estimate obtained by treating the signals from a succession of nominally-identical impacts as a random process, we must note that, strictly, each such sample is a deterministic, and not probabilistic, calculation and should contain no statistical uncertainty. Thus, the main source for low coherence in this instance can only be leakage errors, non-linearity or high noise levels — not the same situation as for random excitation.

As mentioned above, there is a particular requirement in the case of a transient excitation that the signals (in effect, the response signals) must have died away by the end of the sample time. For lightly-damped structures this may well result in rather long sample times being required and this in turn poses a problem because it has a direct influence on the frequency range that can be covered. For example, if it is decided that a sample length of 2 seconds is necessary in order to ensure that the response has died away, and we have a 512-point transform, then the minimum time interval possible between successive points on the digitised time histories will be approximately 4 ms. This, in turn, means that the frequency resolution of the spectrum will be 0.5 Hz and the highest frequency on that spectrum will be as low as 250 Hz and it will not be possible to measure the FRF at higher frequencies than this. Such a restriction may well clash with the demands of the test and quite elaborate action may be required to remove it, for example by using an exponential window or by using a zoom facility as explained in the earlier sections.

Another feature usually employed in transient testing is that of making a whole series of repeat measurements under nominally identical conditions and then averaging the resulting FRF estimates. The idea behind this is that any one measurement is likely to be contaminated by noise, especially in the frequency regions away from resonance where the response levels are likely to be quite low. While this averaging technique does indeed enhance the resulting plots, it may well be that several tens of samples need to be acquired before a smooth FRF plot is obtained and this will diminish somewhat any advantage of speed which is a potential attraction of the method. Further details of both methods may be found in Reference [12] and Reference [35].

3.8.6   Postscript
In this section we have described above the various different types of excitation which can be used for mobility measurements. Each has its

good and bad features and, of course, its advocates and followers. Not surprisingly, no single method is the 'best' and it is probably worth making use of several types in order to optimise the time and effort spent on the one hand, and the accuracy obtained on the other. To this end, it is often found to be useful to make preliminary measurements using a wide frequency range and a transient or random excitation type and to follow this up with accurate sinusoidal-test measurements at a few selected frequency points in the vicinity of each resonance of interest. It is these latter data which are then submitted for subsequent analysis at which stage their greater accuracy is put to full advantage while excessive time has not been wasted on unnecessary points.

3.9 CALIBRATION

As with all measurement processes, it is necessary to calibrate the equipment which is used and in the case of FRF measurements, there are two levels of calibration which should be made. The first of these is a periodic 'absolute' calibration of individual transducers (of force and response) to check that their sensitivities are sensibly the same as those specified by the manufacturer. Any marked deviation could indicate internal damage which is insufficient to cause the device to fail completely, but which might nevertheless constitute a loss of linearity or repeatability which would not necessarily be detected immediately. The second type of calibration is one which can and should be carried out during each test, preferably twice — once at the outset and again at the end. This type of calibration is one which provides the overall sensitivity of the complete instrumentation system without examining the performance of the individual elements.

The first type of calibration is quite difficult to make accurately. As in all cases, the absolute calibration of a transducer or a complete system requires an independent measurement to be made of the quantity of interest, such as force or acceleration, and this can be quite difficult to achieve. The use of another transducer of the same type is seldom satisfactory as it is not strictly an independent measure, except in the case of using a reference transducer which has been produced to have very stable and reliable characteristics and which has previously been calibrated against an accepted standard under strictly controlled conditions. Other means of making independent measurements of displacements are generally confined to optical devices and these are not widely available, while independent methods of measuring force are even more difficult to obtain.

As a result, absolute calibration of transducers is generally undertaken only under special conditions and is most often performed using a reference or standard accelerometer, both for accelerometers and — with the aid of a simple mass — of force transducers.

One of the reasons why the absolute type of calibration has not been further developed for this particular application is the availability of a different type of calibration which is particularly attractive and convenient. With few exceptions, the parameters measured in a modal test are ratios between response and force levels, such as mobility or receptance, and so what is required is the ability to calibrate the whole measurement system. The quantities actually measured in the great majority of cases are two voltages, one from the force transducer and its associated electronics and the other from the response transducer. These voltages are related to the physical quantities being measured by the sensitivities of the respective transducers thus:

$$v_f = E_f f$$
$$v_{\ddot{x}} = E_{\ddot{x}} \ddot{x}$$
(3.12)

As mentioned above, there is some difficulty in determining values for E_f and $E_{\ddot{x}}$ individually but we note that, in practice, we only ever use the measured voltages as a ratio, to obtain the frequency response function:

$$\frac{\ddot{x}}{f} = \left(\frac{v_{\ddot{x}}}{v_f}\right)\left(\frac{E_f}{E_{\ddot{x}}}\right) = E\left(\frac{v_{\ddot{x}}}{v_f}\right)$$
(3.13)

and so what is required in the ratio of the two sensitivities:

$$E = \left(E_f / E_{\ddot{x}}\right)$$

This overall sensitivity can be more readily obtained by a calibration process because we can easily make an independent measurement of the quantity now being measured — the ratio of response to force. Suppose the response parameter is acceleration, then the FRF obtained is inertance (or accelerance) which has the units of (1/mass), a quantity which can readily be independently measured by other means. If we undertake a mobility or inertance measurement on a simple rigid mass-like structure, the result we should obtain is a constant magnitude over the frequency range at a level which is equal to the reciprocal of the mass of the calibration block, a quantity which can be accurately determined by weighing.

Fig. 3.44 shows a typical calibration block in use together with the result from a calibration measurement indicating the overall system calibration factor which is then used to convert the measured values of (volts/volt) to those of (acceleration/force), or whatever frequency

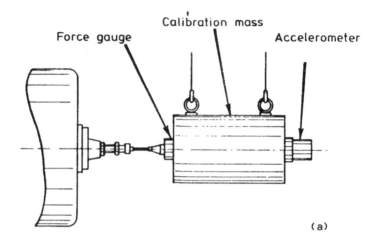

(a)

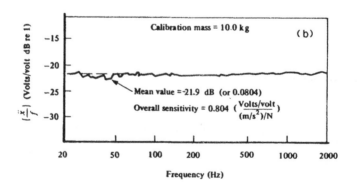

Fig. 3.44 Mass calibration procedure
(a) Measurement setup; (b) Typical measurement

response quantity is to be produced. The scale factor thus obtained should be checked against a corresponding value computed using the manufacturers' stated sensitivities and amplifier gains to make sure that no major errors have been introduced and to see whether either of the transducers has changed its sensitivity markedly from the nominal value. In practice, this check need only be made occasionally as the approximate scale factor for any given pair of transducers will become known and so any marked deviations will be spotted quite quickly.

A calibration procedure of this type has the distinct advantage that it is very easy to perform and can be carried out in situ with all the measurement equipment in just the same state as is used for the FRF measurements proper. In view of this facility, and the possibility of occasional faults in various parts of the measurement chain, frequent checks on the overall calibration factors are strongly recommended: as mentioned at the outset, at the beginning and end of each test is ideal.

3.10 MASS CANCELLATION

It was shown earlier how it is important to ensure that the force is measured directly at the point at which it is applied to the structure, rather than deducing its magnitude from the current flowing in the shaker coil or other similar indirect processes. This is because near resonance the actual applied force becomes very small and is thus very prone to inaccuracy. This same argument applies on a lesser scale as we examine the detail around the attachment to the structure, as shown in Fig. 3.45(a).

Here we see part of the structure, an accelerometer and the force transducer and also shown is the plane at which the force is actually measured. Now, assuming that the extra material (shown by the cross hatching) behaves as a rigid mass, m^*, we can state that the force actually applied to the structure, f_t, is different from that measured by the transducer, f_m, by an amount dependent on the acceleration level at the drive point, $\ddot{x}$, according to:

$$f_t = f_m - m^* \ddot{x} \tag{3.14}$$

Physically, what is happening is that some of the measured force is being 'used' to move the additional mass so that the force actually applied to the structure is the measured force minus the inertia force of the extra mass.

Now, the frequency response quantity we actually require is $A_t(\omega)$ $(= \ddot{X}/F_t)$, although we have measurements of $\ddot{X}$ and F_m only, yielding an indicated inertance, $A_m(\omega)$. This is a complex quantity and if we express it in its Real and Imaginary parts, we can obtain a relationship between $A_m(\omega)$ and $A_t(\omega)$ as follows:

258

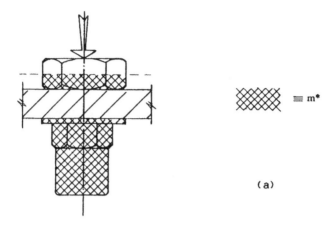

(a)

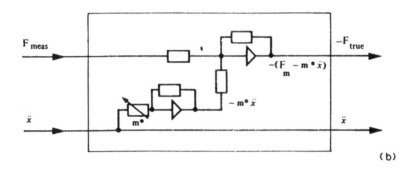

Fig. 3.45 Mass cancellation.

(a) Added mass to be cancelled; (b) Typical analogue circuit

$$\text{Re}(F_t) = \text{Re}(F_m) - m * \text{Re}(\ddot{X})$$
$$\text{Im}(F_t) = \text{Im}(F_m) - m * \text{Im}(\ddot{X})$$

(3.15a)

or

$$\text{Re}(1/A_t) = \text{Re}(1/A_m) - m *$$
$$\text{Im}(1/A_t) = \text{Im}(1/A_m)$$

(3.15b)

Equally, it is possible to perform this process of 'mass cancellation', or 'cancelling the mass below the force gauge', by using an electronic circuit which takes the two signals and carries out the vector addition of equation (3.11) before the signals are passed to the analyser. A suitable circuit using high-gain operational amplifiers is shown in Fig. 3.45(b), containing a variable potentiometer whose setting is selected according to the magnitude of the mass to be cancelled, m^*. A straightforward way of determining the correct setting with a device of this type is to make a 'measurement' with the force transducer and accelerometer connected together but detached from the structure. Then, by adjusting the variable potentiometer until there is effectively zero corrected force signal, f_t, we can determine the appropriate setting for that particular combination of transducers.

Mass cancellation is important when the mass to be cancelled (m^*) is of the same order as the apparent mass of the modes of the structure under test, and this latter is a quantity which varies from point to point on the structure. If we are near an antinode of a particular mode, then the apparent mass (and stiffness) will tend to be relatively small — certainly, only a fraction of the actual mass of the structure — and here mass cancellation may well be important. However, if we now move the same force and response transducers to another position which is near a nodal point of that same mode, now we shall find that the apparent mass (and stiffness) is much greater so that the addition of m^* is less significant and the mass cancellation correction is less urgent. This phenomenon manifests itself by a given structure appearing to have different values for each of its natural frequencies as the excitation and/or response points are varied around the structure.

One important feature of mass cancellation is that it can only be applied to point measurements, where the excitation and response are both considered at the same point. This arises because the procedure described above corrects the measured force for the influence of the additional mass at the drive point. If the accelerometer is placed at another point then its inertia force cannot be subtracted from the measured force since it no longer acts at the same point. It is still possible to correct for that part of m^* which is due to part of the force transducer mass but this is not the total effect. It should also be noted that the transducers' inertia is effective not only in the direction of the excitation but also laterally and in rotation. There are therefore several inertia forces and moments in play, only one of which can usually be compensated for. Nevertheless, there are many cases where this correction provides a valuable improvement to the measurement and should also be considered.

3.11   ROTATIONAL FRF MEASUREMENT
3.11.1   Significance of Rotational FRF Data

It is a fact that 50% of all degrees of freedom are rotations (as opposed to translations) and 75% of all frequency response functions involve rotational DOFs. However, it is relatively rare to find reference to methods for the measurement of rotational FRFs and this reflects the fact that virtually none are made. This situation arises from a considerable difficulty which is encountered when trying to measure either rotational responses or excitations and also when trying to apply rotational excitation, i.e. an excitation moment.

3.11.2   Measurement of Rotational FRFs Using Two or More Transducers

A number of methods have been tried to measure FRF data for the rotational DOFs, with limited success, but these are still in a development stage. However, it is believed that these FRF terms will be of increasing importance for future applications of modal testing and so it is appropriate to include here a brief discussion of some of the aspects of measuring rotational FRFs.

There are basically two problems to be tackled; the first is that of measuring rotational responses and the second is a companion one of generating and measuring the rotational excitations. The first of these is the less difficult and a number of techniques have been developed which use a pair of matched conventional accelerometers placed a short distance apart on the structure to be measured, or on a fixture attached to the structure. Both configurations are illustrated in Fig. 3.46 which

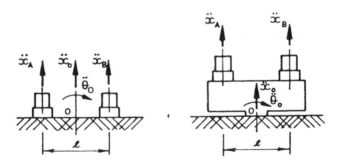

Fig. 3.46   Measurement of rotational response

also shows the DOFs of interest, x_0 and θ_0. The principle of operation of either arrangement is that by measuring both accelerometer signals, the responses x_0 and θ_0 can be deduced by taking the mean and difference of x_A and x_B:

$$x_0 = 0.5(x_A + x_B)$$

$$\theta_0 = (x_A - x_B)/\ell \tag{3.16}$$

This approach permits us to measure half of the possible FRFs — all those which are of the X/F or Θ/F type. The others can only be measured directly by applying a moment excitation and in the absence of any suitable rotational exciters it is necessary to resort to a similar device to the above. Fig. 3.47 shows an extension of the exciting block principle in which a single applied excitation force, F_1 say, can become the simultaneous application of a force F_0 ($=F_1$) and a moment M_0 ($=-F_1.\ell_1$). A second test with the same excitation device applied at position 2 gives a simultaneous excitation force F_0 ($=F_2$) and moment M_0 ($=F_2.\ell_2$). By adding and differencing the responses produced by

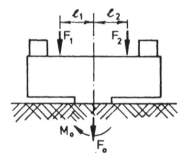

Fig. 3.47 Application of moment excitation

these two separate excitation conditions, we can deduce the translational and rotational responses to the translational force and the rotational moment separately, thus enabling the measurement of all four types of FRF:

$$\frac{X}{F}, \frac{\Theta}{F}, \frac{X}{M} \text{ and } \frac{\Theta}{M}$$

The same principle can be extended to more directions by the use of a multidimensional excitation fixture until the full 6×6 mobility matrix at any given point can be measured. However, it must be noted that the procedures involved are quite demanding, not least because they require the acquisition of subsequent processing of many different measurements made at different times.

Other methods for measuring rotational effects include specially developed rotational accelerometers and shakers but in all cases, there is a major problem that is encountered, which derives from the fact that the prevailing levels of output signal generated by the translational components of the structure's movement tend to overshadow those due to the rotational motions, a fact which makes the differencing operations above liable to serious errors. For example, the magnitude of the difference in equation (3.14) is often of the order to 1-2% of either of the two individual values. When the transducers have a transverse sensitivity of the order of 1-2%, the potential errors in the rotations are enormous. Nevertheless, several applications of these methods have been quite successful.

3.11.3 Measurement of Rotation DOFs Using a Scanning Laser

The scanning laser Doppler velocimeter (LDV) has already been introduced (in section 3.5.8) as a non-contact transducer which has applications to difficult response measurement tasks. It has now been applied to the rotational DOF measurement problem, with promising results.

The original use of the LDV in the present context was to provide a convenient means of measuring the two responses required by the method described in the previous section, and this it was found possible to do. However, advantage can be taken of the scanning capability, described in section 3.5.8.3, to measure the rotational DOFs with greater facility than is possible in the two-point method just discussed. The technique is to use the SLDV to conduct a continuous measurement while performing a small-amplitude linear scan at frequency Ω along a line between the two points of measurement in the previous approach while the structure is undergoing steady-state harmonic vibration at frequency, ω. This produces an amplitude-modulated signal whose three frequency components ($\omega - \Omega$, ω, $\omega + \Omega$) can be used to reveal the translation motion normal to the surface of the test structure and the rotation about an axis in the plane of the surface and perpendicular to the line of the scan. It will be found that the two side-band frequency components are actually complex conjugates of each other, so that data are only available for two separate measurements: in this case, one translation DOF and one rotation DOF, as required.

In fact, the optimum use of this technique is a simple extension of the foregoing method in which the LDV is set to perform a small-diameter *circular* scan around the point of measurement in order to measure three response DOFs: the normal translation plus the two rotations about perpendicular axes in the plane of the surface. The procedure is illustrated in Fig. 3.48(a) and shows the three DOFs at the measurement site, O, as including the normal translational response, $z(t)$, as well as the two rotations about the perpendicular axes, $\theta_x(t)$

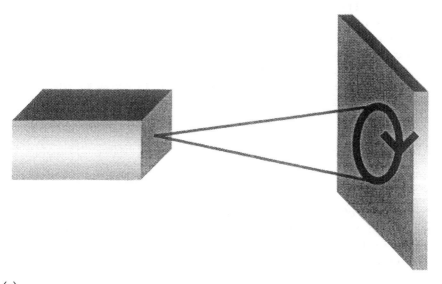

(a)

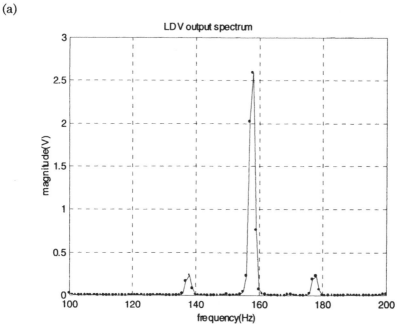

(b)

Fig. 3.48 Circular LDV scan to measure 3DOFs at point.
(a) Setup; (b) Typical RDOF measurement

and $\theta_y(t)$. The point at which the LDV makes its instantaneous measurement is at the end of the beam which is circling around the point, O, at a scanning speed of Ω_z and at a radius of r.

It can be shown that the output of the laser velocimeter, in the case where the structure is vibrating sinusoidally at frequency, ω, and where the small area of the vibrating surface which is covered by the scan is effectively rigid, is once again a signal with three sinusoidal components — V_ω, $V_{\omega-\Omega}$, $V_{\omega+\Omega}$ — at the three frequencies:

- ω
- $\omega - \Omega_z$
- $\omega + \Omega_z$

For each of these three frequency components there is a distinct measurable magnitude and phase and these data can be converted to yield the complex amplitudes — V, Θ_x, Θ_y — of the three response DOFs of interest, $z(t)$, $\theta_z(t)$ and $\theta_y(t)$, using the following formulae:

$$\mathrm{Re}(Z) = \mathrm{Re}(V_\omega) \; ;$$

$$\mathrm{Im}(Z) = \mathrm{Im}(V_\omega)$$

$$\mathrm{Re}(\Theta_x) = \left(\mathrm{Im}(V_{\omega+\Omega}) - \mathrm{Im}(V_{\omega-\Omega})\right)/r \; ;$$

$$\mathrm{Im}(\Theta_x) = \left(\mathrm{Re}(V_{\omega-\Omega}) - \mathrm{Re}(V_{\omega+\Omega})\right)/r$$

$$\mathrm{Re}(\Theta_y) = \left(\mathrm{Re}(V_{\omega+\Omega}) + \mathrm{Re}(V_{\omega-\Omega})\right)/r \; ;$$

$$\mathrm{Im}(\Theta_y) = \left(\mathrm{Im}(V_{\omega-\Omega}) + \mathrm{Im}(V_{\omega+\Omega})\right)/r$$

(3.17)

An example of this application of the SLDV is shown in Fig. 3.48(b) and includes data obtained during a narrow-band random excitation and response measurement, demonstrating the applicability of the method to more general vibration than the simple harmonic case.

It should be noted that the processing of the LDV output which results in the tri-frequency spectra is subject to the various features of signal processing already discussed, and care needs to be taken to avoid or to minimise any such problems. In particular, the problem of leakage encountered in discrete Fourier transform computations can be avoided altogether by selecting a scan rate (frequency) which is periodic with respect to the length of the analysis window. This setting is in addition to those already in place in respect of the frequency(ies) of vibration which are being measured, but since the scanning process is under the user's direct control, arranging for this requirement to be met need not be difficult.

3.12 MEASUREMENTS ON NON-LINEAR STRUCTURES

3.12.1 Introduction to Testing Structures with Non-linear Behaviour

Most of the theory upon which modal testing and FRF measurement is founded relies heavily on the assumption that the test structure's behaviour is linear. By this is meant that (i) if a given loading is doubled, the resulting deflections are doubled and (ii) the deflection due to two simultaneously-applied loads is equal to the sum of the deflections caused when the loads are applied one at a time. In practice, real structures are seldom completely linear although for practical purposes many will closely approximate this state. However, there are many complex structures which do behave in a non-linear way, especially in the vicinity of resonances, usually — but not always — at lower frequencies, and these can give rise to concerns and problems when they are being tested.

Signs of non-linear behaviour include:

" natural frequencies varying with position and strength of excitation;
" distorted frequency response plots, especially near resonances;
" unstable or unrepeatable data.

Probably the most obvious way of checking for the existence of non-linearity is to repeat a particular FRF measurement a number of times using different levels of excitation (and hence response) each time, or using different types of excitation signal (see below). If the resulting curves differ from one such measurement to the next, especially around the resonances, as illustrated in Fig. 3.49, then there is a strong possibility of non-linearity, and this is a check which will work with most types of excitation signal.

If signs of non-linear behaviour are thus detected, it is useful to have a strategy for how to proceed with the modal test because many of the basic relationships used in the various stages of data processing and analysis can no longer be relied upon. It turns out that most types of non-linearity are amplitude-dependent and so if it were possible to measure an FRF curve while keeping the amplitude of response (specifically the displacement response) at a constant level, the behaviour of the structure would be linearised and it would exhibit the characteristics of a linear system. Of course, the data and the model obtained in this way would strictly only apply at that particular vibration level but the fact that the behaviour had been maintained as linear is an important achievement, and one that means that we can extend our use of modal analysis into the realm of non-linear structures, even if only slightly.

266

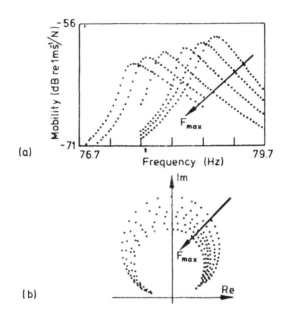

(a)

(b)

Fig. 3.49 Example of non-linear system response for different excitation levels.
(a) Mobility FRF modulus plots; (b) Receptance FRF Nyquist plots

3.12.2 Effects of Different Excitations on Non-linear System Response

While discussing the question of non-linearity, it is interesting to see what form is taken by a single FRF curve for a non-linear system when no particular response level control is imposed. The result depends markedly on the type of excitation signal used as well as the nature of the non-linearity. Fig. 3.50 shows some typical results from measurements made on an analogue computer circuit programmed to exhibit a cubic stiffness characteristic using sinusoidal, random and transient excitations. The stepped-sine test method using the FRA clearly shows the distortion to the frequency response plots caused by the slight non-linearity. However, neither of the other two cases — both of which used the DFT spectrum analyser to decompose the signals — are anything like as effective at detecting the presence of the non-linearity. It can be concluded from this result that, when using a spectrum analyser, the appearance of a normal-looking FRF does not guarantee that the test system is indeed linear: there is some aspect of the signal processing which has the effect of linearising the structure's behaviour. Indeed, true random excitation applies a linearisation procedure to the structure's behaviour which is considered to provide an

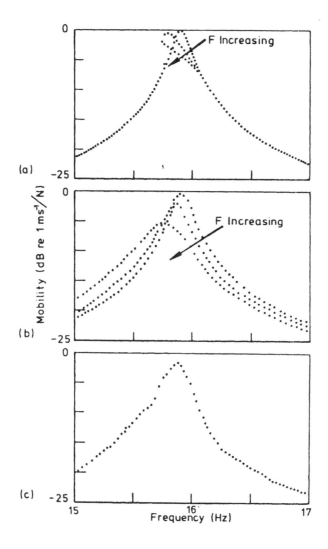

Fig. 3.50 FRF measurements on non-linear analogue system.
(a) Sinusoidal excitation; (b) Random excitation; (c) Transient excitation

optimised linear model for the test structure. Other methods of excitation — sine, periodic, transient — each produce different results, although that from a sinusoidal excitation is the one that can readily be related to the theoretical predictions based on the analysis presented in Section 2.11.

3.12.3 Level Control in Measuring FRF Data on Non-linear Systems

It is widely accepted that one of the best ways of dealing with slight non-linearities in practical structures is to exercise a degree of amplitude control on the vibration levels developed during measurement. This is in recognition of the previously-asserted feature that most practical non-linearities are amplitude-dependent. Some measurements made on a markedly non-linear structure are presented in Fig 3.51, and show the result of using (a) **response level control**,

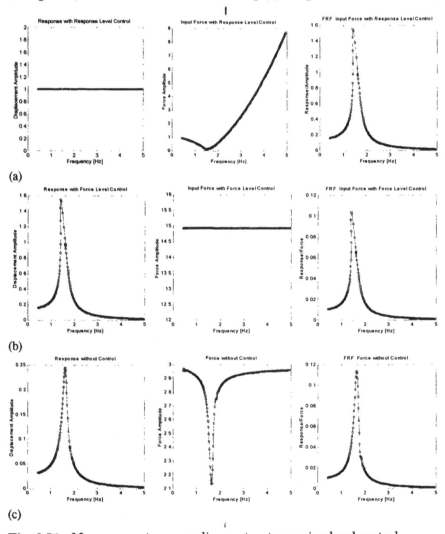

Fig. 3.51 Measurements on non-linear structure using level control.
(a) Response level control; (b) Force level control; (c) No control

(b) **force level control** and (c) **no** level control, in passing through a resonance region. In these methods, the level of excitation must be adjusted at each frequency of measurement until the control parameter is steady at the chosen level. As the structure is non-linear, this can be more difficult to achieve than expected. Results are shown in (a) Nyquist, (b) Bode and (c) Inverse plot formats and it can clearly be seen how the constant-response level test provides the closest to linear behaviour.

We shall discuss in the next chapter a means of detecting and identifying non-linear behaviour from the modal analysis processing of measured FRF data. However, as it is important to detect these effects as early as possible in the test procedure, it is appropriate to consider every opportunity for this task. We shall introduce two such opportunities here at the stage where the FRFs are being measured.

One approach offers the possibility of identifying different types of structural non-linearity from a single FRF curve, always providing that this still contains the necessary evidence (see comments above regarding the use of the Fourier transform methods for measuring FRF data). This method is based on the properties of the Hilbert transform (related to the Fourier transform but different in that it transforms within the frequency domain (or time domain) rather than between the two, which dictates that the Real Part of a frequency response function is related to the corresponding Imaginary Part by a direct Hilbert transform. Thus, if we have measurements of both parts (as is usually the case), we can perform a check on the data by using the measured Real Part to compute an estimate for the Imaginary Part and then comparing this with the actual measured values of the Imaginary Part. A similar comparison can also be made in the other direction using the measured Imaginary Part to estimate the Real Part. Differences between estimated and measured curves are then taken as an indication of non-linearity and the nature of the differences used to identify which type. An example of the procedure is shown in Fig. 3.52 and further details may be found in Reference [37]. However, it must be noted that this procedure will only work effectively if the FRF data have been measured using a sinusoidal excitation and care must be taken to avoid misleading results which can arise if the data relate only to a limited frequency range, thereby restricting the reliability of the Hilbert transform.

The second method is altogether simpler and is based on the features expected to be found in the Inverse FRF plots shown in Chapter 2. In these plots, for an SDOF system, the Real Part of the plotted function is expected to be linear with $(\text{frequency})^2$ — at least if the behaviour is that of a linear system with constant mass and stiffness properties — while the Imaginary Part plot should be linear

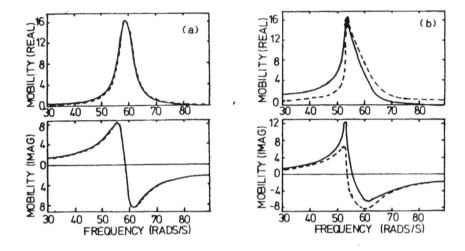

Fig. 3.52 Use of Hilbert transform to detect non-linearity.
(a) Effectively linear system; (b) Non-linear system

with frequency, provided that the damping is constant. If any of these three physical parameters were to be amplitude-dependent (and that is most likely for the stiffness and/or the damping), then one or other, or both, of these plots would be expected to lose their straight-line characteristic. Such a deviation from the expected form is very easy to see by eye and so this can be a very effective first-line detector of non-linear behaviour in the test object — see Fig. 3.53.

In order for either of these approaches to be effective, any non-linearity on the structure must be exercised and so the constant-response-level type of test should **not** be used in this situation: it is better to use force control in order to cover a wide range of amplitudes as the resonance is traversed.

3.12.4   Summary
The techniques required to deal effectively with structures which have any significant degree of non-linearity are beyond the scope of this text, and reference should be made to one of the many detailed works which address this subject. However, it must be noted that the majority of practical structures, while exercising a small degree of non-linear behaviour, **do** respond to testing and modelling by the methods of modal analysis promoted in this book. This short section has sought to demonstrate that slight deviations from the standard modal testing techniques permit a first-level detection of such non-linear behaviour and, even if its characterisation is only very approximate, and not

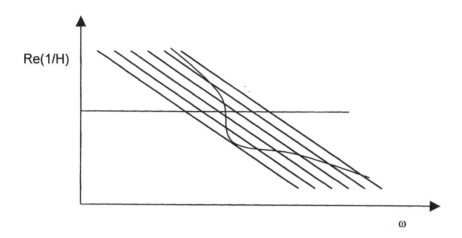

Fig. 3.53 Inverse FRF plots for non-linear systems

satisfying to the mathematician, guide the practitioner towards testing methods which will not suffer unduly from the slight deviations from expected behaviour which such structures will demonstrate.

This represents a pragmatic strategy to the measurement phase when dealing with non-linear structural behaviour in modal testing.

3.13   MULTI-POINT EXCITATION METHODS
3.13.1   Multi-point Excitation in General
During the decade from the mid-1980s to the mid 1990s there was a significant growth in the development and use of multiple-excitation methods for modal testing — so-called MIMO (Multi-input, Multi-output) test methods. The drive for this development was primarily in the need for high-quality test data, and especially for FRF data which possesses a high degree of consistency; a requirement of growing proportions to satisfy the needs of the increasingly sophisticated multi-reference modal analysis algorithms. There are other benefits, also, including:

* the excitation of large structures in a way which more closely simulates their vibration environment in service (than is achieved by a single-point excitation test),
* the facility of detecting and identifying double or repeated modes, and
* the need to complete some tests in the very minimum of on-structure time.

Partly because of the complexity of the procedures, and of the expense involved, the majority of modal tests are still performed using a single-point excitation procedure, but the methods which make use of several simultaneously-applied excitations are today well developed and will be described in outline below. A full exposition of these methods falls outside the scope of this work although it is recognised that they are highly suitable for certain specific applications, such as large aerospace-type structures.

The theory upon which such multi-point excitation methods are based has been presented in Chapter 2 and we shall briefly discuss here the practical implementation of some of the different methods in current use. A more detailed account of these methods, and others of the same type, may be found in Reference [38].

3.13.2   Appropriation or Normal Mode Testing

The first method to be discussed is the oldest, historically, and one which has been used consistently for almost 50 years in the aerospace industry. It is the method known as 'Normal Mode testing' or sometimes 'Appropriation' and is different from all other methods used in modal testing in that it seeks to establish vibration in a pure mode of vibration by careful selection of the locations and magnitudes of a set of sinusoidal excitation forces.

It was shown earlier (equations (2.67) to (2.69)) that it is theoretically possible to generate a monophased response vector by a multi-point, monophased harmonic force vector, and furthermore that these two vectors would be exactly 90° out of phase when the excitation frequency is equal to one of the undamped system's natural frequencies. If we satisfy the conditions posed in equations (2.67) to (2.69) where the force and response vectors are exactly in quadrature with each other, then (2.63) may be written as:

$$i\{X\} = [H_{\text{Re}}(\omega) + iH_{\text{Im}}(\omega)]\{F\} \qquad (3.18)$$

It follows that this equation is valid only if $\det| H_{\text{Re}}(\omega) | = 0$ and this condition provides the basis of a method to locate the undamped system natural frequencies from measured FRF data. A second stage — to find the appropriate force vector — then follows by substitution of the specific frequencies back into equation (3.18).

3.13.3   Multi-phase Stepped-Sine (MPSS) Testing

A modified version of the Normal Mode Test is provided by the MPSS method which is also a steady-state sinusoidal type of test, using several simultaneous exciters applied at different points, but in this case the magnitudes of each force are not so tightly controlled as before.

The basic system is one of an MDOF system excited at a single sinusoidal frequency, ω, by a set of p excitation forces, $\{F\}e^{i\omega t}$, such that there is a set of steady-state responses, described by $\{X\}e^{i\omega t}$. Of course, it is clear that these two vectors are related by the system's FRF properties as:

$$\{X\}_{n \times 1} = [H(\omega)]_{n \times p} \{F\}_{p \times 1} \tag{3.19}$$

However, it is not possible to derive the FRF matrix from a single equation of the form shown in (3.19), not least because there will be insufficient data in two vectors, one of length p and the other of length n, to define completely an $n \times p$ matrix, even if it is symmetric. What is required is to make a series of p' measurements of the same basic type, using different excitation vectors, $\{F\}_i$, each time, choosing these in such a way that when the forcing matrix is assembled from all the p' individual vectors, $[F]_{p \times p'} = [\{F\}_1 \{F\}_2 ... \{F\}_i ... \{F\}_{p'}]$, it is non-singular. This can only be assured if:

- there are at least as many vectors as there are forces, $p' \geq p$;
- the individual force vectors are linearly independent of each other, a condition which requires some care in the selection of a pattern of forces for each part of the test.

A second matrix is also constructed, this time containing the corresponding response vectors: $[X]_{n \times p'} = [\{X\}_1 \{X\}_2 ... \{X\}_i ... \{X\}_{p'}]$ and now these two collections of measured data can be used to determine the required FRF matrix, from:

$$[H(\omega)]_{n \times p} = [X]_{n \times p'} [F]_{p' \times p}^{+} \tag{3.20}$$

where + denotes the generalised inverse of the forcing matrix. In practice, it is often possible to include more than the minimum number of force patterns (p'), especially if the number of exciters used (p) is small, say less than 4 or 5, and thereby to obtain a least-squares solution from the overdetermination of the problem.

3.13.4   Multi-point Random (MPR) Testing
3.13.4.1 Concept

The most popular of the multi-point test methods to evolve over the past decade is that known as 'Multi-point Random' or, simply 'MPR'. In this method, advantage is taken of the incoherence of several uncorrelated random excitations which are applied simultaneously at several different points. By this device, the need to repeat the test several times, as was necessary for the MPSS method, is avoided.

In the case of the multi-point random approach, the purpose of the method is to obtain the FRF data required for modal analysis but in an optimal way. Specifically, the aim of using several exciters simultaneously is to reduce the probability of introducing systematic errors to the FRF measurements, such as can arise when using conventional single shaker methods. Once again, the underlying theory is presented in Chapter 2, this time in equation (2.131). In order to introduce and to explain the concept, we shall consider the simplest form of a multiple excitation as that of a system excited by two simultaneous forces, $f_1(t)$ and $f_2(t)$, where the response $x_i(t)$ is of particular interest. Assuming that we have measured the various auto- and cross-spectral densities of and between the (three) parameters of interest, we can derive expressions for the required FRF parameters, $H_{i1}(\omega)$ and $H_{i2}(\omega)$, dropping the (ω), for simplicity, from:

$$H_{i1} = \frac{(S_{1i} S_{22} - S_{2i} S_{12})}{(S_{11} S_{22} - S_{12} S_{21})} \tag{3.21a}$$

$$H_{i2} = \frac{(S_{2i} S_{11} - S_{1i} S_{21})}{(S_{11} S_{22} - S_{12} S_{21})} \tag{3.21b}$$

These expressions can be used provided that $S_{11} S_{22} \neq |S_{12}|^2$, a condition which is more readily described by the requirement that the two excitation forces must not be fully correlated. Care must be taken in practice to ensure satisfaction of this condition, noting that it is the applied forces, and not the signal sources, which must meet the requirement.

3.13.4.2 General formulation
The simple 2-input, 1-output case described above serves to illustrate the concept. However, in practice, the method is applied using different numbers of exciters and, certainly, several response points simultaneously. The expressions quoted in (3.17) can be extended to these more general cases, as follows:

$$[H_{xf}(\omega)]_{n\times p} = [S_{xf}(\omega)]_{n\times p} [S_{ff}(\omega)]_{p\times p}^{-1} \tag{3.22}$$

where it can be seen that the matrix of spectral densities for the forces must be non singular. As mentioned above, care must be taken in practice to ensure satisfaction of this condition, noting that it is the applied forces, and not the signal sources, which must meet the requirement. This is a requirement which is more readily stated than achieved, largely because of the interaction between the different shakers which is provided by the structure to which they are all attached. Even if the input signals to the exciters' amplifiers are strictly uncorrelated (with each other), it is almost certain that the forces applied to the structure will not be. Worst of all, this feature is most pronounced at or near resonance regions where the local mode shape dominates the dynamic response, and is largely independent of the actual forcing pattern, so that the forces applied by the different exciters are influenced as much by the structure itself as by the inputs to the exciters. Ways around this problem may have to be found if the method is to succeed in practice and the discussion in Section 2.11.3.4 offers some proposals for the best way to proceed: by using the input signals to the exciters as parameters instead of, or as well as, the actual force signals. Thus, we have the alternative version (from [25]):

$$[H_{XF}(\omega)]_{n \times p} = [S_{XV}(\omega)]_{n \times p} [S_{FV}(\omega)]_{p \times p}^{-1} \tag{3.23}$$

which may be easier to use in practice.

3.13.4.3 Coherence in MPR measurements
In a similar way to that in which we defined coherence for the simple two-channel SISO system, we can also make use of the same concepts in this more general case. If, during a MIMO test, we have measured various spectral properties which can be assembled into the three matrices:

$$[S_{FF}(\omega)] \; ; \; [S_{XX}(\omega)] \text{ and } [S_{FX}(\omega)]$$

then we can derive an estimate for the FRF matrix, $[H_1(\omega)]$, using:

$$[H_1(\omega)]^T = [S_{FF}(\omega)]^{-1} [S_{FX}(\omega)] \tag{3.24}$$

and then compute an estimate for the autospectrum of the response from:

$$[\tilde{S}_{XX}(\omega)] = [H_1^*(\omega)][S_{FX}(\omega)] \tag{3.25a}$$

which leads to

$$[\tilde{S}_{XX}(\omega)] = [S_{XF}(\omega)][S_{FF}(\omega)]^{-1} [S_{FX}(\omega)] \tag{3.25b}$$

Now, by comparing the estimated response spectrum, $[\tilde{S}_{XX}(\omega)]$, with the actual measurement, $[S_{XX}(\omega)]$, we obtain a formula for the multiple coherence between the two parameters, $\{f(t)\}$ and $\{x(t)\}$, $[\gamma^2(\omega)]$, as follows:

$$\left[\gamma^2(\omega)\right] = [S_{XX}(\omega)]^{-1}[S_{XF}(\omega)][S_{FF}(\omega)]^{-1}[S_{FX}(\omega)]. \tag{3.26}$$

This whole topic becomes quite complex from this point and the interested reader is directed to one of the specialist texts on the subject, such as Reference [39] for further detailed discussion.

3.13.5   Multiple-reference Impact Tests

Although not falling clearly into the category of multiple-exciter methods, the class of hammer excitation tests referred to as 'Multi-reference Impact Tests' (MRIT) should be included in this section of the book because they do address some of the features which the classical multi-input methods are intended to cover. In the MRIT, typically three response references are measured (often, the x, y and z components at the response measurement location) every time a hammer blow is applied to the structure. As will be seen more clearly when we discuss the modelling process in Chapter 5, the FRF data collected by performing a test in this way will be the equivalent of exciting the structure at three points simultaneously while measuring the response at each of the n points of interest. Thus, in the sense that a multiple-input test is a multi-reference measurement (measuring several parallel columns of the FRF matrix), so too is the MRIT since it provides a multi-reference measurement by including several rows of the same FRF matrix. Its principal advantage over the single-reference measurement made in the same way is its ability to detect double or repeated roots in structures.

3.14   MEASURING FRFs AND ODSs USING THE SCANNING LDV

3.14.1   Types of Scan Using the SLDV

One application of the scanning LDV to the measurement of FRF data has already been introduced in the section in measuring RDOF data. However, that was a special case of a whole range of similar applications in which two major types of measurement are made possible by the capabilities of this transducer. These types of measurement are (i) FRF data at individual frequencies or over specified frequency ranges and (ii) operating deflection shapes at individual frequencies. What all the different measurement procedures have in common is the positive and controlled use of the scanning capability of the SLDV whereby time can be used as a direct measure of spatial position on the structure so that spectral analysis of time-varying signals translates directly into spatial, as opposed to (or as well as) spectral, descriptions of the vibrating structure. Most of the scanning applications described in this section are based primarily on the measurement of a structure which is undergoing steady-state harmonic vibration. However, many of them can be extended to include narrow-band random or periodic vibration, and even transient vibration, under the condition that the width of the frequency band encompassed is smaller than the spacing between the sideband spectral lines that are generated by the scanning process (this spacing is usually determined by the scanning rate employed for the measurement).

There are several different types of scan that are useful for this class of measurement. These include:

- straight line scans (in one dimension),
- circular scans,
- conical scans,
- raster scans,
- Lissajou-type scans,
- arbitrary scans (in two dimensions) — see Fig. 3.54.

The scans can be 'short' or 'long', this classification referring to the scan length relative to the spatial wavelength of any vibration patterns that might be measured. In effect, 'short' scans assume that the vibrating surface is effectively rigid along the length of the scan line while 'long' scans make no such assumption. Straight line scans can be linear (constant velocity of scan with abrupt changes of direction at each end) or sinusoidal (giving smoother changes of direction at each end of a scan).

3.14.2 Straight Line Scans with the SLDV

3.14.2.1 Short straight-line scans (two-axis FRF measurement)
The first use of the SLDV exercising a short scan as part of a modal testing task has already been described in Section 3.11.3: to the measurement of vibration in the normal and one of the in-plane rotational DOFs of a vibrating surface.

3.14.2.2 Long straight-line scans (linear ODS measurement)
The concept of a long scan is similar to that of the short scan except that the order of the displacement pattern that will be measured is greater than the simple case (first order, $z(x,t) = (a_0 + a_1 x)\cos(\omega t + \theta)$) and can be generalised to a polynomial form:

278

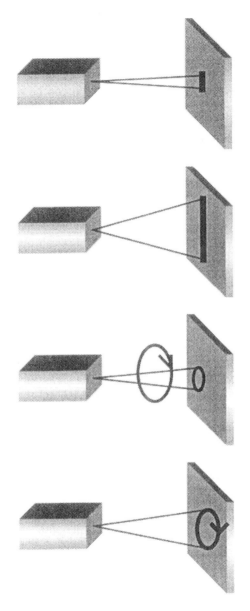

Fig. 3.54 Alternative scans to measure vibration patterns

$$z(x,t) = \left(a_0 + a_1 x + a_2 x^2 + a_3 x^3 \ldots\right)\cos(\omega t + \vartheta) \tag{3.27}$$

(In this example, it is assumed for clarity that the vibration pattern is 'real' in the sense that all points are vibrating with the same phase: the more general case where the vibration pattern is complex can also be addressed by this same approach, but by using Real Part and Imaginary Part expressions in place of the mono-phase version shown in equation (3.27).) The structure is vibrated with a steady-state harmonic excitation at frequency, ω, and the LDV transducer is scanned along the designated line in a sine wave, at frequency, Ω, so that $x = X \sin \Omega t$. The output signal from the LDV will therefore be an amplitude-modulated sine wave that can be described by:

$$v(t) = \sum_n \left(a_n \cos(\omega t + \varphi_n)\cos^n \Omega t\right) \tag{3.28}$$

It should be noted that the LDV signal may also be represented as a spatial Fourier series if a uniform-rate scan is used, but in that case, because of the signal discontinuities at the ends of the scan, there is inevitably an infinite series of Fourier coefficients, and so a sinusoidal scan is preferred. If this output signal is subjected to a conventional Fourier spectral analysis, with the condition that the duration of the analysed time-record is synchronised with the scan length, then a spectrum will be obtained which contains several spectral components, A_0, A_1, A_2,..., A_{-1}, A_{-2},..., these being in a definite pattern based on the central excitation frequency and a series of sidebands as follows:

$$\omega, (\omega + \Omega), (\omega + 2\Omega), \ldots, (\omega - \Omega), (\omega - 2\Omega), \ldots \tag{3.29}$$

It can be shown (see Reference [30]) that there is a simple relationship between the spectral coefficients, A_0, A_1, A_2,..., and the coefficients, a_0, a_1, a_2,..., in the polynomial expression that describes the vibration pattern along the scanned line (the operating deflection shape) which takes the following form:

$$\{a\} = [T]\{A\}$$

where

$$[T] = \begin{bmatrix} 1 & 0 & -2 & 0 & 2 & 0 & \dots \\ 0 & 2 & 0 & -6 & 0 & 10 & \dots \\ 0 & 0 & 4 & 0 & -16 & 0 & \dots \\ 0 & 0 & 0 & 8 & 0 & -40 & \dots \\ 0 & 0 & 0 & 0 & 16 & 0 & \dots \\ 0 & 0 & 0 & 0 & 0 & 32 & \dots \\ \dots & \dots & \dots & \dots & \dots & \dots & \dots \end{bmatrix} \tag{3.30}$$

Special measures, other than simple application of the FFT, are necessary to derive phase, or real and imaginary parts, for the sideband components. An example of applying this procedure to measurement of the ODS along an edge of a rectangular cantilever plate at a fixed frequency of vibration, and for a specific point excitation, is shown in Fig. 3.55(a). It should be noted that this approach yields the vibration pattern as a continuous function of position, instead of the usual series of discrete measured points that result from conventional modal test procedures. One advantage of this result is the possibility of obtaining derivatives of this function (to reveal slopes or curvatures), a process which is notoriously unreliable when using discrete points. However, there is a limitation to be tolerated and that derives from the ever-present speckle noise problem which effectively limits the number of spectral lines that can be measured above the background noise which affects the entire spectrum.

An alternative to this method of measuring the ODS pattern is simply to perform an analogue demodulation of the output signal from the LDV. This can be done using a uniform scan to reveal a measure of the ODS, again as a continuous function of distance along the scan, but this time in discrete format rather than as a continuous series — see Fig. 3.55(b). The demodulation process works even if the ODS is discontinuous, in which case the sine-scan polynomial process will be inaccurate.

3.14.3   Circular Scans with the SLDV

In much the same way that we can make short or long straight line scans, so also is it possible to make circular scans of varying radii by simultaneously scanning in both the x- and y-directions with the SLDV mirrors. If the mirrors are driven with sine and cosine waves at the same scan frequency and with the same scan length, then a circular scan will be produced and this is useful for a number of similar applications.

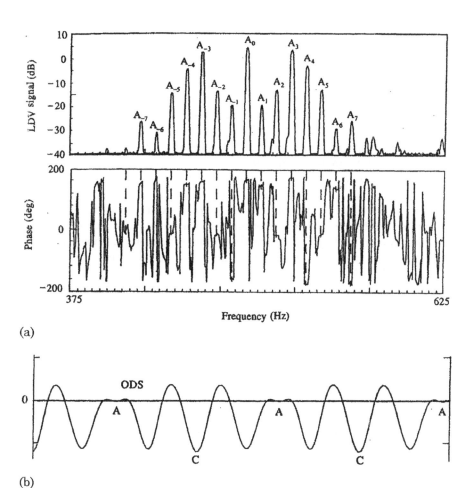

Fig. 3.55 Measurement of operating deflection shape (ODS) on plate using SLDV.
(a) Spectral analysis of measured signal; (b) Demodulation of measured signal

3.14.3.1 Small-radius scans (for 3DOF FRF measurement)

First, we can note that the small-radius scan has already been introduced in Section 3.10.3 as a means of measuring RDOF data. In fact, the use of a small circular scan centred on a point on a vibrating structure leads directly to a 3DOF measurement which includes the responses in the z, θ_x and θ_y directions from a single measurement and thus to three FRFs referred to the single point of excitation (wherever that may have been applied).

3.14.3.2 Large-radius scans (for ODS measurements)

The use of the large-radius scan extends the application of the straight-line scan discussed above to two dimensions with the particular result that the ensuing function which describes the operating deflection shape over the scanned line is one which is made up of trigonometrical terms in place of the simple polynomial of equation (3.27) This format is particularly useful for testing the many structures that have a degree of axisymmetry, such as many of the components that are described in the earlier section on rotating structures (Section 2.8.4). In these cases, the predominant mode shape information of interest is that which describes the nodal diameter components, and that is exactly the information that is contained in the terms in the trigonometrical expression provided by a circular scan. A good example of this feature is illustrated in the measurements made on a radial-flow turbomachine impeller shown in Fig. 3.56. The first plot, Fig. 3.56(a), shows a point FRF made on the rim of the impeller; the second and third plots, Figs. 3.56(b) and (c), show the resulting Fourier analyses of the LDV signal obtained from a circular scan at each of two of the many resonances evident in the FRF plot — at 2138 Hz and 2740 Hz, respectively. It can be shown that, in the event of the scan encountering an n ND mode, the spectral analysis of a time record produced by the SLDV from a single circular scan (at frequency Ω) around the periphery of the impeller will contain components at $(\omega - n\Omega)$ and $(\omega + n\Omega)$ only. Thus we can conclude from the illustrations in Figs. 3.56(b) and (c) that these two resonances suggest mode shapes of 4 ND and a combination of 9 and 10 ND, respectively. The measurement revealing this result took a fraction of the time that would normally be required to measure a sufficient number of points around the rim for a subsequent Fourier analysis to reveal the underlying nodal diameter components.

Furthermore, the possibility of other, higher-order, terms in the mode shape is quickly eliminated by the scan, although much less easily so by the conventional discrete-point approach.

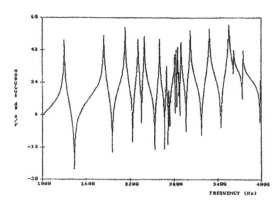

(a)

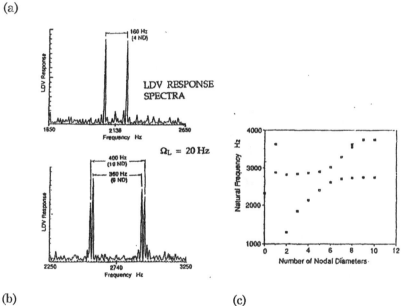

LDV RESPONSE
SPECTRA

$\Omega_L = 20\,\text{Hz}$

(b)   (c)

Fig. 3.56 Measurement of nodal diameter patterns on axisymmetric impeller structure.
(a) Point FRF; (b) Mode 1; (c) Mode 2

3.14.3.3 Conical scan (triaxial FRF measurement)

A variant on the circular scan is the *conical* scan, made using a short-focus lens positioned between the LDV and the test structure, and focused at the point of interest on that structure, as shown in Fig. 3.57(a). With this method, a single scan is capable of yielding the three translational response DOFs (*x, y, z*) at the measurement point,

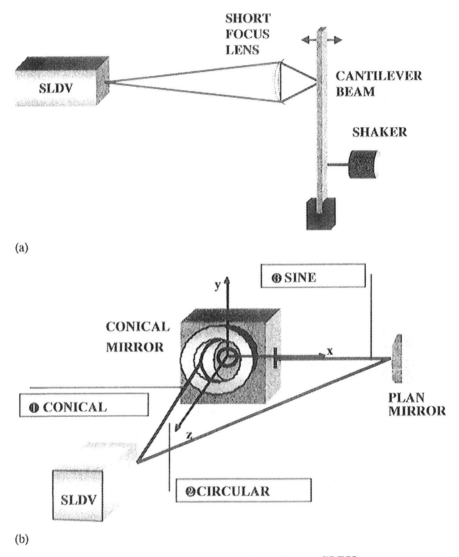

Fig. 3.57 Triaxial measurement at a point using an SLDV.
(a) Conical scan; (b) Circular scan with mirror

thereby providing three FRFs referred to the single point excitation which is assumed (here) as being the source of the steady-state harmonic vibration response.

Two measurements centred on the same point, with conical scan and a small circular scan (which can be achieved simply by moving the lens axially), can yield five of the six DOFs at the measurement point.

An alternative to using the short-focus lens described above is to use a conical mirror; see Fig. 3.57(b). Conical and circular scans may then be obtained with the same setup merely by adjusting the scan radius.

3.14.4 Area Scanning with the SLDV (ODS Measurements of Two-dimensional Surfaces)

It is clear that a more general version of the circular scan is one in which the two scanning axes are driven at different frequencies to each other so as to trace out a moving Lissajou-type of pattern across the scanned surface: here, the two scan rates are different to each other and not related by a simple multiple. In this way, a whole area can be scanned in a relatively short time and a two-dimensional version of the polynomial expression above (equation (3.27)) can be derived. One of the most effective two-dimensional scans has been found to be one in which the scan rate along one axis is much slower than that along the other. Such a scan is illustrated in Fig. 3.58(a) and a typical result obtained by spectral analysis of a measurement on a rectangular plate, vibrating in a steady-state harmonic manner is shown in Fig. 3.58(b). In this spectrum, a complex pattern based on two series of sidebands can be seen, each series being related to one of the two scan directions. In this spectrum, each spectral component is identified by a double index, e.g. $A_{1,2}$, and these components can be converted to a corresponding set of two-dimensional polynomial coefficients, $a_{i,j}$, using the same transformation matrix as before, but in the double format:

$$[a] = [T][A][T]^T$$

where

$$z(x, y) = a_{0,0} + a_{1,0}x + a_{2,0}x^2 \ldots + a_{1,1}xy + a_{1,2}xy^2 \ldots \qquad (3.31)$$

Further details of these advanced techniques can be found in the referenced literature [30], but it can be noted here that most of the methods summarised above are applicable with any type of scanning transducer: at the present time, the scanning LDV is the most immediately available option but it is conceivable that other devices based on different transduction techniques, and perhaps less vulnerable to noise the limitations of the LDV, may be developed in the future.

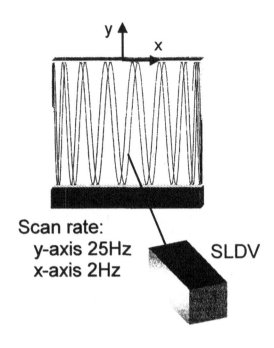

(a)

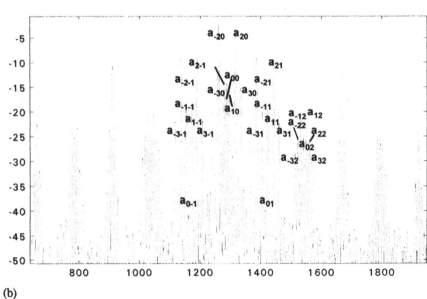

(b)

Fig. 3.58 Area scan for 2D mode shape measurement.
(a) Scan profile; (b) Spectrum of SLDV signal

CHAPTER 4

Modal Parameter Extraction (Modal Analysis) Methods

4.1 INTRODUCTION
4.1.1 Introduction to the Concept of Modal Analysis

Having dealt with the first phase of any modal test — that of measuring the raw data from which the desired mathematical model is to be derived — we now turn our attention to the various stages of analysis which must be undertaken in order to achieve the objective of constructing a model. A major part of this analysis consists of curve-fitting a theoretical expression for an individual FRF (as developed in Chapter 2) to the actual measured data obtained by one of the methods discussed in Chapter 3. The present chapter describes some of the many procedures which are available for this task and attempts to explain their various advantages and limitations: as with all other aspects of the subject, no single method is 'best' for all cases.

In increasing complexity, the methods discussed involve the analysis, or curve-fitting, first of part of a single FRF curve, then of a complete curve encompassing several resonances and, finally, of a set of many FRF plots all on the same structure. In every case, however, the task undertaken is basically the same: to find the coefficients in a theoretical expression for the frequency response function which then most closely matches the measured data. This task is most readily tackled by using the partial fraction series-form for the FRF, as developed in Sections 2.4 to 2.7 for different types of system, although some methods use the rational fraction version and yet others prefer to work in the time domain, based on the impulse response function. The particular advantage of the series-form FRF approach is that the coefficients thus determined are directly related to the modal properties of the system under test, and these are generally the very parameters that are sought.

This phase of the modal test procedure is often referred to as 'modal parameter extraction', or 'modal analysis' because it is the corresponding stage in an experimental study to that called modal analysis in a theoretical study. In both approaches, modal analysis leads to the derivation of the system's modal properties. However, it should be noted that the two processes themselves are quite different: one is a curve-fitting procedure while the other is a root-finding or eigensolution exercise. The various methods used for experimental modal analysis tend to divide into two philosophies: one in which the analysis is essentially automatic — FRF data are supplied as input data and modal parameters are extracted without further involvement of the user — and a second one which is much more interactive, in which the user is expected to participate in various decisions throughout the analysis. Although, theoretically, there should be no need for this latter course of action, it is often found expedient in the light of the imperfect and incomplete data which are inevitably obtained in practical situations with real and complicated engineering structures.

4.1.2   Types of Modal Analysis

A great many of the current curve-fitting methods operate on the response characteristics in the frequency domain — i.e. on the frequency response functions themselves — but there are other procedures which perform a curve-fit in the time domain. These latter methods use the fact that the Inverse Fourier Transform of the FRF is itself another characteristic function of the system — the Impulse Response Function — which represents the response of the system to a single unit impulse as excitation (analogous to the single unit-sinusoid for the FRF). The majority of the following sections are concerned with modal analysis performed directly on the FRF curves but in later sections of the chapter we shall discuss this alternative approach to the problem using the impulse response properties.

Modal analysis methods can be classified into a series of different groups, and it is convenient to do so in order to summarise the essential features of each. First of all, it is appropriate to define the domain in which the analysis is performed:

- Frequency domain (of FRFs)
- Time domain (of IRFs or response histories)

Next, it is appropriate to consider the frequency range over which each individual analysis will be performed, and this divides into two categories, also, depending upon whether a single mode is to be extracted at a time, or several. These two subgroups are referred to as:

- SDOF methods; and
- MDOF methods, respectively.

A further classification relates to the number of FRFs which are to be included in a single analysis (bearing in mind that several similar analyses may well be undertaken to complete the processing for one test). There are three different types of FRF data sets, the differences usually depending upon how the data are collected, or measured, rather than what the list of contents is. The simplest type of FRF measurement is referred to as SISO (Single-input, Single-output), and this describes an individual FRF curve so that an SISO data set is made up of a set of FRFs which have been measured individually, usually sequentially. The second type of data set is referred to as SIMO (Single-input, Multi-output) and this refers to a set of FRFs which have been measured simultaneously at several response points, but all under the same single-point excitation. This describes the FRFs in a column or row of the FRF matrix. The third category is the MIMO type, (Multi-input, Multi-output) in which the responses at several points are measured simultaneously while the structure is excited at several points, also simultaneously: this is the standard format for a multi-exciter test method.

Bearing these classifications in mind, it can be said that modal analysis methods can be divided into two types, those which process one single FRF curve at a time, and those which analyse several curves simultaneously. These are referred to as:

* single-FRF methods; and
* multi-FRF methods, which are sometimes (but not universally) subgrouped into Global methods (which deal with SIMO data sets) and Polyreference (which deal with MIMO data).

Although, in principle, it is possible to use any method on any type of data set, it should be noted that the simpler methods (e.g. SDOF, single-FRF) tend to be very time-consuming for the analyst if applied to large sets of measured data while the more powerful methods (e.g. MDOF, Polyreference) may be rather intolerant of the small inconsistencies present in data amassed by repeated application of the SISO measurement approach. A degree of matching is necessary between the quality of the data obtained and the analysis method to be used.

4.1.3   Difficulties Due to Damping
It should be noted at the outset of a study of this subject that there are a number of problems to be expected in its application in practice. Many of these relate to the difficulties (which have already been mentioned) associated with the reliable modelling of damping effects. In practice, we are obliged to make certain assumptions about what model is to be used for the damping effects, and these assumptions must often be

made at the outset of the modal analysis process and can only be re-visited by starting the analysis over again. Sometimes, significant errors can be incurred in the modal parameter estimates — and not only in the damping parameters — as a result of a conflict between the assumed damping behaviour and that which actually occurs in reality. Naturally, such occurrences must be checked for and remedial action taken if they are detected. Another feature which sometimes gives rise to difficulties in modal analysis and which also derives from the damping modelling is that of real modes and complex modes. In practice, all modes of practical structures are expected to be complex, although in the majority of cases such complexity will be very small and, often, quite negligible. However, there are situations in which complex modes **will** exist and they need to be recognised and observed, especially when the results of a modal analysis are being scrutinised. These conditions have been discussed in some detail in Chapter 2.

The situation with modal complexity is rather analogous to that with non-linearity in which it is true to assert that all structures are likely to be non-linear to some degree — but that in the great majority of cases the degree of non-linearity, and the extent of the errors it causes, can be regarded as negligible. Here, also, vigilance is necessary to detect those few cases where this dismissal is not justified and such a task is usefully performed when assessing the results of the modal analysis stage.

4.1.4   Difficulties of Model Order

There is one further problem area which it is worth raising at the outset, and that concerns the question of model order: exactly how many modes are there in the measured FRF(s), or at least in the frequency range being analysed. This question is one of the most difficult to resolve in many practical situations where a combination of finite resolution and noise in the measured data combine to make the issue very unclear, the more so with the apparently-more-powerful global and polyreference analysis methods. It is a drawback of the power of many modern modal analysis curve-fitters that they are capable of fitting an FRF of almost any order (i.e. number of assumed 'modes') to most data sets. While some of the modes thus identified are 'real' ('genuine' is probably a better word here), many others may be fictitious, or 'computational' modes, introduced by the analysis process in pursuit of the optimum curve-fit which, after all, is the mission that the numerical algorithm is on. Correct differentiation between genuine and fictitious modes remains a critical task in many modal tests, and will be discussed in more detail in those methods which are particularly sensitive to it.

4.1.5   Contents of Chapter 4

In this chapter, we shall seek to explain the significance and basis of the various methods of modal analysis which are in common use, but shall place less emphasis than in other chapters on the details of the various methods. The reason for this is that the modal analyst must rely increasingly on the external provision of the most effective modal analysis software and can no longer expect to write his/her own routines, as was the case only a few years ago. The numerical and computation sophistication of state-of-the-art modal analysis procedures mean that practical implementation of these methods are beyond the means of the modal analyst.

However, it is important that the analyst knows how to use the various methods available, and appreciates the differences embodied in the different choices. He/she also needs to be well aware of the assumptions that are made, often implicitly, in the various approaches so that a proper interpretation can be made of the results obtained in practice, especially on real practical, imperfect (in the sense of not necessarily behaving according to the tidy presumptions of the underlying theory) structures whose behaviour will often not conform exactly to the prescriptions of the theoretical models.

This chapter thus seeks to guide the user through the maze of different approaches and to equip him/her with sufficient understanding and appreciation of the different methods available for him/her to make an informed choice of which one(s) to use, and an intelligent interpretation of the results that are provided by that choice.

The chapter starts with a discussion of various means of checking the reliability of the measured data, for if these are contaminated with errors, especially of a systematic nature, then much of the time spent in analysing the curves may be largely wasted. Then the various methods for extracting the modal parameters of the model, which is deemed to represent the observe behaviour of the test structure, are reviewed, starting with the simplest (although not the fastest) and progressing through successive enhancements to the more powerful and automatic analysis procedures that provide the basis of much routine modal analysis today. Methods based on a frequency-domain presentation of the measured data (FRFs) as well as their time-domain equivalents (the IRFs) are reviewed. Finally, the question of what happens when conventional modal analysis is applied to structures whose behaviour is distinctly non-linear is addressed. This leads to a simple extension of conventional linear modal analysis methods that can give a first-level estimation of the level and nature of non-linear components in the structure.

4.2 PRELIMINARY CHECKS OF FRF DATA

4.2.1 Visual Checks

Before commencing the modal analysis of any measured FRF data, it is always prudent to undertake a few preliminary and/or simple checks in order to ensure that time is not wasted on what subsequently turns out to be obviously bad data. It is not always possible to ascertain from visual inspection of an FRF plot whether it is a valid measurement, but there are certain characteristics which should be observed and these should be checked as soon as possible after the measurement has been made.

4.2.1.1 Low-frequency asymptotes

Most of the checks are made using a log-log plot of the modulus of the measured FRF, whether that be receptance, mobility or — as is usually the format of the raw measurements — accelerance. The first feature to be examined is the characteristic at very low frequencies — below the first resonance, if data extend down that far — since in this region we should be able to see the behaviour corresponding to the support conditions chosen for the test. If the structure is grounded (see Chapter 3), then we should clearly see a stiffness-like characteristic at low frequencies, appearing as asymptotic to a stiffness line at the lowest frequencies, and the magnitude of this should correspond to that of the static stiffness of the structure at the point in question. Conversely, if the structure has been tested in a free condition, then we should expect to see a mass-line asymptote in this low frequency range and, here again, its magnitude may be deduced from purely rigid-body considerations. Deviations from this expected behaviour may be caused by the frequency range of measurements not extending low enough to see the asymptotic trend, or they may indicate that the required support conditions have not in fact been achieved. In the case of a freely-supported structure, there will generally be some rigid body modes at very low frequencies (i.e. considerably lower than the first flexural mode) and these will tend to interrupt the mass-like asymptotic trend.

4.2.1.2 High-frequency asymptotes

A second similar check can be made towards the upper end of the frequency range where it is sometimes found, especially on point mobility measurements, that the curve becomes asymptotic to a mass line or, more usually, to a stiffness line. Such a tendency can result in considerable difficulties for the modal analysis process and reflects a situation where the excitation is being applied at a point of very high mass or flexibility. Although not incorrect, the data thus obtained will often prove difficult to analyse because the various modal parameters to be extracted are overwhelmed by the dominant local effects. Such a situation suggests the use of a different excitation point.

4.2.1.3 Incidence of antiresonances

Another set of checks can be made for systems with relatively clear resonance and antiresonance characteristics. The first of these is a check to satisfy the expected incidence of antiresonances (as opposed to minima) occurring between adjacent resonances. For a point FRF, there must be an antiresonance after each resonance, while for transfer FRFs between two points well-separated on the structure, we should expect more minima than antiresonances. A second check to be made at the same time, is that the resonance peaks and the antiresonance 'troughs' exhibit the same sharpness (on a log-log plot). Failure to do so may well reflect poor measurement quality, either because of a spectrum analyser frequency resolution limitation (see Section 3.7) causing blunt resonances, or because of inadequate vibration levels resulting in poor definition of the antiresonance regions.

4.2.1.4 Overall shape of FRF skeleton

There is another technique which will be described more fully in the next chapter that enables an overall check to be made on the relative positions of the resonances, antiresonances and ambient levels of the FRF curve. Essentially, it is found that the relative spacing of the resonance frequencies (R) and the antiresonance frequencies (A) is related to the general level of the FRF curve, characterised by its magnitude at points roughly halfway between these two types of frequency. Fig. 4.1 shows two example mobility plots, one of which (Fig. 4.1(a)) is mass-dominated and tends to drift downwards with antiresonances occurring immediately before resonances, while the other (Fig. 4.1(b)) is of a stiffness-dominated characteristic which generally drifts upwards and has antiresonances immediately above resonances. There is a procedure for sketching a simple skeleton of mass-lines and stiffness-lines through an FRF curve which confirms whether or not the R and A frequencies are consistent with the general level of the curve: the skeleton should pass through the 'middle' of the actual FRF plot. Fig. 4.1(c) shows a sketch of an apparently plausible FRF which does not satisfy the skeleton check. This may be because of poor data or because the parameter plotted is not in fact that expected (mobility, in this case) but a different format.

4.2.1.5 Nyquist plot inspection

Finally, at a more detailed level, we can assess the quality of the measured data again at the stage of plotting the FRF data in a Nyquist format. Here, each resonance region is expected to trace out at least

294

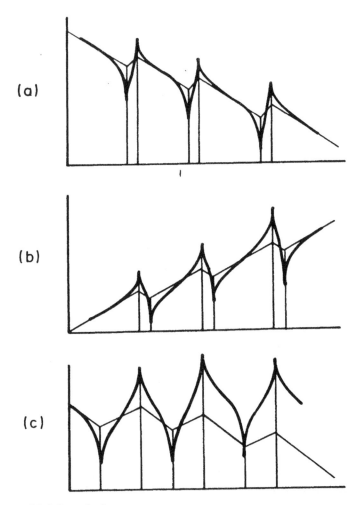

Fig. 4.1   Mobility skeletons.
(a) Mass-dominated characteristics; (b) Stiffness-dominated characteristics; (c) Anomalous characteristics

part of a circular arc, the extent of which depends largely on the interaction between adjacent modes. For a system with well-separated modes, it is to be expected that each resonance will generate the major part of a circle but as the modal interference — or 'overlap' — increases, with closer modes or greater damping levels, it is to be expected that only small segments — perhaps 45° or 60° — will be identifiable. However, within these bounds, the Nyquist plot should ideally exhibit a smooth curve and failure to do so may be an indication of a poor measurement technique, often related to the use of the analyser.

4.2.2   Assessment of Multiple-FRF Data Set using SVD

There are some much more detailed quality assessment checks possible which have become available in recent years as some of the many applications which have been found for the Singular Value Decomposition (SVD). These methods are appropriate for situations where several FRFs have been acquired, sometimes from a single excitation or reference DOF (SIMO data) and other times for data from several references (MIMO data). As mentioned elsewhere in this book, the SVD has proved to be a very useful tool in several aspects of modal testing, and in particular in matters pertaining to the quality, reliability and order of the data we have acquired in our measurements and intend to use in our analysis processes. The first method which is outlined here is described in more detail in [40] and it may be adapted and extended in the foreseeable future as the basic concept becomes more established and the results of the process and their interpretation become more widespread and more familiar. The method is relatively simple to apply: the set of FRFs which are to be assessed are stored in a series of vectors, $\{H_{jk}(\omega)\}$, each of which contains the values for one FRF at all measured frequencies, $\omega = \omega_1, \omega_2, ..., \omega_L$. These vectors are assembled into a matrix:

$$[A]_{L \times np} = [\{H_{11}(\omega)\}_{L \times 1} \{H_{21}(\omega)\}_{L \times 1} ... \{H_{np}(\omega)\}_{L \times 1}] \tag{4.1}$$

where n and p represent the maximum number of measured DOFs and the number of excitation points, respectively, and L represents the number of frequencies at which the FRF data are defined.

Then, an SVD is performed on the matrix, $[A]$, and computing three matrices, $[U]_{L \times L}$, $[V]_{np \times np}$ and $[\Sigma]_{w \times np}$, as normal for this type of process, with the relationship:

$$[A]_{L \times np} = [U]_{L \times L} [\Sigma]_{L \times np} [V]_{np \times np}^{\mathrm{T}} \tag{4.2}$$

The first-level interpretation of these results is as follows:

- the singular values of $[\Sigma]$, σ_1, σ_2, ..., σ_w describe the amplitude information and the number of non-zero singular values represents the order of the system (i.e. the number of independent modes of vibration which effectively contribute to the measured FRFs);
- the columns of $[U]$ represent the frequency distribution of these amplitudes; and
- the columns of $[V]$ represent their spatial distribution.

A further stage of analysis is possible, and that is to create a sub-factor of the decomposition by computing a new matrix, $[P]_{L\times np}$, which is referred to as the 'Principal Response Function (PRF)' matrix, each column of which contains a response function corresponding to one of the original FRFs:

$$[U]_{L\times L}[\Sigma]_{L\times np} = [P]_{L\times np} \tag{4.3}$$

It will be seen that these PRFs have similar properties to (and the same dimensions and units as) the original FRFs, but possess certain advantages which we can exploit both at this preliminary stage, and later, during the analysis. Each PRF is, simply, a particular combination of the original FRFs. As such, each PRF contains all the essential information in those FRFs including — importantly — the eigenvalue information in an explicit form (i.e. a modal analysis of any one of the PRFs will yield the eigenvalue properties for each mode which is visible on that function).

One example of this form of pre-processing is shown in Fig. 4.2 for the case of numerically-simulated test data, and another in Fig. 4.3 for the case of real measured test data. In both cases, a series of three plots are shown:

(a)   the original FRFs, overlaid;
(b)   the singular values, σ_r, plotted in descending order of magnitude; and
(c)   the PRFs, overlaid.

The second plot, (b), has the possibility of conveying the true order of the system because the number of non-zero singular values is equal to this parameter. In most practical cases, the demarcation between non-zero and zero singular values is not sharply-defined, but can often be inferred from close inspection of the plot. The third plot, (c), is perhaps the most useful because it shows the genuine modes distinct from the computational modes. It can be seen that the PRFs tend to fall into two groups: the most prominent are a set of response functions, each of which has a small number of dominant peaks (i.e. resonances), while the second, lower group shows less distinct and clear-cut behaviour. In fact, the first set represent the physical modes of the system while the second set, which are generally at a much lower level of response, represent the noise or computational modes present in the data. In the limit, it is possible to determine whether the noise present on the data permits the separation of these two groups of modes from each other, a requirement which must be satisfied for a successful modal analysis to be achievable. This is established by the need for a clear gap between the two sets of functions: if such a gap is present, then it will be

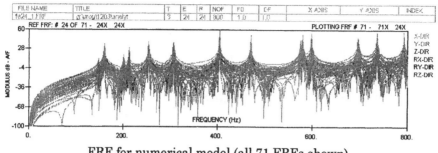

FRF for numerical model (all 71 FRFs shown)

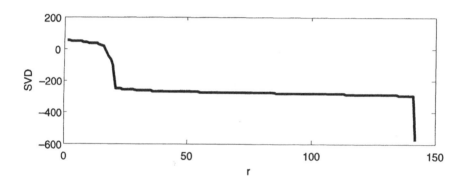

Singular values for numerical data (only 9 FRFs shown)

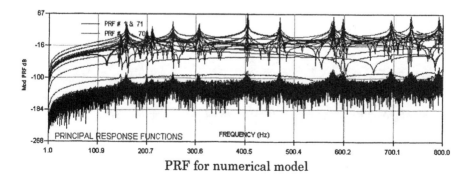

PRF for numerical model

Fig. 4.2   FRF and PRF characteristics for numerical model

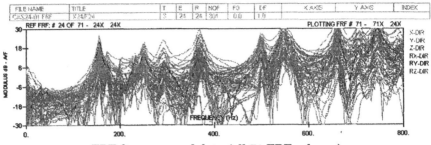

FRF for measured data (all 71 FRFs shown)

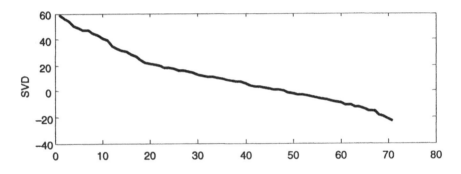

Singular values for measured data (only 9 FRFs shown)

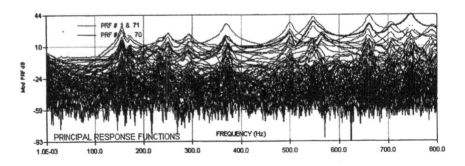

PRF for measured data

Fig. 4.3   FRF and PRF characteristics for measured data

possible to extract the properties of the *m* modes which are active in the measured responses over the frequency range covered; if not, then it may be very difficult, if not impossible, to perform a successful modal parameter extraction. It is important to note here that different results are obtained from the same original data set if the analysis is performed over different frequency bands, see Fig. 4.4. From this example it is seen that there may be distinct advantages of attempting a modal analysis over a limited frequency range (i.e. using less than the full frequency range covered by the measurement) in order to be working with a well-conditioned set of data. Thus, it may be possible to perform a successful modal analysis on a given set of data by doing so in several parts rather than in one single run. Furthermore, it may well be found that by using the PRF curves, instead of the FRFs, a simpler modal analysis is possible for the global properties (the eigenvalues) as a result of the dominance in each PRF of just one or two modes.

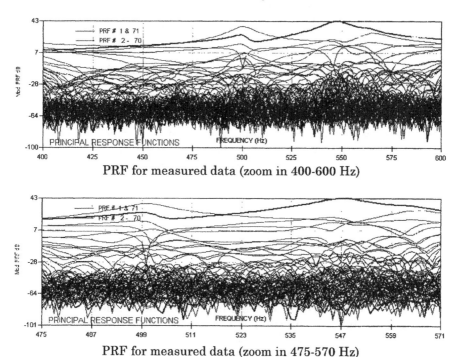

PRF for measured data (zoom in 400-600 Hz)

PRF for measured data (zoom in 475-570 Hz)

Fig. 4.4   PRF characteristics for different frequency ranges

While the ultimate interpretation goal of this approach is to determine whether or not the measured data are of adequate quality for the modal analysis process that will be applied to them, it may also offer the advantage of a pre-analysis processing that renders the measured data more amenable to this important step.

4.2.3   Mode Indicator Functions (MIFs)
4.2.3.1   General
A more complex version of the above idea has been developed in the form of Mode Indicator Functions (MIFs) which are intended for sets of FRF data from multiple references. Such data are available from multiple-excitation measurements such as MPR or MPSS tests, or from multi-reference impact tests (MRIT) and typically consist of an $n \times p$ matrix where n is a relatively large number of measurement DOFs and p is the number of excitation or reference DOFs, typically 3 or 4. In these methods, the frequency-dependent submatrix of FRFs which are available explicitly is subjected to an eigenvalue or singular value decomposition analysis which thus yields a small number (3 or 4) of eigen- or singular values, these also being frequency-dependent. The different versions of the MIF employ slightly different formulations which sometimes result in an eigenvalue decomposition and sometimes in a singular value decomposition, but it should be noted here that the singular values of a rectangular matrix, $[A]$, are the square roots of the eigenvalues of the square matrix, $[A]^H[A]$; hence the close connection between these two types of decomposition.

The original provenance of these mode indicator functions was from the methods of stepped-sine normal mode testing in which a single or pure mode is excited by suitable tuning of a number of separate exciters. These methods rely on guidance as to the optimum selection and magnitude of the set of forces which must be applied in order to achieve the single-mode response condition necessary and such guidance is sought in the form of estimates for the forcing vector, $\{F\}_r$. Nowadays, these same techniques can be used to determine the number of modes present in a given frequency range, to identify repeated natural frequencies and to pre-process the FRF data prior to modal analysis.

4.2.3.2   Complex mode indicator function (CMIF)
One of the most widely-used of these indicator functions is the CMIF, or complex mode indicator function, which is defined simply by the SVD of the FRF (sub) matrix. This decomposition, which is defined as:

$$[H(\omega)]_{nxp} = [U(\omega)]_{n \times n}[\Sigma(\omega)]_{n \times p}[V(\omega)]_{p \times p}^H$$
$$[CMIF(\omega)]_{pxp} = [\Sigma(\omega)]_{pxn}^T[\Sigma(\omega)]_{n \times p}$$

(4.4)

results in singular values and left and right singular vectors, all of which are frequency-dependent. The actual mode indicator values are provided by the squares of the singular values and are usually plotted as a function of frequency in logarithmic form, as shown in Fig. 4.5, with natural frequencies indicated by large values of the first CMIF (the highest of the singular values) and double or multiple modes by simultaneously large values of two or more CMIF values. Associated with the CMIF values at each natural frequency, ω_r, are two vectors, the left singular vector, $\{U(\omega_r)\}_1$, which approximates the mode shape of that mode, and the right singular vector, $\{V(\omega_r)\}_1$, which represents the approximate force pattern necessary to generate a response on that mode only. (Here, it is assumed that only a single mode exists at $\omega = \omega_r$: if two modes exist at this same natural frequency, then there will be two such left vectors and two right vectors which correspond to the two modes, or combinations of them.)

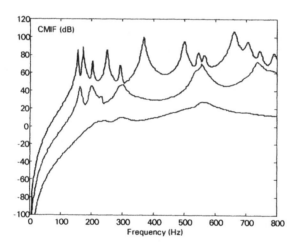

Fig. 4.5 Complex Mode Indicator Function (CMIF)

In addition to identifying all the significant natural frequencies, including double or multiple modes, the CMIF can also be used to generate a set of enhanced FRFs from the formula:

$$\left[EFRF(\omega)\right]_{n \times p} = \left[H(\omega)\right]_{n \times p}\left[V(\omega)\right]_{p \times p} \qquad (4.5)$$

There is one non-trivial EFRF for each mode, r, the result of which is an almost SDOF characteristic response function which is then readily amenable to modal analysis by the simplest of methods. As in the previous case, these modified FRFs are simply linear combinations of the original measured data and, as such, contain no more and no less information than in their original form. However, such an approach lends itself to a very reliable extraction of the global (eigenvalue) properties for the measured FRF data set which can then be re-visited in a second stage to determine the local (mode shape) properties for all the measured DOFs.

It should be noted here that an alternative definition for the CMIF is provided by the eigenvalue decomposition as applied to the given FRF matrix, in the form:

$$[H(\omega)]^H [H(\omega)]\{F(\omega)\} = \lambda(\omega)\{F(\omega)\} \tag{4.6}$$

where the eigenvalues, λ_r, are identical to the squares of the singular values, σ_r^2, above and both are functions of frequency, ω, as before.

4.2.3.3   Other MIFs

A number of other variants on the mode indicator function concept are also in use. Two are worth mentioning here: the MMIF (or multivariate MIF) and the RMIF (real MIF). The first of these is a refinement of the original force appropriation applications and defines the function as a frequency-dependent eigenvalue decomposition in the following form:

$$\beta[[H_R]^T[H_R] + [H_I]^T[H_I]]\{F\} = [H_I]^T[H_I]\{F\} \tag{4.7}$$

The MMIF consists of the eigenvalues, β_r, which result from the eigenvalue solution to (4.7) for each frequency, ω, and these values are plotted as a function of frequency, ω, in the form shown in Fig. 4.6, where it can be seen that the MMIF takes a value between 0 and 1, with the resonance frequencies now identified by minimum values of MMIF, instead of the maximum values for the CMIF.

Lastly, in this introduction to the indicator functions in use, we mention the RMIF, whose definition is a simpler form of the MMIF and slightly different to that of the CMIF as:

$$[H_I]^+[H_R]\{F\} = \lambda\{F\} \tag{4.8}$$

In this version, natural frequencies are identified by zero crossings of the RMIF values (the eigenvalues of the solution to (4.8)) in place of the minima and maxima of the other functions. A full discussion of these and other MIFs can be found in [41].

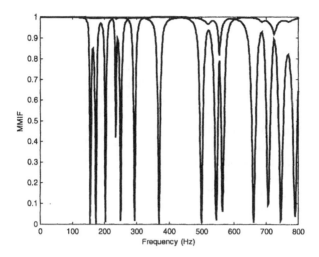

Fig. 4.6   Multivariate Mode Indicator Function (MMIF)

4.3   SDOF MODAL ANALYSIS METHODS
4.3.1   Review of SDOF Modal Analysis Methods
4.3.1.1   General approach

The first type of modal analysis method to be considered here, and to have been developed in the earliest days of modal testing, is that which is described as the 'SDOF approach'. This title does not imply that the system being modelled is reduced to a single degree of freedom: rather, that just one resonance is considered at a time, and that each modal analysis of this type seeks only to extract the properties of one of the system's modes. All the modes in the frequency range of interest are thus analysed, but sequentially, one after the other, rather than simultaneously as has become commonplace in more recent methods.

There are, of course, limitations to such a simple approach, the principal one being that very close modes (modes with close natural frequencies — and 'close' must be defined carefully - cannot easily be separated. This is a restriction which can be minimised by careful application of the method, although it remains quite a lengthy process and demanding of the user in its relatively high level of operator interaction.

There are several implementations of the basic concept of SDOF analysis, ranging from the simple peak-picking method, through the classic circle-fit approach to more automatic algorithms such as the inverse FRF 'line-fit' method and the general least-squares methods developed recently, both of which can be applied in a global sense (SIMO data) as well as on a single FRF (SISO).

Notwithstanding the relatively time-consuming nature of these SDOF methods, and their inherent approximations, they remain extremely useful tools for the modal test engineer. Their simplicity of use, and the facility of 'observing' the analysis in process, render them particularly valuable for the preliminary phases of a modal test, and for situations where rapid estimations for the basic features of the structure's behaviour are required. As will be seen in the following paragraphs, they are amenable to detailed scrutiny of such features as the linearity of the structure's behaviour and even to the type of damping which is present in the test structure. It could even be advocated that no large-scale modal test should be permitted to proceed until some preliminary SDOF analyses have been performed on the first FRF data obtained.

4.3.1.2 The SDOF assumption

Before detailing the actual steps involved in any of the SDOF analysis methods, it is necessary to examine the assumptions which will be made and the basis on which the methods are founded. As the name implies, the method exploits the fact that in the vicinity of a resonance, the behaviour of most systems is dominated by a single mode. Algebraically, this means that the magnitude of the FRF is effectively controlled by one of the terms in the series, that being the one relating to the mode whose resonance is being observed. We can express the assumption as follows. From Chapter 2, we have:

$$\alpha_{jk}(\omega) = \sum_{s=1}^{N} \frac{{}_sA_{jk}}{\omega_s^2 - \omega + i\eta_s\omega_s^2} \qquad (4.9a)$$

This can be rewritten, without simplification, as:

$$\alpha_{jk}(\omega) = \frac{{}_rA_{jk}}{\omega_r^2 - \omega^2 + i\eta_r\omega_r^2} + \sum_{\substack{s=1 \\ \neq r}}^{N} \frac{{}_sA_{jk}}{\omega_s^2 - \omega^2 + i\eta_s\omega_s^2} \qquad (4.9b)$$

Now, the SDOF assumption is that for a small range of frequency in the vicinity of the natural frequency of mode r, the second of the two terms in (4.9b) is approximately independent of frequency, ω, and the expression for the receptance may be written as:

$$\alpha_{jk}(\omega)_{\omega \approx \omega_r} \cong \frac{{}_rA_{jk}}{\omega_r^2 - \omega^2 + i\eta_r\omega_r^2} + {}_rB_{jk} \qquad (4.10)$$

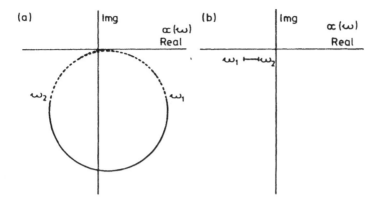

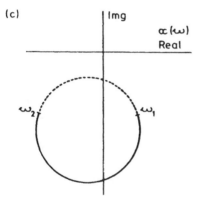

Fig. 4.7   FRF Nyquist plots for 4DOF system.
(a) Contribution of local mode; (b) Contribution of other modes;
(c) Complete FRF

This can be illustrated by a specific example, shown in Fig. 4.7. Using a 4DOF system, the receptance properties have been computed in the immediate vicinity of the second mode and each of the two terms in equation (4.9b) has been plotted separately, in Figs. 4.7(a) and (b) using the Nyquist display. Also shown, in Fig. 4.7(c), is the corresponding plot of the total receptance over the same frequency range. What is clear in this example is the fact that the first term (that relating to the mode under examination) varies considerably through the resonance region, sweeping out the expected circular arc in the Nyquist plot, while the second term, which includes the combined effects of all the other modes, is effectively constant through the narrow frequency range covered. Thus we see from the total receptance plot in Fig. 4.7(c) that this may, in effect, be treated as a circle with the same properties as the modal circle for the specific mode in question but which is displaced from the origin of the Argand plane by an amount determined by the contribution of all the other modes. Note that this is not to say that the other modes are unimportant or negligible — quite the reverse, their influence can be considerable — but rather that their combined effect can be represented as a constant term around this resonance. It is also clear from this discussion, that the subtle differences discussed above concerning the effect of other modes are almost invisible in the other plot, that in which the FRF modulus is plotted against frequency, a result which tends to reduce the effectiveness of performing a modal analysis on data presented in that format.

4.3.2 SDOF Modal Analysis I — Peak-Amplitude Method

We shall begin our study of the various methods available for analysing measured FRF data to obtain the described mathematical models of our test structure by examining the very simplest of SDOF approaches — the so-called 'peak-picking' or 'peak-amplitude' method. In this method it is assumed that **all** the response can be attributed to the local mode and that any effects due to the other modes can be ignored. This is a method which works adequately for structures whose FRF exhibit well-separated modes which are not so lightly-damped that accurate measurements at resonance are difficult to obtain but which, on the other hand, are not so heavily damped that the response at a resonance is strongly influenced by more than one mode. Although this appears to limit the applicability of the method, it should be noted that in the more difficult cases, such an approach can be useful in obtaining initial estimates to the parameters required, thereby speeding up those of the more general curve-fitting procedures, described later, which require starting estimates.

The peak-picking method is applied as follows:

(i) First, individual resonance peaks are detected on the FRF plot (Fig. 4.8(a)), and the frequency of one of the maximum responses taken as the natural frequency of that mode (ω_r).

(ii) Second, the local maximum value of the FRF is noted ($|\hat{H}|$) and the frequency bandwidth of the function for a response level of $|\hat{H}|/\sqrt{2}$ is determined ($\Delta\omega$). The two points thus identified as ω_b and ω_a are the 'half-power points': see Fig. 4.8(b).

(iii) The damping of the mode in question can now be estimated from one of the following formulae (whose derivation is given below in (4.21)):

$$\eta_r = \frac{\omega_a^2 - \omega_b^2}{2\omega_r^2} \cong \frac{\Delta\omega}{\omega_r}$$

$$2\zeta_r = \eta_r$$

(4.11)

(iv)   Last, we may now obtain an estimate for the modal constant of the mode being analysed by assuming that the total response in this resonant region is attributed to a single term in the general FRF series (equation (2.66)). This can be found from the equation

$$\left|\hat{H}\right| = \frac{A_r}{\omega_r^2 \eta_r}$$

or   (4.12a)

$$A_r = \left|\hat{H}\right|\omega_r^2 \eta_r$$

It is appropriate now to consider the possible limitations to this method. First, it must be noted that the estimates of both damping and modal constant depend heavily on the accuracy of the maximum FRF level, $|\hat{H}|$, and as we have seen in the previous chapter on measurement techniques, this is not a quantity which is readily measured with great accuracy. Most of the errors in measurements are concentrated around the resonance region and particular care must be taken with lightly-damped structures where the peak value may rely entirely on the validity of a single point in the FRF spectrum. Also, it is clear that only real modal constants — and that means real modes, or proportionally-damped structures — can be deduced by this method.

   The second most serious limitation will generally arise because the single-mode assumption is not strictly applicable. Even with clearly-separated modes, it is often found that the neighbouring modes do contribute a noticeable amount to the total response at the resonance of the mode being analysed. It is to deal with this problem that the more general circle-fit method, described in the next section, was developed as a refinement of this current approach. The problem is illustrated in Figs. 4.8(c) and (d) where a Nyquist type of plot is used to show two possible FRF characteristics which might equally well give the modulus plot shown in Fig. 4.8(b). The limitation of the method described above becomes evident and it will, in the second example, produce an overestimate of the damping level and an erroneous modal constant.

   However, it is possible to adapt the above procedure slightly,

308

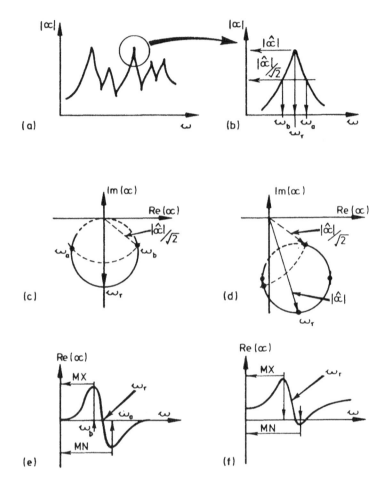

Fig. 4.8 Peak-amplitude method of modal analysis.
(a) Modulus plot; (b) Resonance detail; (c) Nyquist plot of single
mode — 1; (d) Nyquist plot of single mode — 2; (e) Real part of
single mode plot; (f) Imaginary part of single mode plot

without involving the curve-fitting processes to be discussed next, by working with a plot of the Real Part of the receptance FRF, instead of the modulus plot as shown in Figs. 4.8(a) and (b). Figs. 4.8(e) and (f) show plots of the Real Part of the receptance detail previously illustrated in Figs. 4.8(c) and (d). From these it can be seen that the positions and values of the maximum and minimum values of the plot yield good estimates of the locations of the half-power points and of the diameter of the circle in the Nyquist plot. This last quantity is a better indicator of the maximum magnitude of the single term in the FRF series upon which the estimate of the modal constant is based, equation (4.12). Furthermore, a more refined estimate of the natural frequency itself can be derived from the midway point between the maximum and minimum on the Imaginary plot: see Figs. 4.8(c) and (d). Thus we can use

$$A_r = \left(|MX| + |MN|\right)\omega_r^2 \eta_r \qquad\qquad (4.12b)$$

4.3.3   SDOF Modal Analysis II — Circle-Fit Method
4.3.3.1   Properties of the modal circle
We shall now examine the slightly more detailed SDOF analysis method based on circle-fitting FRF plots in the vicinity of resonance. It was shown in Chapter 2 that for the general SDOF system, a Nyquist plot of frequency response properties produced circle-like curves and that, if the appropriate parameter were chosen for the type of damping model, this would produce an exact circle. Further, we saw in the later sections concerned with MDOF systems that these also produce Nyquist plots of FRF data which include sections of near-circular arcs corresponding to the regions near the natural frequencies. This characteristic provides the basis of one of the most important types of modal analysis, that known widely as 'the SDOF circle-fit method'.

We shall base our treatment in this section on a system with structural damping and thus shall be using the receptance form of FRF data as it is this parameter which produces an exact circle in a Nyquist plot for the properties of a simple oscillator (see Section 2.2). However, if it is required to use a model incorporating viscous damping, then it is the mobility version of the FRF data which should be used. Although this gives a different general appearance to the diagrams — as they are rotated by 90° on the complex plane — most of the following analysis and comments apply equally to that choice. Some of the more discriminating modal analysis packages offer the choice between the two types of damping and simply take receptance or mobility data for the circle-fitting according to the selection.

310

Having established the plausibility of observing an individual modal circle from a (measured) FRF plot, we shall now explore some of the properties of the modal circle since these provide the means of extracting the required modal parameters. In the case of a system assumed to have structural damping, the basic function with which we are dealing is:

$$\alpha(\omega) = \frac{1}{\omega_r^2 \left(1 - (\omega/\omega_r)^2 + i\eta_r\right)} \qquad (4.13)$$

since the only effect of including the modal constant $_rA_{jk}$ is to scale the size of the circle (by $|_rA_{jk}|$) and to rotate it (by $\angle_rA_{jk}$). A plot of the quantity $\alpha(\omega)$ is given in Fig. 4.9. Now, it may be seen that for any frequency, ω, we may write the following relationships:

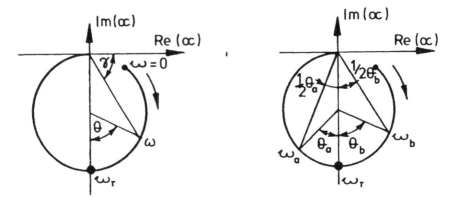

Fig. 4.9   Properties of modal circle

$$\tan\gamma = \frac{\eta_r}{1 - (\omega/\omega_r)^2} \qquad (4.14a)$$

$$\tan(90° - \gamma) = \tan\left(\frac{\theta}{2}\right) = \frac{1 - (\omega/\omega_r)^2}{\eta_r} \qquad (4.14b)$$

from which we obtain:

$$\omega^2 = \omega_r^2\left(1 - \eta_r\tan\left(\frac{\theta}{2}\right)\right) \qquad (4.14c)$$

If we differentiate equation (4.14c) with respect to θ, we obtain:

$$\frac{d\omega^2}{d\theta} = \frac{-\omega_r^2 \eta_r \left(1 - (\omega/\omega_r)^2\right)^2}{2} \frac{1}{\eta_r^2} \tag{4.15}$$

The reciprocal of this quantity — which is a measure of the rate at which the locus sweeps around the circular arc — may be seen to reach a maximum value (maximum sweep rate) when $\omega = \omega_r$, the natural frequency of the mode. This is shown by further differentiation, this time with respect to frequency:

$$\frac{d}{d\omega}\left(\frac{d\omega^2}{d\theta}\right) = 0 \quad \text{when} \quad \left(\omega_r^2 - \omega^2\right) = 0 \tag{4.16}$$

It may also be seen from this analysis that an estimate of the damping is provided by the sweep rate parameter since:

$$\left(\frac{d\theta}{d\omega^2}\right)_{\omega=\omega_r} = -\frac{2}{\omega_r^2 \eta_r} \tag{4.17}$$

The above property proves useful in analysing MDOF system data since, in general, it is not known exactly where is the natural frequency, but if we can examine the relative spacing of the measured data points around the circular arc near each resonance, then we should be able to determine its value.

Another valuable result can be obtained by further inspection of this basic modal circle. Suppose we have two specific points on the circle, one corresponding to a frequency (ω_b) below the natural frequency, and the other to one (ω_a) above the natural frequency. Referring to Fig. 4.9, we can write:

$$\tan\left(\frac{\theta_b}{2}\right) = \frac{1 - (\omega_b/\omega_r)^2}{\eta_r}$$

$$\tag{4.18}$$

$$\tan\left(\frac{\theta_a}{2}\right) = \frac{(\omega_a/\omega_r)^2 - 1}{\eta_r}$$

and from these two equations we can obtain an expression for the damping of the mode:

$$\eta_r = \frac{\omega_a^2 - \omega_b^2}{\omega_r^2 \left(\tan(\theta_a/2) + \tan(\theta_b/2) \right)} \qquad (4.19)$$

This is an exact expression, and applies for all levels of damping. If we are concerned with light damping (say, loss factors of less than 2 to 3 per cent), the expression above simplifies to:

$$\eta_r \cong \frac{2(\omega_a - \omega_b)}{\omega_r \left(\tan(\theta_a/2) + \tan(\theta_b/2) \right)} \qquad (4.20)$$

and if we further restrict our interest, this time to the two points for which $\theta_a = \theta_b = 90°$ (the half-power points), we obtain the familiar formula:

$$\eta_r = \frac{\omega_2 - \omega_1}{\omega_r} \qquad (4.21a)$$

or, if the damping is not light:

$$\eta_r = \frac{\omega_2^2 - \omega_1^2}{2\omega_r^2} \qquad (4.21b)$$

and this is **exact** for any level of damping. The final property relates to the diameter of the circle which, for the quantity specified in equation (4.13), is given by $(1/\omega_r^2 \eta_r)$. When scaled by a modal constant added in the numerator, the diameter will be

$$_r D_{jk} = \frac{\left| _r A_{jk} \right|}{\omega_r^2 \eta_r}$$

and, as mentioned earlier, the whole circle will be rotated so that the principal diameter — the one passing through the natural frequency point — is oriented at an angle $\arg(_r A_{jk})$ to the negative Imaginary axis. (Note that this means that the circle will be in the upper half of the plane if $_r A_{jk}$ is effectively negative, a situation which cannot arise for a point measurement but which can for transfer data.)

We shall complete this section by deriving corresponding formulae to (4.19) to (4.21) for the case of a SDOF system with viscous, rather than structural, damping. Recalling that in this case we should use mobility in place of receptance, we can write

$$Y(\omega) = \frac{i\omega}{\left(k - \omega^2 m\right) + i(\omega c)} \tag{4.22}$$

or

$$\text{Re}(Y) = \frac{\omega^2 c}{\left(k - \omega^2 m\right)^2 + (\omega c)^2}$$

$$\text{Im}(Y) = \frac{i\omega\left(k - \omega^2 m\right)}{\left(k - \omega^2 m\right)^2 + (\omega c)^2}$$

From there, and referring to the sketch in Fig. 4.9, we have

$$\tan\left(\frac{\theta}{2}\right) = \frac{\omega\left(k - \omega^2 m\right)}{\omega^2 c} = \frac{1 - (\omega/\omega_r)^2}{(2\zeta\omega/\omega_r)} \tag{4.23}$$

and using the same procedure as before for ω_b and ω_a (points before and after ω_r, respectively):

$$\tan\left(\frac{\theta_b}{2}\right) = \frac{1 - (\omega_b/\omega_r)^2}{(2\zeta\omega_b/\omega_r)}$$

$$\tan\left(\frac{\theta_a}{2}\right) = \frac{(\omega_a/\omega_r)^2 - 1}{(2\zeta\omega_a/\omega_r)}$$

These expressions yield:

$$\zeta = \frac{\omega_a^2 - \omega_b^2}{2\omega_r\left(\omega_a\tan(\theta_a/2) + \omega_b\tan(\theta_b/2)\right)} \tag{4.24a}$$

or, for light damping,

$$\zeta \cong \frac{\omega_a - \omega_b}{\omega_r\left(\tan(\theta_a/2) + \tan(\theta_b/2)\right)} \tag{4.24b}$$

Finally, selecting the half-power points as those frequencies for which $\theta_a = \theta_b = 90°$, we have:

$$\zeta = \frac{\omega_2 - \omega_1}{2\omega_r} \qquad\qquad (4.24c)$$

4.3.3.2 Circle-fit analysis procedure

Armed with the above insight into the structure of an FRF plot near resonance, it is a relatively straightforward matter to devise an analysis procedure to extract the necessary coefficients in equation (4.11a), and thence the modal parameters themselves. Basing the following comments on the case for structural damping, the sequence is:

(i)   select points to be used;
(ii)   fit circle, calculate quality of fit;
(iii)   locate natural frequency, obtain damping estimate;
(iv)   calculate multiple damping estimates, and scatter;
(v)   determine modal constant modulus and argument.

Step (i) can be made automatic by selecting a fixed number of points on either side of any identified maximum in the response modulus or it can be effected by the operator whose judgement may be better able to discern true modes from spurious perturbations on the plot and to reject certain suspect data points. The points chosen should not be influenced to any great extent by the neighbouring modes and, whenever possible without violating that first rule, should encompass some 270° of the circle. This is often not possible and a span of less than 180° is more usual, although care should be taken not to limit the range excessively as this becomes highly sensitive to the accuracy of the few points used. Not less than six points should be used.

The second step, (ii), can be performed by one of numerous curve-fitting routines and consists simply of finding a circle which gives a least-squares deviation for the points included. Note that there are two possible criteria which can be applied here: one is that which minimises the deviations of points from the nearest point on the circle and the other, which is more accurate, minimises the deviations of the measured points from where they ought to be on the circle. This latter condition is more difficult to apply and so it is the former which is more common. At the end of this process, we have specified the centre and radius of the circle and have produced a quality factor which is the mean square deviation of the chosen points from the circle. 'Errors' of the order to 1 to 2 per cent are commonplace and an example of the process is shown in Fig. 4.10(a).

Step (iii) can be implemented by constructing (here used metaphorically as the whole process is performed numerically) radial lines from the circle centre to a succession of points around the resonance and by noting the angles they subtend with each other. Then, the rate of sweep through the region can be estimated and the frequency at which it reaches a maximum can be deduced. If, as is usually the case, the frequencies of the points used in this analysis are spaced at regular intervals (i.e. a linear frequency increment), then this process can be effected using a finite difference method. Such a procedure enables one to pinpoint the natural frequency with a precision of about 10 per cent of the frequency increments between the points. At the same time, an estimate for the damping is derived using (4.17) although this will be somewhat less accurate than that for the natural frequency. Fig. 4.10(b) shows the results from a typical calculation.

It is interesting to note at this point that other definitions of the natural frequency are sometimes used. Including:

(a)   the frequency of maximum response;
(b)   the frequency of maximum imaginary receptance;
(c)   the frequency of zero real receptance.

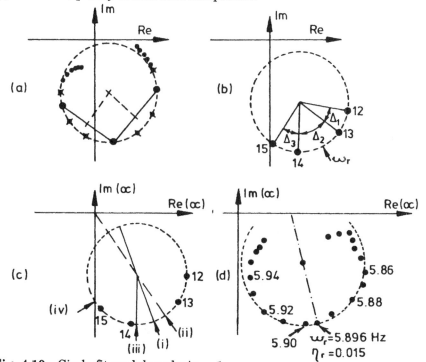

Fig. 4.10   Circle-fit modal analysis — 1.
(a) Circle-fit; (b) Location of natural frequency; (c) Alternative definitions; (d) Damping estimates

316

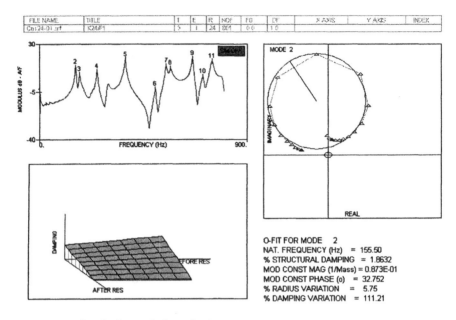

FILE NAME	TITLE		T	E	R	NOF	FO	DF	X AXIS	Y AXIS	INDEX
Ca124-01.trf	X24F1		5	I	24	804	0 0	1 0			

MODE 2

O-FIT FOR MODE   2
NAT. FREQUENCY (Hz)   = 155.50
% STRUCTURAL DAMPING   = 1.8632
MOD CONST MAG (1/Mass) = 0.873E-01
MOD CONST PHASE (o)   = 32.752
% RADIUS VARIATION   =   5.75
% DAMPING VARIATION   = 111.21

Fig. 4.10   Circle-fit modal analysis — 1.
(e) Typical application

These are all indicated on Fig. 4.10(c) and while they seldom make a significant difference to the value of the natural frequency itself, selecting the wrong one can have implications for the values found for the damping factor and for the modal constant (and thus the mode shapes).

Next, for step (iv), we are able to compute a set of damping estimates using every possible combination from our selected data points of one point below resonance with one above resonance using equation (4.19). With all these estimates we can either compute the mean value or we can choose to examine them individually to see whether there are any particular trends. Ideally, they should all be identical and so an indication not only of the mean but also of the deviation of the estimates is useful. If the deviation is less than 4 to 5 per cent, then we have generally succeeded in making a good analysis. If, however, the scatter is 20 or 30 per cent, there is something unsatisfactory. If the variations in damping estimate are random, then the scatter is probably due to measurement errors but if it is systematic, then it could be caused by various effects (such as poor experimental set-up, interference from neighbouring modes, non-linear behaviour, etc.), none of which should, strictly, be averaged out. Thus, if a large scatter of damping estimates is indicated, a plot of their values such as that shown in Fig. 4.10(d) should be examined (see

Section 4.3.3.3, below).

Lastly, step (v) is a relatively simple one in that it remains to determine the magnitude and argument of the modal constant from the diameter of the circle, and from its orientation relative to the Real and Imaginary axes. This calculation is straightforward once the natural frequency has been located and the damping estimates obtained.

Finally, if it is desired to construct a theoretically-regenerated FRF plot against which to compare the original measured data, it will be necessary to determine the contribution to this resonance of the other modes and that requires simply measuring the distance from the 'top' of the principal diameter to the origin, this quantity being the value of $_rB_{jk}$ in equation (4.10). Then, using that equation together with the modal parameters extracted from the circle-fit, it is possible to plot a curve based on the 'model' obtained.

NOTE: if previous estimates for ω_r and η_r are available, steps (iii) and (iv) can be omitted, and only the modal constant derived.

A typical circle-fit analysis of some practical data is presented in Fig. 4.10(e).

4.3.3.3 Interpretation of damping plots

It has been shown how the multiple estimates of damping available from this approach can be used to obtain an average value for the damping factor. Indeed, the variation and distribution of the individual damping estimates, as illustrated by the damping 'carpet' plots shown in Fig. 4.10(c), can serve as a very useful diagnostic of the quality of the entire analysis. As mentioned earlier, good measured data should lead to a smooth plot of these damping estimates and any roughness of the surface can be explained in terms of noise on the original data — see, for example, Fig. 4.11(a). However, any systematic distortion of the plot (the surface should be smooth, flat and level) is almost certainly caused by some form of error in the data, or its analysis, or in the assumed behaviour of the structure. Thus, leakage errors show up as illustrated in Fig. 4.11(b); modal analysis errors, such as in the estimate for the natural frequency, as in Fig. 4.11(c) and non-linearity in the structure's behaviour as in Fig. 4.11(d).

It is possible to accumulate quite quickly a record of standard damping-plot characteristics of structures which are tested frequently, and the faults which are sometimes observed in their measurement catalogued, and this readily detected from inspection of the damping plots produced. They are an effective means of checking the quality of the measured and analysed data.

318

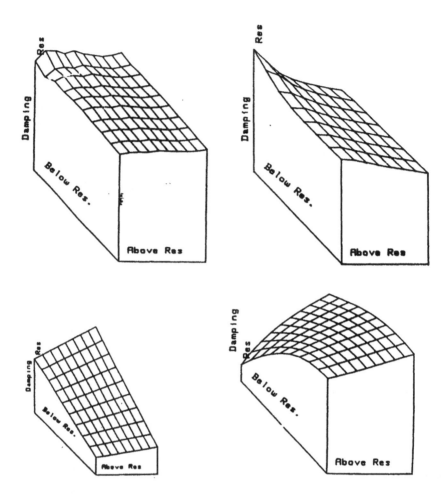

Fig. 4.11 Circle-fit modal analysis — 2: interpretation of damping plots

4.3.4 SDOF Modal Analysis III — Inverse or Line-fit Method
4.3.4.1 Properties of inverse FRF plots

Although the circle-fit method is very widely used, there are alternative procedures available which work within the same general confines and assumptions. We shall discuss one here: a direct alternative to the circle-fit called the 'Inverse' or 'Line-fit' Method.

The original version of this method used the fact that a function which generates a circle when plotted in the complex (Nyquist) plane will, when plotted as a reciprocal, trace out a straight line. Thus, if we were to plot the reciprocal of receptance (not dynamic stiffness in the strict sense) of a SDOF system with structural damping, we would find that in the Argand diagram it produces a straight line as can be seen from inspection of the appropriate expressions for a SDOF system:

$$\alpha(\omega) = \frac{\left(k - \omega^2 m\right) - i(d)}{\left(k - \omega^2 m\right)^2 + d^2} \tag{4.25a}$$

and

$$\frac{1}{\alpha(\omega)} = \left(k - \omega^2 m\right) + i(d) \tag{4.25b}$$

Sketches of these two forms of the FRF are shown in Fig. 4.12 and the procedure which may be used to determine the modal parameters using the inverse FRF is as follows. First, a least-squares best-fit straight line is constructed through the data points and an estimate for the damping parameter is immediately available from the intercept of the line with the Imaginary axis. Furthermore, an indication of the reliability of that estimate may be gained from the nature of the deviations of the data points from the line itself — if these are randomly scattered above and below the line, then we probably have typical experimental errors, but if they deviate in a systematic fashion, such as by being closer to a curve than a straight line, or to a line of other than zero slope, then there is a source of bias in the data which should be investigated before making further use of the results.

Then, a second and independent least-squares operation is performed, this time on the deviation between the Real part of the measured data points and that of the theoretical model. Resulting from this, we obtain estimates for the mass and stiffness parameters in the theoretical model to complete the description.

It should be noted that this approach is best suited to systems with real modes (effectively assumed in the analysis) and to relatively well-separated modes as corrective action is required in the event that the FRF is not locally dominated by a single mode. However, the method is relatively insensitive to whether or not data are measured exactly at the natural frequency (at which point the Real part of the inverse receptance is zero) as the straight line can readily be obtained with data points which are well away from resonance.

4.3.4.2 General inverse analysis method

More recently, another application of the same basic concept was developed which has led to a robust and simple SDOF modal analysis

320

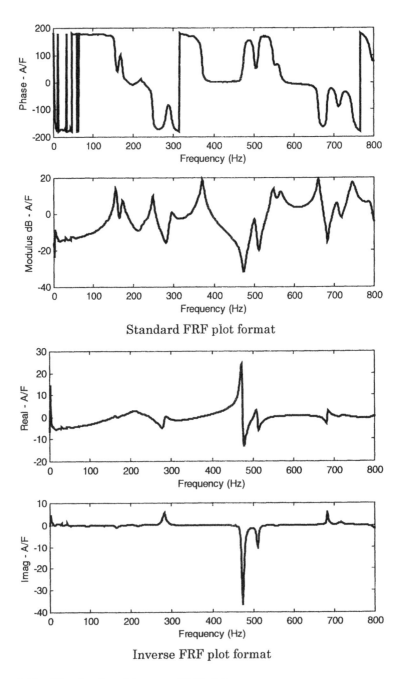

Standard FRF plot format

Inverse FRF plot format

Fig. 4.12   Standard and inverse FRF plots

approach which is found to be more usable than the circle-fit method while retaining its essential features, including the diagnostic potential of multiple estimates. This method is the 'Line-fit' method of the title and is summarised below.

In this version, use is made of the inverse FRF $(H_{jk}^{-1}(\omega))$ when plotted in a different way to that just described: two plots are presented, one of the Real Part of $(H_{jk}^{-1}(\omega))$ vs. (frequency, $\omega)^2$, and the other of the Imaginary Part of $(H_{jk}^{-1}(\omega))$ vs. (frequency, $\omega)^2$. It has already been shown that if a purely SDOF system FRF is plotted in this way, then both plots demonstrate straight lines, and separately reveal useful information about the mass, stiffness and damping properties of the measured system, including an indication as to whether the damping is structural (Imaginary Part constant with frequency, ω) or viscous (Imaginary Part linear with frequency, ω).

However, this simplicity of form is not present when the FRF data contain the effects of more than a single mode, as is usually the case, because it can readily be seen that, although the inverse of the FRF for a truly SDOF system has the advantageous properties that result in this feature of the plots:

$$(H_{jk}^{-1}(\omega)) = (k - \omega^2 m) + i(\omega c \text{ or } d)$$

the corresponding inverse in the case of more than one mode present is not similarly convenient:

$$(H_{jk}^{-1}(\omega)) = \frac{1}{\Sigma\{(k - \omega^2 m) + i(\omega c \text{ or } d)\}}$$

which means that in reciprocal form the modal series does not apply:

$$(H_{jk}^{-1}(\omega)) \neq \sum \frac{1}{\{(k - \omega^2 m) + i(\omega c \text{ or } d)\}}$$

Thus it becomes clear that if the early advantages of the inverse plot are to be available in the more general case, then some modification to the basic formulation must be found. This is done as follows.

We start with the basic formula for SDOF analysis, quoted earlier in (4.2), which is:

$$\alpha_{jk}(\omega)_{\omega \approx \omega_r} \cong \frac{{}_r A_{jk}}{\omega_r^2 - \omega^2 + i\eta_r \omega_r^2} + {}_r B_{jk}$$

and we note that the presence of the $_r B_{jk}$ term — the contribution of modes other than the one of current interest — 'spoils' the inverse plot. The trick is to define a new FRF term, $\alpha'_{ik}(\omega)$, which is simply the difference between the actual FRF and the value of the FRF at one fixed frequency in the range of interest - a frequency which is referred to as the 'fixing frequency' and which is denoted by Ω . Thus we have:

$$\alpha'_{jk}(\omega) = \alpha_{jk}(\omega) - \alpha_{jk}(\Omega)$$

from which the inverse FRF parameter that we shall use for the modal analysis, $\Delta(\omega)$, can be defined:

$$\Delta(\omega) = (\omega^2 - \Omega^2)/\alpha'_{jk}(\omega)$$

$$= \text{Re}(\Delta) + i\text{Im}(\Delta)$$

It can be seen that these two quantities, $\text{Re}(\Delta(\omega))$ and $\text{Im}(\Delta(\omega))$, are simply related to the variable frequency, ω , as:

$$\text{Re}(\Delta) = m_R \omega^2 + c_R \quad ; \quad \text{Im}(\Delta) = m_I \omega^2 + c_I$$

and, also, that

$$m_R = a_R(\Omega^2 - \omega_r^2) - b_r(\omega_r^2 \eta_r);$$
$$m_I = -b_R(\Omega^2 - \omega_r^2) - a_r(\omega_r^2 \eta_r);$$
$$_r A_{jk} = a_R + ib_r$$

So, the first step of our analysis procedure can be made, as follows:

(i) using the FRF data measured in the vicinity of the resonance, ω_r, choose one of the measured points as the datum (or 'fixing') frequency, Ω_j, and then calculate the possible values of $\Delta(\omega)$ using the remaining measured data points;

(ii) plot these values on Re vs. (frequency)^2 and Im vs. (frequency)^2 plots and compute the best-fit straight line in each case so as to determine $m_R(\Omega_j)$ and $m_I(\Omega_j)$ for that particular value of the fixing frequency, Ω_j.

Typical results of this stage of the process using data from a practical example are shown in Fig. 4.13, where the essentially-straight line characteristic of both Real and Imaginary Parts can be seen.

Now, it can be shown that both these straight-line slopes, m_R and m_I, are simple functions of Ω and that we can write:

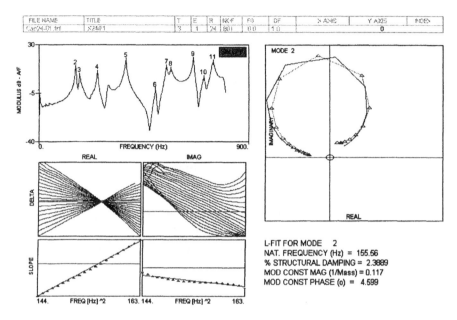

Fig. 4.13 Line-fit modal analysis: typical results.
(a) Plots of $\mathrm{Re}(\Delta(\omega))$ and $\mathrm{Im}(\Delta(\omega))$; (b) Slopes from (a)

$$m_R = n_R\Omega^2 + d_R \quad \text{and} \quad m_I = n_I\Omega^2 + d_I$$

where

$$n_R = a_r \quad ; \quad n_I = -b_r$$

$$d_R = -b_r(\omega_r^2\eta_r) - a_r(\omega_r^2) \quad ; \quad d_I = b_r(\omega_r^2) - a_r(\omega_r^2\eta_r)$$

Now, let:

$$p = n_I/n_R \quad \text{and} \quad q = d_I/d_R \tag{4.26}$$

and, noting that

$$\eta_r = \frac{(q-p)}{(1+pq)} \quad ; \quad \omega_r^2 = \frac{d_R}{(p\eta_r - 1)n_R}$$

$$a_r = \frac{\omega_r^2(p\eta_r - 1)}{(1+p^2)d_R} \quad ; \quad b_r = -a_r p \tag{4.27}$$

we now have sufficient information to extract estimates for the four parameters for the resonance which has been analysed:

$$\omega_r, \eta_r, \text{and } {}_r A_{jk} = a_r + ib_r$$

using the second stage of our process, which consists of the following steps:

(iii)   plot graphs of $m_R(\Omega)$ vs. Ω^2 and of $m_I(\Omega)$ vs. Ω^2, using the results from the repeated application of step (i), above, each time using a different one of the measurement points as the fixing frequency, Ω_j ;

(iv)   determine the slopes of the best-fit straight lines through these two plots, n_R and n_I ,and their intercepts with the vertical axis, d_R and d_I ;

(v)   using these four quantities and the formulae in equation (4.27), determine the four modal parameters required for that mode.

The FRF data shown earlier in Fig. 4.13(a) and relating to the first part of the analysis are again used to illustrate this second part, in Fig. 4.13(b). Once again, the straight-line feature of each of the plots can be seen and the modal parameters deduced from the second level of curve-fitting are shown at the foot of the graphs. This is a typical line-fit analysis.

Certain features of the line-fit type of modal analysis are worth mentioning. First is the fact that the straight-line curve-fitting which is required is simpler and quicker than that based on circles or more complicated forms. Second, the straight-line feature means that it is easier to spot discrepancies from the expected form, such as poor data, inappropriate damping models or non-linear behaviour. Third, the inverse nature of the functions being used mean that the most important data in determining the best-fit lines are those points which are furthest from the resonance point. This is significant because, as we have seen in the earlier chapters, the points closest to resonance are often the most difficult to measure accurately: both the exciter-structure interaction dynamics, and various signal processing effects, combine to make the resonance region the most vulnerable to error.

Thus, a modal analysis method which places more weight to points slightly away from the resonance region is likely to be less sensitive to these measurement difficulties.

4.3.5 Other SDOF Methods

We have now introduced the main SDOF methods in current use but it is worth mentioning another approach which is sometimes used in certain circumstances and, anyway, will find a role in our subsequent discussion of modal analysis on slightly non-linear systems. As we have seen, each resonance is defined by four modal parameters:

$$\omega_r, \eta_r, \text{and } _rA_{jk} = a_r + ib_r$$

and since each individual FRF data point contains three quantities:

$$\omega, \text{Re}(H(\omega)), \text{Im}(H(\omega))$$

one might imagine that the four modal parameters could possibly be determined using as few as two complex FRF data points. This idea forms the basis of another SDOF modal analysis approach, although because of its sensitivity to errors in the data, it must be used with considerable caution. The usual application is to take several pairs of FRF data points, usually each pair straddling the resonance, and to compute the modal parameters using the limited data provided by the two points. This is repeated several times and the resulting estimates of the modal parameters averaged to yield a mean value for each. If the measured FRF data are of very high quality, then this can be a useful and valid approach, but if not, then one of the preceding methods — each of which take advantage of the over-determination of the actual problem — is to be recommended.

4.3.6 Residuals
4.3.6.1 Concept of residual terms

At this point we need to introduce the concept of residual terms, necessary in the modal analysis process to take account of those modes which we do not analyse directly but which nevertheless exist and have an influence on the FRF data we use. Usually, it is necessary to limit the frequency range of measurement and/or analysis for practical reasons and this inevitably means that we cannot identify the properties of modes which exist outside this range. However, the influence of such modes is present in the measured FRF data and we must take account of it somehow. (Note: it should be observed that the topic discussed here is not related to the 'residue' quantities used in some analyses as an alternative definition to our 'modal constant'.)

The first occasion on which the residual problem is encountered is generally at the end of the analysis of a single FRF curve, such as by the repeated application of an SDOF curve-fit to each of the resonances in turn until all modes visible on the plot have been identified. At this point, it is often desired to construct a 'theoretical' curve, based on the modal parameters extracted from the measured data, and to overlay this on the original measured data to assess the success of the curve-fit process. (A more appropriate description of the calculated curve is 'regenerated', since it does not come from a purely theoretical analysis of the system, and we shall use this terminology subsequently.) When the regenerated curve is compared with the original measurements, the result is often disappointing, as illustrated in Fig. 4.14(a). However, by the inclusion of two simple extra terms — the 'residuals' — the modified regenerated curve is seen to correlate very well with the original experimental data, as shown in Fig. 4.14(b). The origin of these residual terms may be explained as follows.

If we regenerate an FRF curve from the modal parameters we have extracted from the measured data, we shall use a formula of the type:

$$H_{jk}(\omega) = \sum_{r=m_1}^{m_2} \frac{{}_r A_{jk}}{\omega_r^2 - \omega^2 + i\eta_r \omega_r^2} \qquad (4.28)$$

in which we have shown the limits in the modal series as m_1 and m_2 to reflect the fact that, in general, we do not always start below the first mode $(r = 1)$ and we seldom continue to the highest mode $(r = N)$. However, just because we choose to limit our frequency range of measurement and analysis does not mean that the measured FRF data is unaffected by modes which lie outside this range. Indeed, the equation which most closely represents the measured data is:

$$H_{jk}(\omega) = \sum_{r=1}^{N} \frac{{}_r A_{jk}}{\omega_r^2 - \omega^2 + i\eta_r \omega_r^2} \qquad (4.29a)$$

which may be rewritten, without loss of generality, as:

$$H_{jk}(\omega) = \sum_{r=1}^{m_1-1} + \sum_{r=m_1}^{m_2} + \sum_{r=m_2+1}^{N} \left(\frac{{}_r A_{jk}}{\omega_r^2 - \omega^2 + i\eta_r \omega_r^2} \right) \qquad (4.29b)$$

In this equation, we shall refer to the first of the three terms as that for the 'low-frequency' modes; to the third term as that for the 'high-

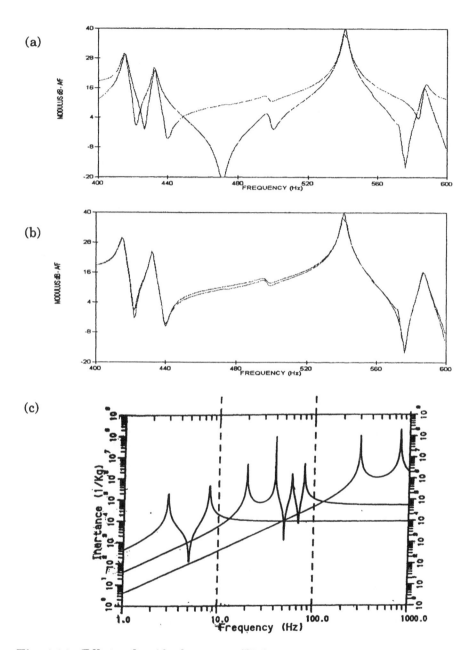

Fig. 4.14 Effects of residual terms on FRF regeneration.
(a) Measured and regenerated without residuals; (b) Measured
and regenerated with residuals; (c) Numerical simulation of
contributions of low-, medium-, and high-frequency modes

frequency' modes while the second term is that which relates to the modes actually identified. Fig. 4.14(c) illustrates plots of typical values for each of the three terms individually, and the middle one is that which is computed using only the modal data extracted from the modal analysis process, such as demonstrated in a particular case in Fig. 4.14(d). It is usually the case that we are seeking a model of the structure which is accurate within the frequency range of our tests (it would be unreasonable to expect to be able to derive one which was representative beyond the measured frequency range) and so we need to find a way of correcting the regenerated plot within the central frequency range to take account of the low-frequency and high-frequency modes. From the sketch, it may be seen that within the frequency range of interest, the first term tends to approximate to a mass-like behaviour, while the third term, for the high-frequency modes, approximates to a stiffness effect. Thus, we have a basis for the residual terms and shall rewrite equation (4.29b):

$$H_{jk}(\omega) \cong -\frac{1}{\omega^2 M_{jk}^R} + \sum_{r=m_1}^{m_2}\left(\frac{{}_rA_{jk}}{\omega_r^2 - \omega^2 + i\eta_r\omega_r^2}\right) + \frac{1}{K_{jk}^R} \qquad (4.30)$$

where the quantities M_{jk}^R and K_{jk}^R are the residual mass and stiffness for that particular FRF and, it should be noted, for that particular frequency range (if we extend or limit the range of analysis, the residual terms will also change).

4.3.6.2 Calculation of residual mass and stiffness terms

The way in which residual terms are calculated is relatively straightforward and involves an examination of the FRF curve at either end of the frequency range of interest. First, we compute a few values of the regenerated FRF curve at the lowest frequencies covered by the tests, using only the identified modal parameters. Then, by comparing these values with those from actual measurements, we estimate a mass residual constant which, when added to the regenerated curve, brings this closely into line with the measured data. Then, the process is repeated at the top end of the frequency range, this time seeking a residual stiffness. Often, the process is more effective if there is an antiresonance near either end of the frequency range and this is used as the point of adjustment. The procedure outlined here may need to be repeated iteratively in case the addition of the stiffness residual term then upsets the effectiveness of the mass term, and so on, but if the frequency range encompassed is a decade or greater, such interaction is generally minor.

Finally, it should be noted that often there is a physical significance to the residual terms. If the test structure is freely-supported, and its rigid body modes are well below the minimum frequency of measurement, then the low-frequency or mass residual term will be a direct reflection of the rigid body mass and inertia properties of the structure and, as such, is amenable to direct computation using simple dynamics. At the other extreme, the high-frequency residual can represent the local flexibility at the drive point. It can be seen, from inspection of the expression (4.29b), that the magnitude of this stiffness residual will vary according to the type of frequency response function considered. If we are concerned with a point measurement, then all the modal constants in the series $r = m_2, N$ will be positive, and as the denominator will always have the same sign, all the contributions from the high-frequency modes will be additive, resulting in the maximum possible magnitude for the residual. On the other hand, for a transfer FRF, we find that the terms in the series will be of varied sign, as well as magnitude, and so the total expression will tend to be less than for a point FRF and, in some cases, will tend to be very small, if not negligible. This characteristic should be borne in mind when computing residual terms.

4.3.6.3 Residuals as pseudo modes

Sometimes, it is convenient to treat the residual terms as if they were modes, simply to minimise the complexity of the data base which has to be stored at the end of each modal analysis. Instead of representing each residual effect by a constant — a residual mass, or residual stiffness — each can be represented by a pseudo mode, that is to say a mass and stiffness quantity for each of the two residuals. For the low-frequency residual effects, this pseudo mode has a 'natural frequency' which is below the lowest frequency on the measured FRF; for the high-frequency residual effects, that pseudo mode has a 'natural frequency' which is above the highest frequency on the measured range. These pseudo modes can be conveniently included in the list of modes which have been extracted by modal analysis of that FRF curve and used, with the genuine modes, to compute the regenerated curve required for comparison with the measured original. In effect, this is the same as saying that the low-frequency residual effect is not exactly a straight mass-line, as depicted in Fig. 4.14(c), nor the high-frequency term a straight stiffness line, but both have a characteristic curve typical of the response of a SDOF system away from its natural frequency. The method of computing these pseudo mode parameters is identical to that described above for the simpler case where there is just one coefficient for each term: here there are two, but since each measured FRF point contains three quantities, both pseudo modes can easily be defined using the classical two-point correction method described above.

However, it is customary to supply the residual correction step in the analysis process with suggested values for the two pseudo modes' natural frequencies and in this case the two processes become identical: just one coefficient term to be found for each residual.

In fact, using pseudo modes instead of simple residual mass and stiffness terms is a more accurate way of representing the out-of-range modes. There is one warning, however, and that is to point out that these pseudo modes are **not** genuine modes and that although they do represent the out-of-range modes' contributions to a given FRF, they cannot be used to deduce the corresponding contributions of these same modes for any other FRF curve. This is an important point to which we shall return later.

4.3.7   Refinement of SDOF Modal Analysis Methods

In the circle-fit and other SDOF modal analysis methods discussed above, an assumption was made that near the resonance under analysis, the effect of all the other modes could be represented by a constant. There will be several situations where this assumption is not strictly valid and where the SDOF analysis will be inadequate as a result. Such situations will arise whenever there are neighbouring modes **close** to the one being analysed: 'close' being loosely defined as a situation where the separation between the natural frequencies of two adjacent modes is less than the typical damping level, both measured as percentages. This is often cited as a failing restriction for SDOF methods. However, by building on the results obtained with a direct analysis of this type, we can usually remove that restriction and thereby make a more precise analysis of the data. The means of introducing this refinement to this important class of analysis method is as follows.

We can write the following expression for the receptance FRF in the frequency range of interest:

$$H_{jk}(\omega) = \sum_{s=m_1}^{m_2} \frac{{}_s A_{jk}}{\omega_s^2 - \omega^2 + i\eta_s\omega_s^2} + \frac{1}{K_{jk}^R} - \frac{1}{\omega^2 M_{jk}^R} \qquad (4.31a)$$

which we can arrange into two terms as:

$$H_{jk}(\omega) = \left(\frac{{}_r A_{jk}}{\omega_r^2 - \omega^2 + i\eta_r\omega_r^2} \right)$$

$$+ \left(\sum_{\substack{s=m_1 \\ \neq r}}^{m_2} \frac{{}_s A_{jk}}{\omega_s^2 - \omega^2 + i\eta_s\omega_s^2} + \frac{1}{K_{jk}^R} - \frac{1}{\omega^2 M_{jk}^R} \right) \qquad (4.31b)$$

In the previous methods, the second term was assumed to be a constant throughout the curve-fit procedure to find the modal parameters for mode r. However, if we have some (good) estimates for the coefficients which constitute the second term, for example by having already completed an SDOF analysis, we may remove the restriction on the analysis. Suppose we take a set of measured data points around the resonance at ω_r and denote these as $H_{jk}^m(\omega)$ and then, at each frequency for which we have a measured FRF value, we can compute the magnitude of the second term in (4.31b) and subtract this from the measurement. The resulting adjusted data points should then conform to a true single-degree-of-freedom behaviour as demonstrated by:

$$H_{jk}^m(\omega) - \left(\sum_{\substack{s=m_1 \\ \neq r}}^{m_2} \frac{{}_s A_{jk}}{\omega_s^2 - \omega^2 + i\eta_s\omega_s^2} + \frac{1}{K_{jk}^R} - \frac{1}{\omega^2 M_{jk}^R} \right)$$

$$= \frac{{}_r A_{jk}}{\omega_r^2 - \omega^2 + i\eta_r\omega_r^2} \qquad (4.32)$$

and we can use the same technique as before to obtain improved estimates to the modal parameters for mode r. This procedure — sometimes referred to as 'SIM', referring to the simultaneous recognition of several modes — can be repeated iteratively for all the modes in the range of interest as many times as is necessary to obtain convergence to acceptable answers. It is often found that on 'normal' FRF data, where most of the modes are relatively weakly coupled, the improvement in the modal parameters is quite small — see Fig. 4.15 for an example — but in cases where there is stronger coupling, the enhancement can be significant.

4.4 MDOF MODAL ANALYSIS IN THE FREQUENCY DOMAIN (SISO)

4.4.1 General Approach

There are a number of situations in which the SDOF approach to modal analysis is simply inadequate or inappropriate and for these there exist

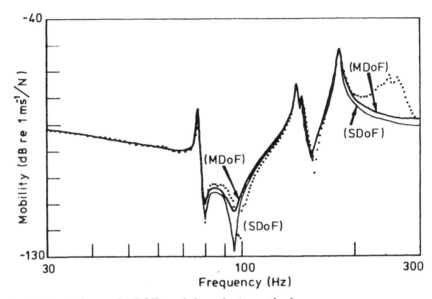

Fig. 4.15 Enhanced SDOF modal analysis method

several alternative methods which may generally be classified as multi-degree-of-freedom (or MDOF) modal analysis methods. The particular cases which demand a more elaborate treatment than that afforded by the SDOF concept are those with closely-coupled modes, where the single mode approximation is inappropriate, and those with extremely light damping, for which measurements at resonance are inaccurate and difficult to obtain. By closely-coupled modes we mean those systems for which either the natural frequencies are very closely spaced, or which have relatively heavy damping, or both, in which the response even at resonance is not dominated by just one mode (or term in the FRF series). A particular problem arises frequently with structures that possess modes which have very close natural frequencies, even identical in some cases, and for these a correct modal analysis can only be achieved using not only an MDOF approach (simultaneous extraction of several modes at a time), but also one which analyses several curves at the same time. These we shall discuss in the next section, confining our interest here to the first part of this generalisation: namely, to the simultaneous extraction of several modes' properties in one calculation.

For these cases, and for all others where a very high degree of accuracy is demanded, we look to a more exact modal analysis than that described in the previous sections. However, as a word of caution: we should be wary of using over-refined numerical analysis procedures on measured data which itself has a finite accuracy.

There are many individual algorithms available for this task and we shall not attempt to describe them all in detail. It will suffice to distinguish the different approaches and to explain the bases on which they operate. It is seldom necessary (and is often impossible) for the modal analyst to have an intimate knowledge of the detailed workings of the numerical processes but it is important that s/he is aware of the assumptions which have been made, and of the limitations and implications. Also, in the happy event that s/he has several different algorithms at his/her disposal, s/he must always be able to select the most appropriate for each application.

In this section we shall outline three different methods of frequency-domain MDOF curve-fitting and shall, as before, sometimes use the hysteretically-damped system as our example, and other times the viscous damping model. However, as we engage the more sophisticated techniques of numerical analysis, we need be less concerned with the detailed differences between viscous and hysteretic damping models. Mathematically, the difference is simply that in one version the imaginary parts of the FRF expression are constant while in the other (viscous) they are frequency-dependent. The various methods all share the feature of permitting a curve-fit to the entire FRF measurement in one step and the three approaches considered here are:

(i)   a general approach to multi-mode curve-fitting;
(ii)   a method based on the rational fraction FRF formulation, and
(iii)   a method particularly suited to very lightly-damped structures.

One final note before we examine the methods in detail: the comments in Section 4.3.6 concerning residuals apply in exactly the same way to these cases for MDOF analysis. Indeed, as we are proposing to consider the entire curve in one step, rather than an isolated mode, it is essential to incorporate the residual terms from the outset. If we do not do so, then the modal parameters which result from the modal analysis will probably be distorted in order to compensate for the influence the out-of-range modes in the measured data.

4.4.2   Method I: General Curve-Fit Approach: Non-Linear Least-Squares (NLLS)

Many of the longer-established modal analysis techniques were devised in the days of less powerful computation facilities than we enjoy today, and may, as a result, seem somewhat pedestrian by current numerical analysis standards. However, they often have the advantage that they permit the user to retain more direct contact with the processes being used than would be the case in a more powerful, and automatic algorithm. The secret is to maintain a balance between computational sophistication on the one hand (which might just be thwarted by the relatively poor quality of the input data), and the lengthy and often tedious practices which involve the user in many of the decisions.

We shall start our study of this important area by describing the basics of a general MDOF curve-fit philosophy, detailed implementation of which has been perfected by many workers and made widely available in the form of the Non-linear Least-squares (NLLS) Method.

We shall denote the individual FRF measured data as:

$$H_{jk}^m(\Omega_\ell) = H_\ell^m \tag{4.33a}$$

while the corresponding 'theoretical' values are denoted by

$$H_{jk}(\Omega_\ell) = H_\ell = \sum_{s=m_1}^{m_2} \frac{{}_sA_{jk}}{\omega_s^2 - \Omega_\ell^2 + i\eta_s\omega_s^2} + \frac{1}{K_{jk}^R} - \frac{1}{\Omega_\ell^2 M_{jk}^R} \tag{4.33b}$$

where the coefficients ${}_1A_{jk}$, ${}_2A_{jk}$, ..., ω_1, ω_2, ..., η_1, η_2, ..., K_{jk}^R and M_{jk}^R are all to be determined. We can define an individual error as ε_ℓ where:

$$\varepsilon_\ell = \left(H_\ell^m - H_\ell\right) \tag{4.34}$$

and express this as a scalar quantity:

$$E_\ell = \left|\varepsilon_\ell^2\right| \tag{4.35}$$

if we further increase the generality by attaching a weighting factor w_ℓ to each frequency point of interest, then the curve-fit process has to determine the values of the unknown coefficients in (4.33) such that the total error:

$$E = \sum_{\ell=1}^{p} w_\ell E_\ell \tag{4.36}$$

is minimised. This is achieved by differentiating the expression in (4.36) with respect to each unknown in turn, thus generating a set of as many equations as there are unknowns, each of the form:

$$\frac{dE}{dq} = 0 \quad ; \quad q = {}_1A_{jk}, {}_2A_{jk}, ..., \text{etc.} \tag{4.37}$$

Unfortunately, the set of equations thus formed are not linear in many of the coefficients (all the ω_s and η_s parameters) and thus cannot be solved directly. It is from this point that the differing algorithms choose their individual procedures, making various simplifications and assumptions in order to contain the otherwise very large computational task to within reasonable proportions. Most use some form of iterative solution, some linearise the expressions in order to simplify the problem and almost all rely heavily on good starting estimates. For further details the reader is referred to various papers such as [42] and [43]. An example of a curve-fit carried out using this NLLS method is shown in Fig. 4.16.

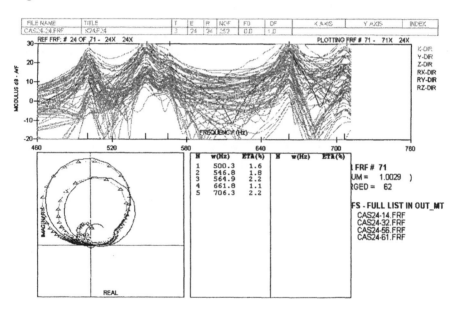

Fig. 4.16 Application of NLLS MDOF modal analysis

4.4.3 Method II: Rational Fraction Polynomial Method (RFP)

The method which has emerged as one of the 'standard' frequency-domain modal analysis methods is that known as the 'Rational Fraction Polynomial' (or RFP) method. This method is a special version of the general curve-fitting approach outlined above, but it is based on a different formulation for the theoretical expression used for the FRF which employs the rational, rather than the partial, fraction formula:

$$H(\omega) = \sum_{r=1,N} \frac{A_r}{(\omega_r^2 - \omega^2 + 2i\omega\omega_r\zeta_r)}$$

or

(4.38)

$$H(\omega) = \frac{\left(b_0 + b_1(i\omega) + b_2(i\omega)^2 \ldots + b_{2N-1}(i\omega)^{2N-1}\right)}{\left(a_0 + a_1(i\omega) + a_2(i\omega)^2 \ldots + a_{2N}(i\omega)^{2N}\right)}$$

It will be seen that in this formulation we have adopted the viscous damping model as this is the norm for this method. Also, we see that in the rational fraction version, the unknown coefficients which will be sought from the curve-fitting process are not the modal properties direct, as is the case in the former, partial fraction, version, but a series of polynomial coefficients, a_0, a_1, ..., a_{2N}, b_0, b_1, ..., b_{2N-1}, which are clearly related to the modal parameters but which will require a further stage of processing before these required quantities are yielded by the analysis method.

The particular advantage offered by this approach is the possibility of formulating the curve-fitting problem as a linear set of equations, thereby making the solution amenable to a direct matrix solution. The basis of the method is as follows.

We shall denote each of our measured FRF data points by $\tilde{H}_k$, where $\tilde{H}_k = \tilde{H}(\omega_k)$, and define the error between that measured value and the corresponding value derived from the curve-fit expression as

$$e_k = \frac{\left(b_0 + b_1(i\omega_k) + b_2(i\omega_k)^2 \ldots + b_{2m-1}(i\omega_k)^{2m-1}\right)}{\left(a_0 + a_1(i\omega_k) + a_2(i\omega_k)^2 \ldots + a_{2m}(i\omega_k)^{2m}\right)} - \tilde{H}_k$$

(4.39a)

leading to the modified, but more convenient, version actually used in the analysis:

$$e'_k = \left(b_0 + b_1(i\omega_k) + b_2(i\omega_k)^2 \ldots + b_{2m-1}(i\omega_k)^{2m-1}\right)$$
$$- \tilde{H}_k\left(a_0 + a_1(i\omega_k) + a_2(i\omega_k)^2 \ldots + a_{2m}(i\omega_k)^{2m}\right)$$

(4.39b)

In these expressions, only m modes are included in the theoretical FRF formula: the true number of modes, N, is actually one of the unknowns to be determined during the analysis.

If the measured FRF (or that section which is being analysed) is defined by a total of L individual frequency points, then a set of L linear equations can be written which relate the measured FRF values,

$\tilde{H}_k$ ($k=1,L$) to the unknown polynomial coefficients, a_j, ..., b_k, ..., and can be expressed in the form:

$$\begin{bmatrix} [Y] & [X] \\ [X]^T & [Z] \end{bmatrix}_{L\times(4m+1)} \begin{Bmatrix} \{b\} \\ \{a\} \end{Bmatrix}_{(4m+1)\times1} = \begin{Bmatrix} \{G\} \\ \{F\} \end{Bmatrix}_{L\times1} \tag{4.40}$$

where all the elements in the constituent matrices and vectors are known measured quantities. For completeness, the composition of these matrices is given below, by extending the analysis from equation (4.39b) above which can be rewritten as follows:

$$e'_k = \left\{1 \ (i\omega_k) \ (i\omega_k)^2 \ ... (i\omega_k)^{2m-1}\right\} \begin{Bmatrix} b_0 \\ b_1 \\ ... \\ b_{2m-1} \end{Bmatrix}$$

$$-\tilde{H}_k\left\{1 \ (i\omega_k) \ (i\omega_k)^2 \ ... (i\omega_k)^{2m-1}\right\} \begin{Bmatrix} a_0 \\ a_1 \\ ... \\ a_{2m-1} \end{Bmatrix} - \tilde{H}_k(i\omega_k)^{2m} a_{2m} \tag{4.41a}$$

and, when L such equations are combined, in matrix form, we obtain:

$$\{E'\}_{L\times1} = [P]_{L\times2m} \{b\}_{2m\times1} - [T]_{L\times(2m+1)} \{a\}_{(2m+1)\times1} - \{W\}_{L\times1} \tag{4.41b}$$

Solution for the unknown coefficients is achieved by minimising the error function, J, which is defined by

$$J = \{E*\}^T\{E\} \tag{4.42}$$

and this leads to equation (4.40), above, in which:

$$[Y] = \text{Re}\left([P*]^T[P]\right); \ [X] = \text{Re}\left([P*]^T[T]\right); \ [Z] = \text{Re}\left([T*]^T[T]\right);$$
$$\{G\} = \text{Re}\left([P*]^T\{W\}\right); \ \{F\} = -\text{Re}\left([T*]^T\{W\}\right) \tag{4.43}$$

While it is possible to obtain a solution from this equation, both the matrices $[P]$ and $[T]$ are found to be ill-conditioned and further refinement of the method is necessary to obtain a reliable numerical implementation. Use is generally made of orthogonal polynomials to transform the equations to a form which is better conditioned in numerical application. For further details, the reader is referred to a more specialist reference, such as [44].

Once the solution has been obtained for the coefficients, a_0, a_1, ..., b_0, ..., etc., then the second stage of the modal analysis can be performed in which the required modal parameters are derived. This is usually done by solving the two polynomial expressions which form the numerator and denominator of equation (4.38): the denominator to obtain the natural frequencies, ω_r, and damping factors, ζ_r and the numerator to determine the complex modal constants, A_r.

All the foregoing analysis presumes that the order of the model, m, is known and in general that is not the case. As mentioned earlier, this quantity is one of the parameters sought from the analysis. What is often done in methods of this type is to repeat the analysis using different assumed values for the order, m, and to compare the results of successive runs. For each run, there will be properties found for as many modes as prescribed by the chosen model order. Some of these will be genuine modes while others will be fictitious, or computational, modes and we need to separate the former from the latter. Various strategies may be adopted for this important phase of the analysis, amongst which are included:

- measuring the difference between the original FRF curve and that regenerated using the modal properties derived;
- measuring the consistency of the various modal parameters for different model order choices and eliminating those which vary widely from run to run.

An example of this latter check is shown in Table 4.1 for a simple case involving measurements around just a single mode of vibration of the test structure.

Mode	ω_r [Hz]	η_r	$\|A_r\|$ [1/kg]	$<A_r$
1	70.54 (10%)	0.004 (20%)	1.720 (40%)	145.21° (40%)
2	110.03 (0.03%)	0.010 (1%)	0.998 (3%)	−176.00° (2%)
3	111.01 (0.02%)	0.009 (1.2%)	1.001 (2.5%)	179.44° (1%)
4	149.83 (7%)	0.007 (15%)	0.401 (53%)	90.32° (57%)

Table 4.1 Averages and variances of modal parameters using RFP

Other checks which are often used to establish the reliability of the various modal parameter estimates, and thus to identify which modes to retain and which to reject in the final list, are:

- to run and re-run the analysis for a given model order using different subsets of the data contained within the chosen frequency range (i.e. not to use all the data points in one single large calculation but instead to perform several smaller calculations);
- to re-run the analysis for slightly different frequency range selections all of which embrace the modes of interest.

In all of these checks, interest is concentrated on the repeatability of the various modal properties: modes which reappear for all choices of data and model condition are believed to be genuine, while those which vary from run to run are more likely to have computational features due to the curve-fitting requirements as their origin, rather than physical ones which derive from the system's vibration modes. A further example of the use of the RFP method is given in Fig. 4.17.

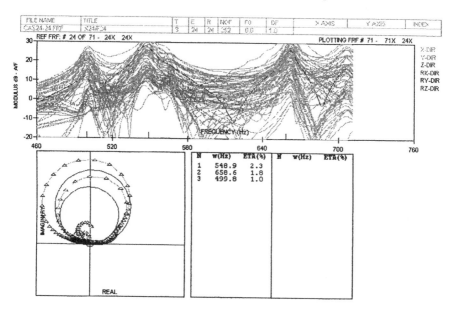

Fig. 4.17 Application of RFP MDOF modal analysis to single resonance

4.4.4   Method III: Lightly-Damped Structures
It is found that some structures do not provide FRF data which respond very well to the above modal analysis procedures mainly because of difficulties encountered in acquiring good measurements near resonance. This problem is met on very lightly-damped structures, such as is the case for many components of engineering structures when

treated individually. For such structures, also, it is often the case that interest is confined to an undamped model of the test structure since the damping in a complete structural assembly is provided mostly from the joints and not from the components themselves. Thus, there is scope for an alternative method of modal analysis which is capable of providing the required modal properties — in this case, natural frequencies and (real) modal constants only — using data measured away from the resonance regions. Such a method, which is very simple to implement, is described below.

The requirements for the analysis are as follows:

(i)   measure the FRF over the frequency range of interest;
(ii)   locate the resonances (obvious for this type of structure) and note the corresponding natural frequencies (which will thus be measured with an accuracy equal to the frequency resolution of the analyser);
(iii)   select individual FRF measurement data points from as many frequencies as there are modes, plus two, confining the selection to points away from resonance;
(iv)   using the data thus gathered, compute the modal constants (as described below);
(v)   construct a regenerated curve and compare this with the full set of measured data points.

The theory behind the method is quite simple and will be presented for the ideal case of all modes being included in the analysis. If, as discussed earlier, the frequency range chosen excludes some modes, these are represented as two additional modes with natural frequencies supposed to be at zero and at a very high frequency respectively and two additional FRF data points are taken, usually one from close to either end of the frequency range covered.

For an effectively undamped system, we may write:

$$H_{jk}(\omega) = \sum_{r=1}^{m} \frac{{}_rA_{jk}}{\omega_r^2 - \omega^2} \tag{4.44}$$

which, for a specific value, measured at frequency Ω_ℓ, can be rewritten in the form:

$$\tilde{H}_{jk}(\Omega_\ell) = \left\{ \left(\omega_1^2 - \Omega_\ell^2\right)^{-1} \quad \left(\omega_2^2 - \Omega_\ell^2\right)^{-1} \quad \ldots \right\} \begin{Bmatrix} {}_1A_{jk} \\ {}_2A_{jk} \\ \vdots \end{Bmatrix} \tag{4.45}$$

If we collect a total of $(2m+1)(n+1)m$ such individual measurements, these can be expressed by a single equation:

$$\begin{Bmatrix} \tilde{H}_{jk}(\Omega_1) \\ \tilde{H}_{jk}(\Omega_2) \\ \vdots \\ \vdots \end{Bmatrix} = \begin{bmatrix} (\omega_1^2 - \Omega_1^2)^{-1} & (\omega_2^2 - \Omega_1^2)^{-1} & \cdots \\ (\omega_1^2 - \Omega_2^2)^{-1} & (\omega_2^2 - \Omega_2^2)^{-1} & \cdots \\ \vdots & \vdots & \vdots \\ \vdots & \vdots & \vdots \end{bmatrix} \begin{Bmatrix} {}_1A_{jk} \\ {}_2A_{jk} \\ \vdots \end{Bmatrix}$$
(4.46a)

or

$$\{\tilde{H}_{jk}(\Omega)\} = [R]\{A_{jk}\}$$
(4.46b)

from which a solution for the unknown modal constants $_rA_{jk}$ in terms of the measured FRF data points $\tilde{H}_{jk}(\Omega_\ell)$ and the previously identified natural frequencies, ω_r, may be obtained:

$$\{A_{jk}\} = [R]^{-1}\{\tilde{H}_{jk}(\Omega)\}$$
(4.47)

An example of the application of the method is shown in Fig. 4.18 for an aerospace structure while further details of its finer points are presented in Reference [45]. The performance of the method is found to depend upon the points chosen for the individual FRF measurements, and these should generally be distributed throughout the frequency range and, wherever possible, should include as many antiresonances as are available. This last feature has the particular advantage that at an antiresonance, the theoretical model will exhibit a zero response: hence it is possible to supply such a nil value for the appropriate data in equation (4.47). In the limit, this means that only one FRF data point may be required from the measurements, all the others being set to be identically zero, even though from the measurements on a real structure their values would be extremely small, but finite.

4.5 GLOBAL MODAL ANALYSIS METHODS IN THE FREQUENCY DOMAIN
4.5.1 General Approach
In the next chapter we shall be discussing how results from the modal analysis of several FRF plots for a given structure may be further processed in order to yield the full modal model. At the present time, having performed modal analysis on each of the individual frequency responses, we have found the natural frequencies and damping factors but we do not yet have the mode shapes explicitly — only combinations

342

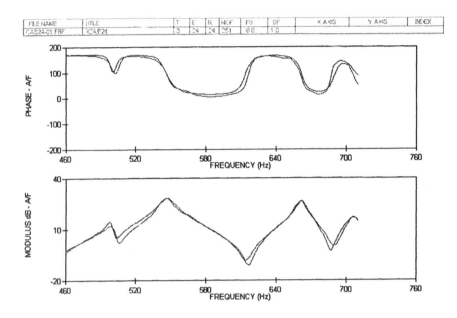

FILE NAME	TITLE		T	E	R	NK/F	FO	DF	X AXIS	Y AXIS	INDEX
CAS24-01.FRF	XC4/F24		3	24	24	251	0.0	1.0			

Fig. 4.18 Example of IDENT modal analysis for a lightly-damped structure

of the individual eigenvector elements as modal constants. A further stage of processing is required — here referred to as 'modelling' — in order to combine the various individual results obtained thus far. This process is described in Chapter 5. However, that phase is somewhat anticipated by some of the more recent curve-fitting procedures which are not confined to working with individual FRF curves but which are capable of performing a multi-curve fit. In other words, they fit several FRF curves simultaneously, taking due account of the fact that the properties of all the individual curves are related by being from the same structure. In simple terms, all FRF plots on a given testpiece should indicate the same values for natural frequency and damping factor of each mode. In practice, this does not happen exactly, unless they are constrained to be identical, and in the next chapter we discuss ways of dealing with this apparently unsatisfactory result. However, in a multi-curve fit, the constraints are imposed ab initio and such methods have the advantage of producing a unique and consistent model as direct output. One of the first published methods was that by Goyder [46], using an extension of the frequency-domain method. Other methods, such as the Ibrahim Time-Domain[47] and one referred to as 'Polyreference' [48] are also available.

Another way in which a set of measured FRF curves may be used collectively, rather than singly, is by the construction of a single composite Response Function. We recall that

$$H_{jk}(\omega) = \sum_{r=1}^{n} \frac{{}_r A_{jk}}{\omega_r^2 - \omega^2 + i\omega_r^2 \eta_r} \qquad (4.48)$$

and note that if we simply add several such FRFs, thus:

$$\sum_j \sum_k H_{jk}(\omega) = \sum_j \sum_k \left(\sum_{r=1}^{N} (...) \right) = HH(\omega) \qquad (4.49)$$

The resulting function $HH(\omega)$ will have the frequency and damping characteristics of the structure appearing explicitly, just as does any individual FRF, although the coefficients Σ_r (which replace the modal constants) are now very complicated combinations of the mode shape elements which depend heavily upon which FRFs have been used for the summation. Nevertheless, the composite function $HH(\omega)$ can provide a useful means of determining a single (average) value for the natural frequency and damping factor for each mode where the individual functions would each indicate slightly different values.

As an example, a set of mobilities measured in a practical structure are shown individually in Fig. 4.19(a) and their summation shown as a single composite curve in Fig. 4.19(b). The results from analysis of the separate curves produce estimates for the natural frequency and damping factor for each mode, and these can be used to derive a mean value for each modal parameter. Also available are the unique values for each parameter produced by analysis of the single composite curve.

A similar property applies to the impulse response functions for use with time-domain, rather than frequency-domain, analysis methods.

Both frequency-domain and time-domain methods are amenable to the expansion to multi-curve analysis. The techniques are simply quantitative extensions of their single curve counterparts and have as disadvantages first, the computation power required is unlikely to be available in an on-line mini computer, and secondly that there may be valid reasons why the various FRF curves exhibit slight differences in their characteristics and it may not always be appropriate to average out all the variations. Throughout the whole procedure of mobility measurements and modal analysis, we invoke the averaging process many times — to smooth rough FRF curves, to reduce the effects of measurement noise, to remove discrepancies and anomalies between different modal properties — and we should always remember that

344

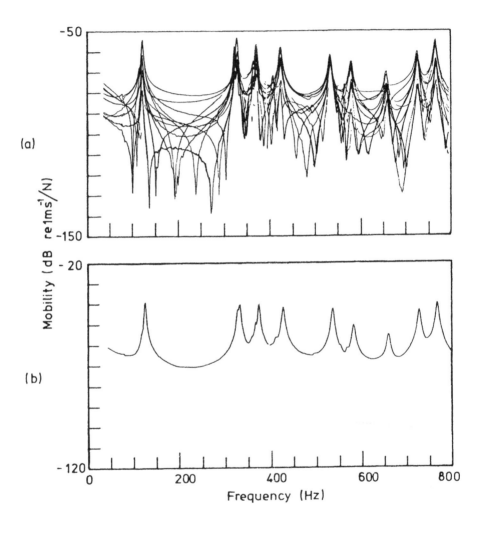

Fig. 4.19 Set of measured FRFs.
(a) Individual curves; (b) Composite curve

averaging is a valid means of removing random variations but it is not an appropriate way of treating systematic variations. We do not always make that distinction.

4.5.2   Global Rational Fraction Polynomial Method (GRFP)

In an earlier section, the basic RFP method was described in the context of use for a single FRF curve. This method is well suited to the more general application to multi-FRF data, both of the SIMO and the MIMO types. The extension beyond the basic form described above is achieved by exploiting the fact that if we take several FRFs from the same structure, then the numerator polynomials will be the same in every case, and so the number of unknown coefficients in a problem where there are n measured FRFs and m modes of vibration is of the order of $(n+1)(2m+1)$, while if the same FRFs were to be analysed individually, instead of in a global approach, the number of coefficients which would be obtained would be almost twice this number, at $2n(2m+1)$ and these would require further rationalising (see Chapter 5). Thus, the global RFP method is simply an extension of that presented in Section 4.3.4 and further details for those readers wishing to develop their own codes may be found in specialist references, such as [44].

4.5.3   Global SVD Method

There have been several implementations of algorithms designed to provide a global modal analysis in the frequency domain. The interested reader is referred to more specialist texts for a comprehensive review but it can be observed that details of these methods are of interest primarily to code developers, and less so to practitioners of modal analysis. This latter group have a definite need to appreciate the major features of such methods, and of the assumptions that may be embedded in them, but the full details tend to be concerned heavily with numerical efficiency and this, while of benefit to the modal analyst, is not a specialisation that he/she needs to acquire directly.

Nevertheless, it is perhaps appropriate to summarise one of these methods which we shall refer to below as the 'Global SVD' method. A set of FRFs with a single reference (such as are contained within a column from the complete FRF matrix) can be referred to the underlying modal model of the structure — assumed to have viscous damping — by the equation:

$$\{H(\omega)\}_k = \begin{Bmatrix} H_{1k}(\omega) \\ H_{2k}(\omega) \\ \vdots \\ H_{nk}(\omega) \end{Bmatrix}_{n\times 1}$$

$$= [\Phi]_{n\times N} \left[(i\omega - s_r)\right]^{-1}_{N\times N} \{\phi_k\}_{N\times 1} + \{R_k(\omega)\} \qquad (4.50)$$

where $\{R_k(\omega)\}$ is a vector containing the relevant residual terms to account for the existence of unknown out-of-range modes. Using the substitution:

$$\{g_k(\omega)\} = \left[(i\omega - s_r)\right]^{-1}_{N\times N} \{\phi_k\}_{N\times 1}$$

equation (4.50) can be re-written as:

$$\{H(\omega)\}_k = [\Phi]\{g_k(\omega)\} + \{R_k(\omega)\}$$

also   (4.51)

$$\{\dot{H}(\omega)\}_k = [\Phi][s_r]\{g_k(\omega)\} + \{\dot{R}_k(\omega)$$

Next we can write the following expressions:

$$\{\Delta H(\omega_i)\}_k = \{H(\omega_i)\}_k - \{H(\omega_{i+c})\}_k$$

$$\{\Delta H(\omega_i)\}_k \approx [\Phi]_{P\times N} \{\Delta g_k(\omega_i)\}_{N\times 1} \qquad (4.52)$$

$$\{\Delta \dot{H}(\omega_i)\}_k \approx [\Phi]_{P\times N} [s_r]\{\Delta g_k(\omega_i)\}_{N\times 1}$$

If we now consider data at several different frequencies, $i = 1, 2, 3, \ldots,$ L, we can write:

$$[\Delta H_k]_{n\times L} = [\Phi]_{n\times N} [\Delta g_k]_{N\times L}$$

$$[\Delta \dot{H}_k]_{n\times L} = [\Phi]_{n\times N} [s_r]_{N\times N} [\Delta g_k]_{N\times L} \qquad (4.53)$$

Eliminating $[\Delta g_k]$, we can construct an eigenvalue problem:

$$\left([\Delta \dot{H}_k]^T - s_r [\Delta H_k]^T\right)\{z\}_r = \{0\}$$

where

$$[z] = [\Phi]^{+T} \tag{4.54}$$

If we solve equation (4.54) using the SVD, we can determine the rank of the FRF matrices and thus the correct number of modes to be identified (m), leading to the appropriate eigenvalues: s_r; $r = 1, 2, ..., m$. Then, in order to determine the mode shapes, the modal constants can be recovered from:

$$\begin{Bmatrix} H_{jk}(\omega_1) \\ H_{jk}(\omega_2) \\ ... \\ H_{jk}(\omega_{1L}) \end{Bmatrix}_{L \times 1} = \begin{bmatrix} (i\omega_1 - s_1)^{-1} & (i\omega_1 - s_2)^{-1} & ... \\ (i\omega_2 - s_1)^{-1} & (i\omega_2 - s_2)^{-1} & ... \\ ... & ... & ... \\ (i\omega_L - s_1)^{-1} & ... & (i\omega_L - s_m)^{-1} \end{bmatrix}_{L \times m}$$

$$\begin{Bmatrix} {}_1 A_{jk} \\ {}_2 A_{jk} \\ ... \\ {}_m A_{jk} \end{Bmatrix}_{m \times 1} \tag{4.55}$$

Using this approach, it is possible to extract a consistent set of modal parameters for the model whose FRFs have been supplied. By way of an illustrative example, a case study has been prepared using actual test data shown in Fig. 4.20(a) measured on a simple but real structure: the box shown in Fig. 4.20(b). The tables shown below provide illustration of typical modal analysis results obtained by two approaches: first, using the SDOF methods described in Section 4.3 and second, using the global method outlined above. For a few typical modes in the middle range, the results are shown from both analyses. First, in Table 4.2(a) are shown the spread of different values obtained for the natural frequencies and damping factors from some 71 measured FRFs. The next stage in this type of analysis is to reduce these multiple estimates to a single value for each modal property and this has been done in the first two columns of Table 4.2(b). Also shown in this table, in the right hand pair of columns, are the set of unique modal properties obtained from a single analysis of all the FRFs using the global method outlined above.
This latter analysis can generally be performed in a fraction of the time that it takes to undertake the former. However, there are complications with the more advanced method in that a decision has to be made as to the number of modes that should be included in the

348

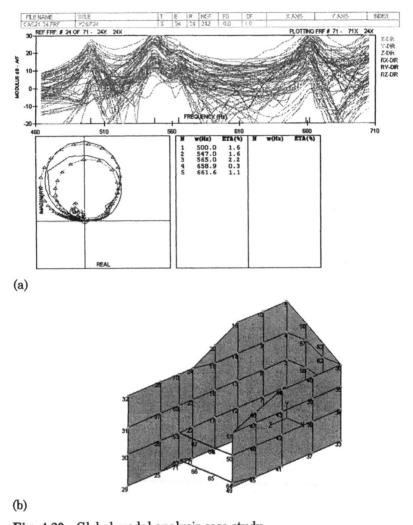

(a)

(b)

Fig. 4.20   Global modal analysis case study.
(a) Set of measured FRFs; (b) Test structure

Mode	(a) SDOF Nat. Freq. *(Hz)* Min (mean) max	(a) *Var.*	(b) GLOBAL *(Hz)*	(a) SDOF Damping *(%)* min (mean) max	(a) *Var.*	(b) GLOBAL *(%)*
1	155.3 **(155.91)** 156.9	*0.20*	**155.97**	2.24 **(3.07)** 3.79	*0.37*	3.09
2	171.4 **(172.08)** 175.5	*0.30*	**172.97**	2.58 **(3.78)** 7.55	*0.79*	4.30
3	197.7 **(203.91)** 209.4	*1.46*	**203.22**	3.72 **(6.86)** 13.22	*2.03*	7.20
4	227.7 **(233.07)** 238.4	*1.48*	**234.31**	2.23 **(4.32)** 17.97	*2.26*	1.70
5	246.9 **(249.66)** 252.5	*0.91*	**250.31**	1.88 **(3.24)** 5.90	*0.75*	4.12

Table 4.2   (a) Multiple SDOF, and (b) Global modal analyses

analysis (a feature of almost all MDOF analysis procedures). While it may be very obvious in many cases, such as the example used here, there are also times when the decision is far less readily arrived at and recourse to a basic SDOF analysis can often help to confirm this decision. One advantage that the single-curve, SDOF, analysis methods offer is an insight into the consistency or reliability of the measured FRF data. Indications of these qualities are given by the uniformity of the individual modal estimates across the whole set of measured data. While the mean values of the SDOF set, and the single values obtained for the global analysis may appear to be close, the former are complemented by a variance which indicates the reliability of each of the mean values: a feature absent from the global estimates.

4.6   MDOF MODAL ANALYSIS IN THE TIME DOMAIN
4.6.1   Introduction
The earliest MDOF modal analysis methods were based on a curve-fitting of the system's response properties presented as time-domain data, rather than the frequency-domain versions that we have studied up until now. These early methods are based on an algorithm devised by the 18th century French mathematician, Prony, and the reference lists of many early papers in the development of modern modal analysis methods refer to the originating work by that author, [49]. This family of methods, known as Complex Exponential Methods, permit a semi-automatic analysis of discrete data derived from the inverse Fourier transform of the FRF: the so-called IRF, or Impulse Response Function (discussed earlier in Chapter 2). The basic tenet is that any IRF, or other free vibration response function, can be expressed by a series of complex exponential components (see equation 2.142), the properties of each of which contain the eigenvalue and eigenvector properties of one mode.

Nowadays, the greater convenience of interpretation of the structure's response properties in the frequency domain, by inspection of FRFs rather than the time-domain IRFs, has led to a greater emphasis and utilisation being made of the frequency domain methods. Nevertheless, the time-domain methods deserve a place in the legacy of modal analysis technology and are still used in a number of specialist applications, especially in cases of structures with very low natural frequencies for which the time required to acquire several cycles' worth of data can become a practical problem.

4.6.2   Complex Exponential Method for Single FRF
4.6.2.1   Theory of method
As mentioned earlier in this chapter, there are a number of alternative modal analysis methods which work on measured data in a time-

domain format, rather than the more familiar frequency-domain versions. Most of the available methods derive from a technique known as the '**complex exponential**' method, although there are several variants and refinements which have been introduced in order to make the numerical procedures more efficient and suitable for small computers. The basis of the method, whose principal advantage is that it does not rely on initial estimates of the modal parameters, is outlined below.

As the method uses the time-domain version of system response data, in the form of the Impulse Response Function, its present application is limited to models incorporating viscous damping only. (It will be recalled that the hysteretic damping model presents difficulties for a time-domain analysis.)

Our starting point is the expression for the receptance FRF of a general MDOF system with viscous damping, which may be written as:

$$
\alpha_{jk}(\omega) = \sum_{r=1}^{N} \frac{{}_rA_{jk}}{\omega_r\zeta_r + i\left(\omega - \omega_r\sqrt{1-\zeta_r^2}\right)} + \frac{{}_rA_{jk}^*}{\omega_r\zeta_r + i\left(\omega + \omega_r\sqrt{1-\zeta_r^2}\right)} \tag{4.56a}
$$

or

$$
\alpha_{jk}(\omega) = \sum_{r=1}^{2N} \frac{{}_rA_{jk}}{\omega_r\zeta_r + i(\omega - \omega_r')} \quad ; \quad
\begin{aligned}
\omega_r' &= \omega_r\sqrt{1-\zeta_r^2} \\
\omega_{r+N}' &= -\omega_r' \\
{}_{(r+N)}A_{jk} &= {}_rA_{jk}^*
\end{aligned} \tag{4.56b}
$$

From classical theory, we can obtain the corresponding Impulse Response Function (IRF) by taking the Inverse Fourier Transform of the receptance:

$$
h_{jk}(t) = \sum_{r=1}^{2N} {}_rA_{jk}\, e^{s_r t} \quad ; \quad s_r = -\omega_r\zeta_r + i\omega_r' \tag{4.57}
$$

If the original FRF has been measured, or obtained, in a discrete form, and is thus described at each of a number of equally-spaced frequencies, the resulting IRF (found via the Inverse Fourier Transform of the FRF) will similarly be described at a corresponding number of equally-spaced time intervals ($\Delta t = 1/\Delta f$). We may conveniently define this data set as follows:

$$h_0, h_1, h_2, \ldots, h_q = h(0), h(\Delta t), h(2\Delta t), \ldots, h(q\Delta t) \tag{4.58}$$

From this point, it is convenient to omit the jk subscript and to use an abbreviated notion, as follows

$$_r A_{jk} \rightarrow A_r \quad ; \quad e^{s_r \Delta t} \rightarrow V_r \tag{4.59}$$

so that equation (4.57) becomes:

$$h(t) = \sum_{r=1}^{2N} A_r e^{s_r t} \tag{4.60}$$

Thus, for the ℓth sample, we have:

$$h_\ell = \sum_{r=1}^{2N} A_r V_r^\ell \tag{4.61a}$$

which, when extended to the full data set of q samples, gives:

$$
\begin{aligned}
h_0 &= A_1 + A_2 + \ldots + A_{2N} \\
h_1 &= V_1 A_1 + V_2 A_2 + \ldots + V_{2N} A_{2N} \\
h_2 &= V_1^2 A_1 + V_2^2 A_2 + \ldots + V_{2N}^2 A_{2N} \\
&\vdots \quad\quad \vdots \quad\quad \vdots \quad\quad \vdots \\
h_q &= V_1^q A_1 + V_2^q A_2 + \ldots + V_{2N}^q A_{2N}
\end{aligned}
\tag{4.61b}
$$

Provided that the number of sample points q exceeds $4N$, this equation can be used to set up an eigenvalue problem, the solution to which yields the complex natural frequencies contained in the parameters V_1, V_2, etc. via a solution using the Prony method.

Taking (4.61), we now multiply each equation by a coefficient, β_i, to form the following set of equations:

$$
\begin{aligned}
\beta_0 h_0 &= \beta_0 A_1 + \beta_0 A_2 + \ldots + \beta_0 A_{2N} \\
\beta_1 h_1 &= \beta_1 A_1 V_1 + \beta_1 A_2 V_2 + \ldots + \beta_1 A_{2N} V_{2N} \\
\beta_2 h_2 &= \beta_2 A_1 V_1^2 + \beta_2 A_2 V_2^2 + \ldots + \beta_2 A_{2N} V_{2N}^2 \\
&\vdots \quad\quad \vdots \quad\quad \vdots \quad\quad \vdots \\
\beta_q h_q &= \beta_q A_1 V_1^q + \beta_q A_2 V_2^q + \ldots + \beta_q A_{2N} V_{2N}^q
\end{aligned}
\tag{4.62}
$$

Adding all these equations gives

$$\sum_{i=0}^{q} \beta_i h_i = \sum_{j=1}^{2N} \left(A_j \sum_{i=0}^{q} \beta_i V_j^i \right) \qquad (4.63)$$

What are the coefficients β_i? These are taken to be the coefficients in the equation

$$\beta_0 + \beta_1 V + \beta_2 V^2 + ... + \beta_q V^q = 0 \qquad (4.64)$$

for which the roots are $V_1, V_2, ..., V_q$.

We shall seek to find values of the β coefficients in order to determine the roots of (4.64) — values of V_r — and hence the system natural frequencies. Now, recall that q is the number of data points from the Impulse Response Function, while N is the number of degrees of freedom of the system's model (constituting N conjugate pairs of 'modes'). It is now convenient to set these two parameters to the same value, i.e. to let $q \equiv 2N$.

Then from (4.64) we can see that

$$\sum_{i=0}^{2N} \beta_i V_r^i = 0 \quad \text{for } r = 1, 2N \qquad (4.65)$$

and thus that every term on the right hand side of (4.65) is zero so that

$$\sum_{i=0}^{2N} \beta_i h_i = 0 \qquad (4.66)$$

thus we shall rearrange (4.66) so that

$$\sum_{i=0}^{2N-1} \beta_i h_i = -h_{2N} \quad \text{by setting } \beta_{2N} = 1 \qquad (4.67)$$

and this may be written as

$$\{h_0 \quad h_1 \quad h_2 \quad ... \quad h_{2N-1}\} \begin{Bmatrix} \beta_0 \\ \beta_1 \\ \vdots \end{Bmatrix} = -h_{2N} \qquad (4.68)$$

Now, we may repeat the entire process from equation (4.58) to equation (4.68) using a different set of IRF data points and, further, we may choose the new data set to overlap considerably with the first set — in fact, for all but one item — as follows

$$\{h_1 \quad h_2 \quad h_3 \quad ... \quad h_{2N}\} \begin{Bmatrix} \beta_0 \\ \beta_1 \\ \vdots \end{Bmatrix} = -h_{2N+1} \qquad (4.69)$$

Successive applications of this procedure lead to a full set of $2N$ equations:

$$\begin{bmatrix} h_0 & h_1 & h_2 & ... & h_{2N-1} \\ h_1 & h_2 & h_3 & ... & h_{2N} \\ \vdots & \vdots & \vdots & ... & \vdots \\ \vdots & \vdots & \vdots & ... & \vdots \\ h_{2N-1} & h_{2N} & h_{2N+1} & ... & h_{4N-2} \end{bmatrix} \begin{Bmatrix} \beta_0 \\ \beta_1 \\ \vdots \\ \vdots \\ \beta_{2N-1} \end{Bmatrix} = - \begin{Bmatrix} h_{2N} \\ h_{2N+1} \\ \vdots \\ \vdots \\ h_{4N-1} \end{Bmatrix}$$

or

$$[h]_{2N \times 2N} \{\beta\}_{2N \times 1} = - \{\tilde{h}\}_{2N \times 1} \qquad (4.70a)$$

from which we can obtain the unknown coefficients:

$$\{\beta\} = - [h]^{-1} \{\tilde{h}\} \qquad (4.70b)$$

With these coefficients, we can now use (4.64) to determine the values V_1, V_2, ..., V_{2N} from which we obtain the system natural frequencies, using the relationship

$$V_r = e^{s_r \Delta t}$$

We may now complete the solution by deriving the corresponding modal constants, A_1, A_2, ..., A_{2N} using equation (4.61). This may be written as

354

$$\begin{bmatrix} 1 & 1 & \cdots & 1 \\ V_1 & V_2 & \cdots & V_{2N} \\ V_1^2 & V_2^2 & \cdots & V_{2N}^2 \\ \vdots & \vdots & \cdots & \vdots \\ V_1^{2N-1} & V_2^{2N-1} & \cdots & V_{2N}^{2N-1} \end{bmatrix} \begin{Bmatrix} A_1 \\ A_2 \\ A_3 \\ \vdots \\ A_{2N} \end{Bmatrix} = \begin{Bmatrix} h_0 \\ h_1 \\ h_2 \\ \vdots \\ h_{2N-1} \end{Bmatrix}$$

or

$$[V]\{A\} = \{h\} \tag{4.71}$$

4.6.2.2 Use of complex exponential method

The foregoing method is generally employed in the following way. An initial estimate is made for the number of degrees of freedom and the above mentioned analysis is made. When completed, the modal properties thus found are used in (4.56) to compute a regenerated FRF curve which is then compared with the original measured data. At this stage, the deviation or error between the two curves can be computed.

The whole procedure is then repeated using a different number of assumed degrees of freedom (2N) and this error again computed. A plot of error vs. number of DOF will generally produce a result of the form shown in Fig. 4.21 in which there should be a clearly-defined reduction in the error as the 'correct' number of degrees of freedom is attained. The inclusion of a larger number than this critical value will cause the creation of a number of 'computational' modes in addition to the genuine 'physical' modes which are of interest. Such additional modes serve to account for the slight imperfection inevitably present in measured data and are generally easily identified from the complete list of V_r and A_r modal properties by their unusually high damping factors and/or small modal constants.

4.6.3 Global Analysis in the Time Domain
4.6.3.1 Ibrahim time-domain method (ITD)

Directly following the previous section on the complex exponential method, it is appropriate now to introduce the Ibrahim Time-Domain (ITD) technique as an example of a multi-curve time-domain analysis, [47]. Although the ITD method is not strictly a curve-fitting procedure in the sense that all the preceding ones are and, indeed, it would fit more readily into the next chapter, we shall present it here as a logical extension of the complex exponential idea.

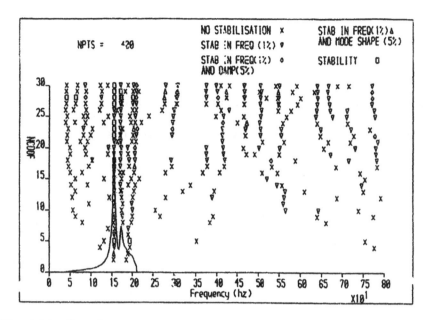

Fig. 4.21 Complex exponential modal analysis: evolution of repeated application

The basic concept of the ITD method is to obtain a unique set of modal parameters — natural frequencies, damping factors and mode shapes — from a set of free vibration measurements in a single analysis. In other words, we shall not be curve-fitting or analysing a single FRF (or equivalent) at a time, as has been the case hitherto, but we shall be processing all the measured data at once. Another feature of the method is that it can be used with any measured free vibration data, whether or not the excitation forces are available. In the event that these data are known, then it is possible to derive fully-scaled eigenvector properties, otherwise only the unscaled mode shapes will be available along with the modal frequency and damping parameters. Perhaps the most likely way in which the method will be applied in a modal testing context is by measuring a set of FRF properties — based on a selection as prescribed in the next chapter — and then by using these to obtain a corresponding set of Impulse Response Functions. These may then be used as the free response data required by the ITD method with the knowledge that the magnitude of the excitation which produced them (a unit impulse) is implicit.

The theory of the method is as follows. As before, it is based on the free vibration solution of a viscously-damped MDOF system and takes as its starting point the assumption that we may write any individual response as:

$$x_i(t_j) = \sum_{r=1}^{2m} (\psi_{ir}) e^{s_r t_j} \tag{4.72}$$

where i represents the coordinate and j the specific time increment at which the response is measured; s_r is the rth root or complex eigenvalue of the system's characteristic equation (see equation (2.73)) and $\{\psi\}_r$ is the corresponding eigenvector, with (ψ_{ir}) being the ith element in that vector. At this point, the eigenvectors are unscaled. Also, we are assuming that the total number of degrees of freedom available in our model is m. In fact, this is not necessarily the same as the number of degrees of freedom of the system (N): m is the number of degrees of freedom which are necessary to represent the measured data and this may be possible with many less than the full set if, for example, the response is confined to a limited frequency range, as will usually be the case in practice.

Now, if we measure the response at several points on the structure — $i = 1, n$ — and at several instants in time — $\ell = 1, q$ — then we can construct a matrix equation of the form:

$$\begin{bmatrix} x_1(t_1) & x_1(t_2) & \dots & x_1(t_q) \\ x_2(t_1) & x_2(t_2) & \dots & x_2(t_q) \\ \vdots & \vdots & \dots & \vdots \\ x_n(t_1) & x_n(t_2) & \dots & x_n(t_q) \end{bmatrix} =$$

$$= \begin{bmatrix} 1^{\psi_1} & 2^{\psi_1} & \dots & 2m^{\psi_1} \\ 1^{\psi_2} & 2^{\psi_2} & \dots & 2m^{\psi_2} \\ \vdots & \vdots & \dots & \vdots \\ 1^{\psi_n} & 2^{\psi_n} & \dots & 2m^{\psi_n} \end{bmatrix} \begin{bmatrix} e^{s_1 t_1} & \dots & e^{s_1 t_q} \\ e^{s_2 t_1} & \dots & e^{s_2 t_q} \\ \vdots & \dots & \vdots \\ e^{s_{2m} t_1} & \dots & e^{s_{2m} t_q} \end{bmatrix} \tag{4.73a}$$

or, in simpler form:

$$[x] = [\Psi][\Lambda] \tag{4.73b}$$

Here,

- $[x]$ is an $n \times q$ matrix of free response measurements from the structure, and is known;
- $[\Psi]$ is an $n \times 2m$ matrix of unknown eigenvector elements; and

- [Λ] is a $2m \times q$ matrix depending on the complex eigenvalues (as yet unknown) and the response measurement times (which are known).

A second, similar, equation is then formed by using a second set of measured response data, each item of which relates to a time which is exactly Δt later than for the first set. Thus we have:

$$x_i(t_\ell + \Delta t) = \sum_{r=1}^{2m} (\psi_{ir}) e^{s_r(t_\ell + \Delta t)} \tag{4.74a}$$

or

$$\hat{x}_i(t_\ell) = \sum (\hat{\psi}_{ir}) e^{s_r(t_\ell)} \quad ; \quad (\hat{\psi}_{ir}) = (\psi_{ir}) e^{s_r \Delta t} \tag{4.74b}$$

which leads to a second set of equations:

$$[\hat{x}] = [\hat{\Psi}][\Lambda] \tag{4.75}$$

Next, remembering that the number of assumed modes (m) is a variable (we do not yet know how many modes are required to describe the observed motion), we can arrange that $n = 2m$, so that the matrices $[\Psi]$ and $[\hat{\Psi}]$ are square. It may be seen that these two matrices are closely related and we can define a matrix $[A]$, often referred to as the 'system matrix', as:

$$[A][\Psi] = [\hat{\Psi}] \tag{4.76}$$

From (4.73b) and (4.75), we find

$$[A][x] = [\hat{x}] \tag{4.77}$$

and this provides us with a means of obtaining $[A]$ from the measured data contained in $[x]$ and $[\hat{x}]$. If we have selected the number of time samples (q) to be identical to the number of measurement points (n), (which we have now set to be equal to $2m$), then $[A]$ can be obtained directly from equation (4.77). However, it is customary to use more data than the minimum required by setting q to a value greater than $2m$. In this case, use of equation (4.77) to determine $[A]$ will be via the pseudo-inverse process which yields a least-squares solution for the matrix, using all the data available. In this case, an expression for $[A]$ is:

$$[A] = [\hat{x}][x]^T \left([x][x]^T\right)^{-1} \tag{4.78}$$

Returning now to equation (4.74), we can see that individual columns in $\{\psi\}_r$ are simply related to the corresponding ones in $\{\hat{\psi}\}_r$ by the relationship:

$$\{\hat{\psi}\}_r = \{\psi\}_r e^{s_r \Delta t} \tag{4.79}$$

Thus, using (4.76) we can write:

$$[A]\{\psi\}_r = \left(e^{s_r \Delta t}\right)\{\psi\}_r \tag{4.80}$$

which will be recognised as a standard form of a set of equations whose solution is obtained by determining the eigenvalues of the matrix $[A]$. It must be noted immediately that these eigenvalues are **not** the same as those of the original equations of motion (since equation (4.80) is not an equation of motion) but they are closely related and we shall see that it is a straightforward process to extract the system's natural frequencies, damping factors and mode shapes from the solution to equation (4.80). The eigenvalues of $[A]$ are the particular values of $(e^{s_r \Delta t})$ and so if we have λ_r as one of these eigenvalues, then we can determine the corresponding complex natural frequency of the system (s_r) from:

$$e^{s_r \Delta t} = \lambda_r = a_r + i b_r = e^{-\omega_r \zeta_r \Delta t} e^{i\omega'_r \Delta t} = c_r e^{i\theta_r}$$

$$c_r = \left(a_r^2 + b_r^2\right)^{1/2} \quad ; \quad \theta_r = \tan^{-1}\left(-\frac{b_r}{a_r}\right) \tag{4.81a}$$

from which we can derive the natural frequency (ω_r) and viscous damping factor (ζ_r) using

$$\omega_r \zeta_r = -\ln\left(\frac{a_r^2 + b_r^2}{2\Delta t}\right)$$

$$\omega'_r = \omega_r \sqrt{1 - \zeta_r^2} = \tan^{-1}\left(\frac{b_r/a_r}{\Delta t}\right) \tag{4.81b}$$

Corresponding to each eigenvalue there is an eigenvector and this can be seen to be identical to the mode shape vector for that mode. No further processing is required in this case. It should, however, be noted that the mode shapes thus obtained are generally unscaled and, as such, are inadequate for regenerating FRF curves. Only if the original free vibration response data were derived from a FRF-IRF procedure can scaled (mass-normalised) eigenvectors be obtained.

4.6.3.2   Use of the ITD method

As with other similar procedures, the ITD method requires the user to make some decisions and judgements concerning which of his/her measurements to use for the calculation (there will generally be an excess of measured data). For example, there will often be more response points than there are genuine modes to be identified, but the method will always produce one mode per two response points. Examination of the set of modal properties, and especially the modal damping factors, will generally indicate which are genuine structural modes and which are 'fictitious' modes caused by noise or other irregularities in the measured data. Various techniques are proposed for a systematic examination of the results as the number of assumed degrees of freedom is increased. Generally, these methods look for a 'settling down' of the dominant modes of vibration, or for a marked reduction in the least-squares error between the original measured data and that regenerated on the basis of the identified modal properties. Further details of the method, and its sensitivity to various parameters are provided in Reference [47].

4.7   MODAL ANALYSIS OF NON-LINEAR STRUCTURES
4.7.1   Introduction

We have seen in earlier chapters that slight nonlinearities in the system behaviour can result in distortions in the measured FRF curves, and that the results produced by sinusoidal, random and transient excitations are all different. At this stage, it is interesting to investigate the consequences of these effects on the modal analysis process and this can effectively be achieved by using the SDOF circle- and line-fit methods described in Section 4.3. We shall use these methods, and in particular the graphical display of the damping estimates it produces, to examine:

(a)   theoretically-generated FRF data;
(b)   data measured on an analogue computer circuit with controlled non-linear effects; and
(c)   tests on practical built-up structures.

We shall then proceed to develop an extension to the aforementioned methods in order to provide a first-level non-linear modal analysis technique, which provides a practical tool for dealing with this important class of structure.

4.7.2 Application of Linear Modal Analysis to Non-linear Structures

4.7.2.1 SDOF modal analysis

For the first case, using theoretical data, we show some results for a system with cubic stiffness nonlinearity, Fig. 4.22(a), and one with coulomb friction (as well as viscous) damping, Fig. 4.22(b). In both cases, a striking and systematic variation in the damping estimates is seen, with a different trend for each of the two cases. It is clear from these results that taking the average of several such damping estimates is pointless — the resulting mean value being heavily dependent upon which estimates are included rather than on how many.

Turning next to some measured data obtained using the standard analysis instrumentation (FRA and FFT) but from measurements made on an analogue computer programmed to behave as a SDOF system with an added cubic stiffness nonlinearity, Fig. 4.23(a) shows a typical result obtained from a sinusoidal excitation measurement while Fig. 4.23(b) gives the corresponding result for narrow band random excitation set to generate approximately the same vibration level. The first result clearly demonstrates the trend predicted by theory while the second one shows no signs of such effects and, indeed, does not suggest the existence of any non-linearity in the tested system. However, this result is not unexpected in view of the processing to which the measured data have been subjected — the formulae used to compute the FRF assume and rely on superposition — and it is found that the FRF produced by a DFT analysis is effectively that for a linearisation of the actual tested system.

The next results are taken from a series of measurements on a complete built-up structure and refer to a specific mode and to a specific FRF — mode 3 in mobility $Y_{7,12}$ — measured using sinusoidal excitation. The first two plots, in Figs. 4.24(a) and (b) result from measurements made during two different constant-excitation level tests (one at a low level and the other at a higer level) and show clear signs of systematic nonlinearly (although for a more complex form than the specific cubic stiffness type examined above). The second two examples, in Figs. 4.24(c) and (d) refer to exactly the same resonance but this time result from tests made at two different constant-response levels. Here, the data from both measurements are clearly very linear although the numerical values of the modal parameters differ for the two different cases. In this test, the behaviour of the structure has been consciously linearised, under known conditions, along the lines proposed in the discussion on measurement techniques, Section 3.11. Thus we see that

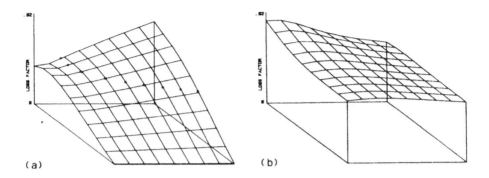

Fig. 4.22   Linear SDOF modal analysis applied to computed FRF of non-
linear system.
(a) Cubic stiffness effect; (b) Coulomb friction damping effect

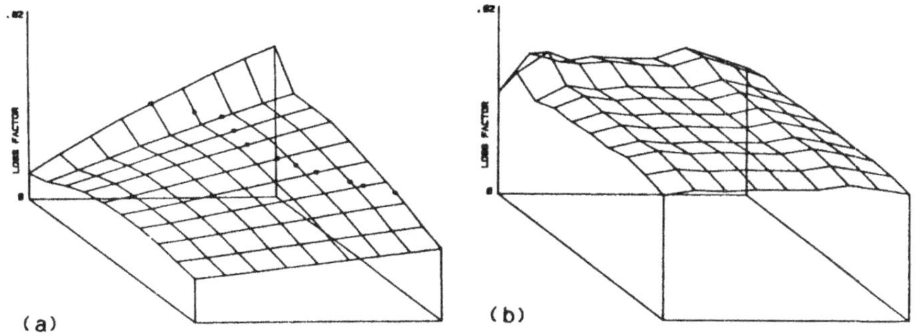

Fig. 4.23   Linear SDOF modal analysis applied to measured FRF on
electronic circuit.
(a) Measured with sinusoidal excitation; (b) Measured with
random excitation

362

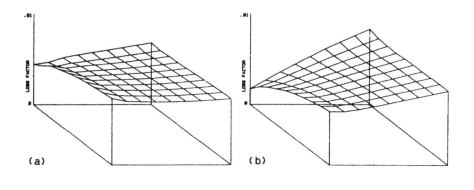

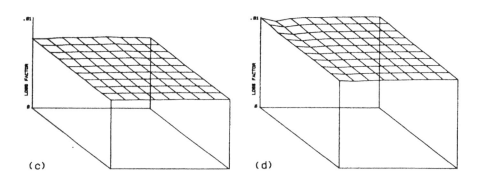

Fig. 4.24 Linear SDOF modal analysis applied to FRFs measured on aerospace structure.
(a) Low-level, constant-force; (b) High-level, constant force; (c) Low-level, constant response; (d) High-level, constant response

SDOF analysis methods, properly used, can indicate the existence of non-linear effects in a test structure excited by sinusoidal excitation (or equivalent). We have also seen that these effects can be induced or suppressed by the exercise of input or output level control, the latter case permitting the construction of a valid linearised model of the structure.

One final observation on this topic: it has been seen on more than one occasion that the FRFs from non-linear structures, when analysed by most SDOF (or other) modal analysis procedures, can reveal a degree of modal complexity which is much greater than might be expected, or even justified. Thus, in many situations, a high modal complexity indicated by the modal analysis process may be an early indication of non-linearity, and, indeed, may be the first clear evidence of such behaviour.

4.7.2.2 MDOF modal analysis

The second illustration of the results of using conventional (i.e. linear) modal analysis procedures on FRF data which has been measured on a non-linear structure relates to the more general MDOF modal analysis methods. As with the preceding examples, the effects which are described below are only encountered if the excitation method used to measure the FRF data is such that the non-linearities are activated and influence the shape of the curves. This situation clearly applies if sinusoidal excitation is used, but is also present when periodic or transient excitation types are employed. It is not the case, however, if the excitation is random.

In those situations where a measured FRF is distorted, as is frequently found for non-linear structures in the region of a resonance peak, the most likely result of applying an MDOF modal analysis curve-fitting procedure is for the analysis to 'find' several modes in the immediate vicinity of the resonance frequency, and to furnish the properties of these modes as the result of the curve-fit. Such a result is illustrated in Fig. 4.25 where the FRF measured in the region of one mode of a non-linear system is shown in (a) together with the curve regenerated using the best SDOF analysis which could be obtained. The second plot, (b), shows the measured curve again, but here with the curve which is regenerated using the results from an MDOF modal analysis (using the RFP method). Clearly, a better fit is obtained than for (a) but the cost of this success is the creation of three modes in the narrow range of the resonance, one of which has negative damping and is thus unstable! It can be seen that two of the three modes 'found' are, in fact, not genuine physical modes at all but are computation modes that are created in order to achieve a better fit between the original data and the regenerated curve. The fact that this can only be achieved

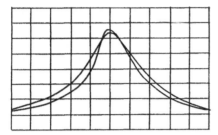 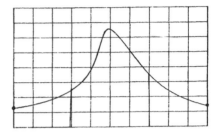

Fig. 4.25 Linear MDOF modal analysis applied to electronic circuit.
(a) Best single-mode fit; (b) Best-fit result; (c)

by the introduction of such artificial modes, especially ones with unrealistic properties, is an indication of the conflict between data from a non-linear behaviour and a model which is entirely linear in its representation.

The example shown here is typical of many such cases where conventional modal analysis is used on data from non-linear structures and so vigilance must be exercised in interpreting the modal properties indicated by such processes.

4.7.3 Extension of SDOF Line-Fit Modal Analysis Method for Non-linear Structures

In the face of results such as those shown above, both of measured FRF data and of the results from applying modal analysis processes to them, the question frequently arises as to what can be done by way of modal analysis for structures which are non-linear to some limited degree. Not surprisingly, this questions has been addressed by several workers in the non-linear dynamics field and some very advanced and sophisticated analysis procedures have been proposed. The results from such studies are generally very advanced material and are not appropriate for a basic text such as this. However, there are some simple quasi-linear analysis procedures which can be developed as slight extension from our linear modal analysis methods which are perhaps useful as a first-level study of some classes of non-linear structure.

The basic approach assumes that if we have steady-state vibration with amplitude, X_0, then the non-linear element(s) in the structure (which is almost always a spring or a damper) can be considered to have an effective or equivalent linear stiffness (or damping) coefficient which can be expressed as:

$$\widetilde{k}(X_0) \quad \text{or} \quad \widetilde{c}(X_0)$$

and that each such component (as is the case for a linear structure) would be characterised not by a single stiffness, or damping factor, but by a range of values, depending upon the amplitude of vibration prevalent at any instant. This characteristic is illustrated in Fig. 2.46. In this case, it would be helpful to be able to identify and to define the values of these equivalent linear stiffnesses or damping factors, and that is what the following method seeks to do.

4.7.3.1 Basic approach

The basic method which is summarised below is an extension of the SDOF methods of modal analysis described earlier in this chapter. It is limited to the analysis of effectively single-degree of freedom behaviour in the vicinity of an isolated mode of vibration and is not well suited to more complex situations, such as close modes. The basis of the approach can be explained graphically from the following plots. Consider first the conventional modulus plot for an FRF of a linear SDOF system with constant coefficients, as shown in Fig. 4.26(a). If we extend this to a non-linear system for which, say, the stiffness is amplitude-dependent, then we can construct a series of such FRF plots for vibration of that system in which the amplitude of the response is maintained at a constant level throughout the small frequency range of the resonance measurement, as shown in Fig. 4.26(b). Each such curve shows the behaviour of a linear system because, if the response amplitude is kept constant throughout the test, the effective stiffness is constant and the system is effectively linear. However, a slightly different system is seen for each different response amplitude. If we were to conduct the measurement under conditions where the response amplitude was allowed to vary, such as results from tests where the excitation force is maintained at a constant level, for each frequency of measurement, then the resulting FRF curve is quite different and can be envisaged as a collection of points drawn from different curves of the first set, depending upon how much response the unit excitation generates at each frequency, as shown in Fig. 4.26(c).

Although not essential to the numerical analysis procedure itself, it is helpful for visualisation purposes to re-visit the so-called 'inverse' format of plotting FRF data, introduced first in Chapter 2, and to replot the above curve in that format because the effect of the non-linearity is to provide a plot which is not a straight line in one or other (or both) of the two plots. A system with a non-linear stiffness, measured at various response levels, gives rise to a non-straight Real Part plot (see Fig. 4.27(b)), while non-linearity on the damping element causes a strongly-distorted Imaginary Part plot: Fig. 4.27(c).

366

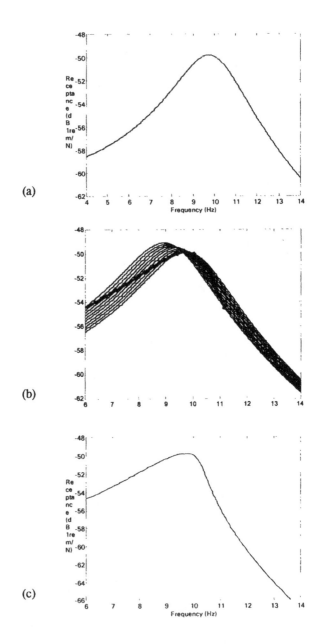

Fig. 4.26 Plots of non-linear system FRF data.
(a) Linear system with constant coefficients; (b) Constant response level FRF for non-linear system; (c) Constant excitation level FRF for non-linear system

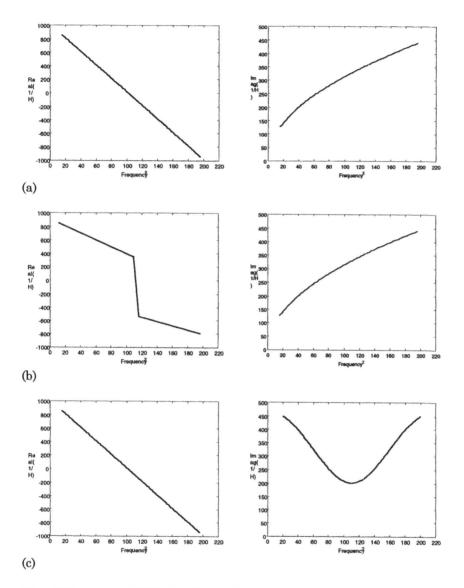

(a)

(b)

(c)

Fig. 4.27 Inverse FRF plots for non-linear systems.
(a) Inverse FRF plots for linear system; (b) Inverse FRF plot for system with non-linear stiffness; (c) Inverse FRF plot for system with non-linear damping

The method of analysis, therefore, is to extract values for the effective stiffness and damping from each individual point on the FRF curve as it passes through the resonance region (which is generally the only region where the amplitudes of vibration are large enough to activate any significant non-linearity, anyway). As can be seen from the plots in Figs. 4.26 and 4.27, each frequency point can be identified by the stiffness of the constant-stiffness straight line on which it lies, as can the damping from the Imaginary Part plots. What must be done in addition to this basic identification stage is to retain the information on the actual amplitude of the system response as required by this representation so that the extracted equivalent stiffness and/or damping can be referred to the relevant amplitude level, X_0. This demands simply that the measured response level information is retained, rather than deleted as is usually the case once the ratio of response/excitation force has been computed. Of course, the measurement of the FRF must be made using excitation levels that cause the response to span a range of amplitudes that are of interest in the final identification and this can be achieved either by controlling the excitation level or by allowing the level to 'float' as this usually has a similar result.

This method of analysis involves a number of assumptions. most of which are frequently quite reasonable, including one which takes the mass to be constant, and never the source of non-linearity.

4.7.3.2 Application of method and examples
In this section we show some examples of the application of the aforementioned method, SDOF-NL. The simplest application assumes real modes (i.e. that the modal constants are always real) and we can develop plots of effective stiffness and/or effective damping as a function of response amplitude which has two branches: one as the resonance is approached (and the response level is rising with increasing frequency) and a second time as the response falls as the resonance is passed. If these two branches trace out effectively the same characteristic (see Fig. 4.28(a)), then it can be surmised that a reasonable result has been achieved, and that the underlying assumptions have been valid.

For more complex situations, where the modal constant might have a non-negligible degree of complexity, then it is possible to refine this method by taking pairs of points from the FRF data set, both relating to the same response amplitude but one below the resonance and one above. These two data points, comprising four quantities, can be used to extract the four modal parameters, which lead directly to the effective stiffness and the effective damping which relate to the response level of those particular pair of data points. This process can be repeated for each available pair of data points.

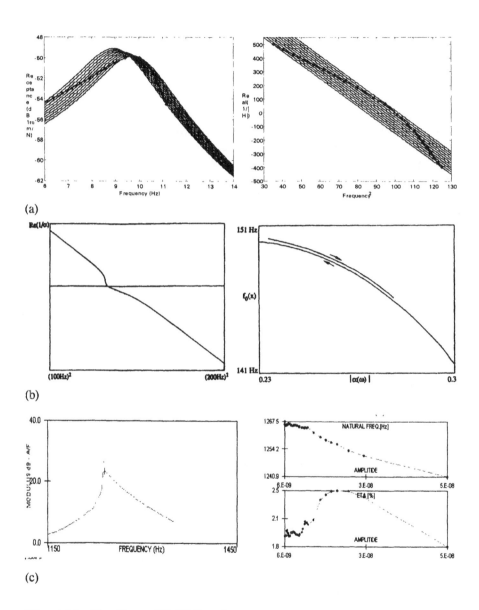

Fig. 4.28 Non-linear SDOF line-fit modal analysis method.
(a) Basis of analysis method; (b) Application to aerospace
structure; (c) Application to structure with added friction
damping

Examples of the application of this method are shown in
Figs. 4.28(b) and (c). The first case is for a structure which contains a

significant non-linearity in the stiffness characteristics while the second presents results taken from measurements of the internal damping in a spinning rotor, which has clear indications of both non-linear stiffness and non-linear damping.

4.8   CONCLUDING COMMENTS

In this chapter we have sought to describe the area of modal analysis which in many ways, is the least under the direct control of the modal test engineer. In the task of extracting modal model parameters from measured test data, the analyst must rely on the skill of others who have coded the various analysis algorithms since these are generally beyond the ability of the practitioner. Against this, or perhaps because of it, the analyst must develop the various skills which enable him or her to select the most appropriate analysis procedure for each case, and to make the best interpretation of what is often ambiguous and sometimes confusing output from these analysis methods.

In this chapter we have first sought to impress upon the user the need for accuracy and reliability in the measured data that is the source of a modal analysis. If these data are not of high quality, the resulting modal model cannot be expected to be any better, and will probably be worse. Thus, attention must be paid at the initial phases to ascertain and to assure the necessary quality of the raw data. Questions as to the correct order for the model (how many modes need to be included in the modal model?) and the most appropriate model for damping (viscous or structural; proportional or non-proportional?) are often foremost amongst these early interpretations.

A hierarchy of different types of modal analysis procedure have been catalogued, from the simple SDOF one-mode-at-a-time for a single response function, through MDOF methods which reveal several modes at a time, to global analysis methods where several modes are extracted simultaneously from several FRFs. These methods are largely carried out today in the frequency domain, but time-domain alternatives exist for some cases and these have also been reviewed.

This chapter does not claim to be exhaustive by any means: nor have the most advanced and powerful methods been discussed. This is an area in the subject which is the subject of constant improvement and enhancements and the interested reader is directed to more specialist works (such as [6] and [50] for more detailed discussions and descriptions.

CHAPTER 5

Derivation of Mathematical Models

5.1 INTRODUCTION
5.1.1 The Role of Different Types of Model

At this stage we have described all the main tools available to the modal analyst. Now we consider how these may be marshalled in order to achieve the primary objective — namely, that of deriving a mathematical model to describe the dynamic behaviour of the test structure. It will be recalled from the earlier parts of the book that we found it convenient to classify the various types of model which can be constructed and also to consider the many different applications for which these models might be required. The subject is sufficiently broad that no single type of model is suitable for all cases and so the particular combination of measurement and analysis steps will vary according to the application. Thus we arrive at a most important aspect of the modelling process: the need to decide exactly which type of model we should seek before setting out on the acquisition and processing of experimental data.

It will be recalled that three main categories of system model were identified, these being the Spatial Model (of mass, stiffness and damping properties), the Modal Model (comprising the natural frequencies and mode shapes) and the Response Model (in our case, consisting of a set of frequency response functions). In addition to this grouping, we have also seen that there exist Complete Models of each type (a theoretically-ideal situation, although in reality these models are usually only approximately correct) and the more realistic Incomplete Models, which consist of something less than a full description of the structure (but whose available parameters are usually quite accurate). In almost all practical cases, we are obliged to consider these incomplete models.

While the relative sequence of these three types of model has previously been stated as Spatial-Modal-Response for a theoretical analysis and, conversely, Response-Modal-Spatial for an experimental study, we now view them in a different order, according to the facility with which each may be derived from the test data. This viewpoint ranks the models: Modal, Response and then Spatial and directly reflects the quantity of the completeness of data required in each case.

A modal model (albeit an incomplete one) can be constructed using just one single mode, and including only a handful of degrees of freedom, even though the structure has many modes and many DOFs. Such a model can be built up by adding data from more modes but it is not a requirement that all the modes should be included nor even that all the modes in the frequency range of interest be taken into account. Thus such a model may be derived with relatively few, or equally, with many data.

The response type of model in the form of a FRF matrix, such as the mobility matrix, also needs only to include information concerning a limited number of points of interest — not all the DOFs need be considered. However, in this case it is generally required that the model be valid over a specified frequency range and here it is necessary that all the modes in that range be included, and moreover, that some account be taken of those modes whose natural frequencies lie outside the range of interest to allow for the residual effects. Thus, the response type of model demands more data to be collected from the tests.

Lastly, a representative spatial model can only really be obtained if we have measured most of the modes of the structure and if we have made measurements at a great many of the DOFs it possesses. This is generally a very demanding requirement to meet and, as a result, the derivation of a spatial model from test data is very difficult to achieve successfully.

5.1.2   Contents of Chapter 5

This chapter is organised with the following structure: first, we shall describe and explain just what data must be measured and analysed in order to construct a suitable model; also, what checks can be made to assess the reliability of the finished product before it is used for its ultimate objective. Next, we shall discuss a number of techniques that are now available for what we shall term 'refining' the model which is obtained from the tests. There are usually a number of features that are present in real test-derived models that are difficult to reconcile with the cleaner, simpler and tidier world of the analytical model and it is these that may need 'refining'. For example, it is common practice to extract complex mode shapes from the test data obtained on most real structures but the corresponding analytical models are almost always undamped so that their modes are real. Closing the gap between these two otherwise disparate types of mode is one of the tasks we may wish to undertake. Then, there are a number of awkward features associated with the usually gross incompleteness of test-derived models, certainly as compared with their analytical counterparts. As a result, we may wish to expand our experimental models, or, alternatively, reduce the theoretical ones so that the two models which are to be compared are at least of the same order. Lastly, we shall explore some of the properties of the models which can be derived by the means described here. There are a number of properties of which the analyst must be keenly aware and some which can be of considerable assistance in various applications for which the models are destined. Some of the more important of these properties will be discussed and explained.

5.2   MODAL MODELS
5.2.1   Requirements to Construct Modal Models
A modal model of a structure is one which consists of two matrices: one containing the natural frequencies and damping factors (the 'eigenvalues') of the modes included, and a second one which describes the shapes of the corresponding modes (the 'eigenvectors'). Thus, we can construct such a model (albeit an incomplete one) with just a single mode and, indeed, a more complete model of this type is assembled simply by adding together a set of these single-mode descriptions.

The basic method of deriving a modal model is as follows. First, we note that from a single FRF curve, $H_{jk}(\omega)$, it is possible to extract certain modal properties for the rth mode by modal analysis so that we can determine:

$$H_{jk}(\omega) \to \omega_r, \eta_r, {}_rA_{jk} \quad ; \quad r = 1, m \tag{5.1}$$

Now, although this gives us the natural frequency and damping properties directly, it does not explicitly yield the mode shape: only a modal constant which is formed from the mode shape data. In order to extract the 'individual' elements, ϕ_{jr}, of the mode shape matrix, $[\Phi]$, it is necessary to make a series of measurements of specific frequency response functions including, especially, the point FRF at the excitation position. If we measure H_{kk}, then by using (5.1) we see that analysis of this curve will yield not only the natural frequency properties, but also the specific elements in the mode shape matrix corresponding to the excitation point, ϕ_{kr}, from:

$$H_{kk}(\omega) \to \omega_r, \eta_r, {}_rA_{jk} \to \phi_{kr} \quad ; \quad r = 1, m \tag{5.2}$$

If we then measure an associated transfer FRF using the same excitation position, such as H_{jk}, we are able to deduce the mode shape element corresponding to the new response point (ϕ_{jr}) using the fact that the relevant modal constants may be combined with those from the point measurement:

$$\left(\phi_{jr}\right) = \frac{{}_r A_{jk}}{\left(\phi_{kr}\right)} \tag{5.3}$$

Hence we find that in order to derive a modal model referred to a particular set of n coordinates, we need to measure and analyse a set of n FRF curves, all sharing the same excitation point (or the same response point, in the event that it is the excitation which is varied) and thus constituting one point FRF and $(n-1)$ transfer FRFs. In terms of the complete FRF matrix, this corresponds to a requirement to measure the individual functions which lie in one column (or one row, since the FRF matrix is generally symmetric), see Fig. 5.1(a). In practice, however, this requirement is the barest minimum of data which will provide the required model and it is prudent to measure rather more than a single column. Often, several additional elements from the FRF matrix would be measured to provide a check, or to replace poor data, and sometimes measurement of a complete second column or row might be advised in order to ensure that one or more modes have not been completely missed by an unfortunate choice of exciter location, see Fig. 5.1(b). It will be clear from inspection of the aforementioned expressions that if the exciter were placed at a nodal point of one of the modes (e.g. if ϕ_{kr} were zero for one mode) then there would be no indications at all of the existence of that mode because every modal constant would be zero for that mode, irrespective of whether the other elements, ϕ_{jr}, are zero or not. Thus, excitation at a node of any mode must be avoided and it may require more than one measurement to confirm that this unwanted condition has not been unwittingly introduced.

Once all the selected FRF curves have been measured and individually analysed, using the most appropriate methods from Chapters 3 and 4, there remains a further stage of processing to be done. Using any of the single-curve modal analysis methods outlined in Chapter 4, we shall find ourselves in possession of a set of tables of modal properties containing rather more data than we are seeking. In particular, we shall have determined many separate estimates for the natural frequency and damping factor of each mode of interest as these parameters are extracted afresh from each FRF curve in the measured set. In theory, all such estimates should be identical but in practice they seldom are, even allowing for experimental errors, and we must find a way to reduce them to the single value for each property which theory demands. A similar situation arises for the mode shape

$$\begin{bmatrix} H_{11} & H_{12} & \cdots & H_{1i} & \cdots & \boxed{H_{1j}} & \cdots & H_{1n} \\ H_{21} & H_{22} & \cdots & H_{2i} & \cdots & H_{2j} & \cdots & H_{2n} \\ \vdots & \vdots & & \vdots & & \vdots & & \vdots \\ H_{i1} & H_{i2} & \cdots & H_{ii} & \cdots & H_{ij} & \cdots & H_{in} \\ \vdots & \vdots & & \vdots & & \vdots & & \vdots \\ H_{j1} & H_{j2} & \cdots & H_{ji} & \cdots & H_{jj} & \cdots & H_{jn} \\ \vdots & \vdots & & \vdots & & \vdots & & \vdots \\ H_{n1} & H_{n2} & \cdots & H_{ni} & \cdots & H_{nj} & \cdots & H_{nn} \end{bmatrix}$$

(a)

$$\begin{bmatrix} \boxed{H_{11}} & H_{12} & \cdots & H_{1i} & \cdots & H_{1j} & \cdots & H_{1n} \\ H_{21} & \boxed{H_{22}} & \cdots & H_{2i} & \cdots & H_{2j} & \cdots & H_{2n} \\ \vdots & \vdots & & \vdots & & \vdots & & \vdots \\ H_{i1} & H_{i2} & \cdots & H_{ii} & \cdots & H_{ij} & \cdots & H_{in} \\ \vdots & \vdots & & \vdots & & \vdots & & \vdots \\ H_{j1} & H_{j2} & \cdots & H_{ji} & \cdots & H_{jj} & \cdots & H_{jn} \\ \vdots & \vdots & & \vdots & & \vdots & & \vdots \\ H_{n1} & H_{n2} & \cdots & H_{ni} & \cdots & H_{nj} & \cdots & H_{nn} \end{bmatrix}$$

(b)

Fig. 5.1 Frequency response function matrix: selection of FRFs to measure.
(a) Minimum data requirement; (b) Typical data selection

parameters also in the event that we have measured more than the minimum FRF data, i.e. if we have measured more than one row or column from the FRF matrix.

The simplest procedure is simply to average all the individual estimates to obtain mean values, $\tilde{\omega}_r$ and $\tilde{\eta}_r$. In practice, not all the estimates should carry equal weight because some would probably derive from much more satisfactory curve fits than others and so a more refined procedure would be to calculate a weighted mean of all the estimates, taking some account of the reliability of each. In fact, it is possible to attach a quality factor to each curve-fit parameter extraction in most of the methods described in Chapter 4 and these quality factors serve well as weighting functions for an averaging process such as that just suggested.

It is important to note that if we choose to accept a mean or otherwise revised value for the natural frequency and damping factor of a particular mode, then in some cases the values assumed for the modal constants should be revised accordingly. For example, we noted in the circle-fitting procedure that the diameter of the modal circle is given by:

$$_r D_{jk} = \frac{_r A_{jk}}{\omega_r^2 \eta_r} \tag{5.4}$$

and so if we choose to revise ω_r and η_r, taking all the measured/analysed data into account, then we should also revise the value of $_r A_{jk}$ since there is no reason to modify the circle diameter itself. Thus, we obtain a corrected set of modal constants, and so mode shape elements for each curve analysed using:

$$_r \tilde{A}_{jk} = {}_r A_{jk} \frac{\tilde{\omega}_r^2 \tilde{\eta}_r}{\omega_r^2 \eta_r} \tag{5.5}$$

where the ~ indicates a revised value.

As mentioned towards the end of Chapter 4, there exist a number of more advanced curve-fitting methods which obviate the need for the above stage of the process by the simple device of performing their analysis of the complete set of FRF curves in a single step. Whatever method is used to reduce the analysed data to their final form, this must consist of the two matrices which constitute a modal model, namely:

$$\left[\omega_r^2 (1 + i\eta_r) \right]_{m \times m} \quad , \quad \left[\Phi \right]_{n \times m}$$

Finally, mention should be made of a simplified form of the modal model which can be obtained rather more quickly than that obtained by following the above procedure. This alternative approach requires first the measurement and full analysis of one FRF, preferably the point mobility, in order to determine values for the natural frequencies and damping factors. Then the analysis of the subsequent FRF curves consists simply of measuring the diameters of the modal circles, omitting the stages which yield further natural frequency and damping estimates. Such a procedure is acceptable in those cases where one has full confidence in the first or reference FRF measurement, or where accuracy is not of paramount concern.

5.2.2 Double Modes or Repeated Roots

One of the areas of difficulty that can arise when constructing models of practical structures is that which occurs when the structure has two or more modes with the same natural frequency. Many structures possess double modes: most structures that are axisymmetric (discs, wheels, cones, ...), or cyclically periodic (such as gear wheels, or bladed assemblies, ...) have most of their modes in such pairs of effectively-identical natural frequencies. In practice, when a structure has two modes that are so close in frequency that it is impossible to detect more than one 'root' when analysing a measured response function in the vicinity of a resonance, then we risk being unable to derive a true model for the structure. All we can define in these circumstances is a single equivalent mode which is, in fact, a combination of the two actual modes that prove difficult to identify individually.

It might be thought that if the consequence of having a double mode is that it is not possible to detect the fact, then it is perhaps not necessary to go to extra lengths to identify and then to include both modes of the pair in the mathematical model we are constructing — why not use just one equivalent mode? However, this is not the case and discovering the existence of multiple modes, and of identifying all the individual modes' properties, is indeed essential for most applications of the resulting model. It is worth recalling a simple case study from a previous section to illustrate and emphasise the significance of this subtlety. Consider a simple disc-like structure which is being excited at a point A on its rim and at a frequency which is at the natural frequency of its 2 nodal-diameter (2ND) mode: Fig. 5.2.(a). The lines show the nodal patterns present at this condition. According to our understanding of structures in general, if the excitation were moved around the disc rim to be placed at one of the nodal points, B, then the mode we see in Fig. 5.2.(a) would no longer be excited, and there would be no resonance: Fig. 5.2.(b). However, experience or intuition suggests to us that if we were to perform the simple experiment just described, then what we would see would be the resonance persisting, even when excited at the new drive point, and a different mode shape would be evident as sketched in Fig. 5.2.(c). Indeed, this is exactly what does happen in these circumstances and the only way that this can be explained is by recognising the existence of a second mode at the same frequency as the first one seen in Fig. 5.2.(a). The first mode shape has not 'moved': another mode has become visible by exciting at a second drive point. If we extracted only the single mode evident from the data which result from excitation at point A, and constructed a mathematical model based on that result, we would simply be unable to use the resulting model to predict the result we obtain with the second excitation. Thus, when a structure has

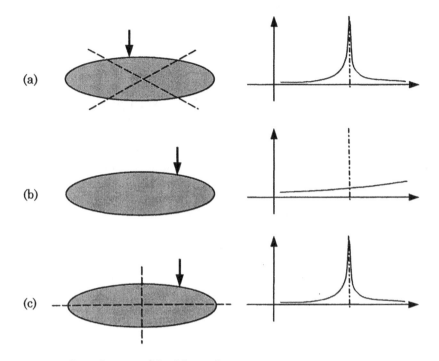

Fig. 5.2 Significance of double modes.
(a) Excitation at resonance of 2ND mode; (b) Excitation moved to node point; (c) New pattern of nodal lines

double modes, it is clearly very important that we can detect the fact, and that we can identify all the modes which are present.

The only way that double or repeated modes can be detected and identified in a modal test is by using data from more than one reference (i.e. by exciting at more than one drive point, or measuring the response at more than one point, if a moving exciter is used). This means that we must measure FRF data from more than a single row or column — in fact, from as many rows/columns as there are repeated roots (double roots, or modes, require two rows/columns; triple modes require data from three rows/columns, etc.).

5.2.3 Constructing Models for NSA Structures
In Chapter 2 we introduced some of the theoretical features of those structures which are classified as Non-Self-Adjoint (NSA). Principle amongst this class of structure are those with non-symmetric mass or stiffness or damping matrices, such as occurs in many structures with rotating components. This class of structure requires additional data to be acquired and analysed before a modal model can be assembled, over

and above those data which have been identified in the preceding section. Even with the qualifications just described in connection with repeated roots, and the dangers of choosing a nodal point as the reference excitation or response measurement point, there is additional information to be obtained in the case of NSA systems.

In the simpler cases, described above, we were able to take advantage of the symmetry of the system matrices (mass, stiffness, FRF, ...) and, as a result, we were obliged to measure just a single row **or** column of the FRF matrix. In the case of NSA structures, the option of choosing between a row or a column of the FRF matrix is not available and we are required to measure and analyse the elements in both a row **and** its associated column. Clearly, this places a significant burden on the measurement phase, especially as it is often very inconvenient to excite a structure at many different locations; the more so if that structure is rotating. Nevertheless, unless we have some additional information, or constraints, that is what is required. A mathematical 'explanation' for this additional requirement can be given by noting that in this class of system there are two types of eigenvector -- left-hand and right-hand. Thus, there are twice as many eigenvector elements to identify than for a conventional structure (which only has one set of eigenvectors). In effect, the FRF data in the column reveal the elements in the right-hand vectors and those in the row lead to the elements in the left-hand eigenvectors.

There is one special case that might apply from time to time: for some systems, the loss of symmetry in the system matrices arises solely because of gyroscopic effects and in the event that only these NSA features are present, then the resulting matrices are not completely asymmetric but are skew-symmetric. For such a system, the left-hand and right-hand eigenvectors are simply the complex conjugates of each other and so knowledge of one set automatically reveals the other, and we can revert to the acquisition and analysis of just one row or one column of the FRF matrix in this special, but realistic, situation.

5.2.4   Quality Checks for Modal Models

As with other stages in the modal testing procedures, it is appropriate at the conclusion of the modelling phase to seek and to implement any checks that can help indicate the reliability of the obtained results. There are two relatively straightforward such checks that can be recommended for this phase of the process. It is possible, of course, to regenerate FRFs from the modal data base that has been constructed and which is, in effect, the modal model. Often, such regeneration of the measured FRFs is an integral part of the quality checks on the modal analysis phase of the procedure. However, it is also possible to use the same data base to synthesise FRFs that have not (yet) been measured and so to do just that, and then to measure the corresponding functions on the structure, can provide a powerful and convincing demonstration of the validity of the modal model which has been assembled from the primary measured data.

A second, more demanding but yet more convincing, demonstration of the validity of the modal model is to use it to predict how the dynamic properties of the test structure will change if it is subjected to a small structural modification, such as can be occasioned by adding a small mass at a selected point. The means of doing this check are described in the next Chapter (Section 6.4): suffice it here to mention that this is a most demanding test of the model but one that, once passed, validates the model convincingly.

5.3   REFINEMENT OF MODAL MODELS
5.3.1   Need for Model Refinement

One of the features of most test-derived models is the likelihood of a degree of incompatibility between such a model and others (such as an analytical model) with which it may be compared. The most obvious difference is likely to be the fact that the elements in the mode shapes contain complex numbers (for the test-derived modes) while those which are produced by theoretical analysis are usually confined to undamped systems and are therefore composed of real mode shape data. Objective comparison between complex mode shapes and real mode shapes is hindered by this fundamental incompatibility and often demands some refinement of one or other of the two sets so that like can be compared with like.

A second obvious incompatibility lies in the almost universal difference in the order or size of the models derived from tests, on the one hand, and theoretical analysis on the other. Usually the former are of relatively small order, typified by the number of measured DOFs, n, while the latter are generally very much more detailed, comprising a mode with the full set of DOFs, N, which is usually one or more orders of magnitude greater than n. Again, when comparisons are to be made between a measured model and its theoretical counterpart, or even between two models of different sizes in general, this order incompatibility presents obstacles to meaningful interpretation and there is a desire to refine one or other model to bring them both to the same size. This can be achieved either by reducing the larger, theoretical, model down to the order determined by the number of measured DOFs, or by expanding the smaller model up by a form of interpolation so that it is described by the same number of DOFs as the complete analytical model. Both of these approaches are possible, as is the conversion of complex modes to real, but it must be recognised at the outset that none of these refinement processes is exact: they each involve approximations that mean that a compromise has been made in order to achieve the greater degree of compatibility which is desired. As such, the methods tend to fall into a more advanced category of process which demand more rigorous attention than can reasonably be presented in the introductory presentation offered below.

5.3.2   Complex-to-Real Conversion

As mentioned above, this first type of model refinement is sought when trying to compare two sets of modes: complex ones, as derived from analysis of test data, and real ones, such as are traditionally produced from theoretical analyses in the absence of a detailed knowledge of the nature, extent and distribution of damping. Such knowledge is rarely available at the modelling stage with the result that the great majority of mathematical models produced by conventional FE modelling are undamped, and have real mode shapes. This results in a requirement for a method which permits us to deduce the associated real modes for a structure for which we know the actual complex modes. Expressed another way, we wish to be able to determine what would be the mode shapes of the tested structure if, by some means, we could remove the damping but leave everything else the same. There are several methods proposed for this task of complex-to-real conversion and we shall mention three here.

5.3.2.1   Simple method

The first is a very simple method which can be applied to many practical situations without much effort, either computational or theoretical, and is the one most widely practised. This simple method is to convert the mode shape vectors from complex to real by taking the modulus of each element and by assigning a phase to each of 0 or 180 degrees. Which of these two alternative phase angles is assigned to each element is decided by a simple 'inspection' and is based on the concept that if the actual measured phase angle is within $\pm$ 10° of 0° or 180° then that phase angle is automatically set to the relevant polarity. This simple and reasonable criterion is then extended to all phase angles so that, in principle, any phase angle of a complex mode shape element which lies between 0° and +90°, or between 270° and 360°, is set to 0°, while those elements with phase angles between 90° and 270° are set to 180°. This procedure can become difficult to apply in borderline cases where there are elements with small moduli and poorly-determined phase angles but reference to a suitable graphical presentation of such a complex mode, as described below in respect of the so-called 'starburst' format, can assist in making the necessary decisions (see Figs. 5.3 (e) and (f)). This visual check will also serve to resolve a small number of cases that are sometimes encountered in which a mode shape is effectively real but has every element rotated by a non-trivial phase angle that is applied equally to every element. In these cases, the typical mode shape element is found to have a phase angle which is a few degrees either side of this uniform offset value or that plus 180° (rather than either side of 0° or 180°, as described above). Such mode shapes are sensibly real, even though the corresponding vectors are populated with complex numbers: it is the **relative** phase angles between the different elements in the vector which determine whether the mode is truly complex, or effectively real.

5.3.2.2 Multi-point excitation (Asher's method)

The second method for extracting real modes from complex is provided by a numerical simulation of the multi-point or 'appropriation' testing technique introduced in Chapter 2. In this method, the test-derived mathematical model based on complex modes is used to synthesise the response that would be produced by the application of several simultaneous harmonic forces (as is actually done in multi-point excitation tests) in order to establish what those forces would need to be in order to produce a mono-modal response vector. There are a number of different algorithms available for the task of finding such a force distribution, as mentioned in Chapter 2, but here we can explore the application of these methods to this specific task of determining the modes that the structure would possess if the damping could be artificially removed. The physical basis of this approach (which is referred to in the literature as 'Asher's Method' [51]) is quite simple: if this optimum set of excitation forces for a given mode can be found, then they represent the forces that are actually being generated by the damping in the system at resonance of that mode. At a resonance condition, it is known that the stiffness forces exactly balance the inertia forces, and the excitation forces exactly balance the damping forces, and so if we can establish this mono-mode response condition we have effectively isolated the damping effects and can then proceed to deduce the dynamic properties of the structure with these effects removed. The problem is that any vector of excitation forces that is determined in this process is likely to be approximate, owing to the inevitable fact that the number of DOFs included in the test-derived model (and at which excitation forces can be numerically applied) is limited and approximations will inevitably follow. Nevertheless, a judicious choice of included DOFs (see Chapter 6) can minimise these effects and can yield useful results in this task.

While not seeking to present a rigorous analysis of this method, it is appropriate to introduce the main features of that analysis: readers who wish to explore the application are referred to the appropriate literature for more detail [51]. We have already introduced the basic equations for this method — in Chapter 2 — and the reader is referred back to equations (2.68)-(2.70), which show that for a generally-damped system, subject to a mono-phased harmonic multi-excitation force vector, $\{F\}$, it is possible to obtain a similarly mono-phased response vector, $\{X\}$, by satisfying the following relationships:

$$\left(\left(-\omega^2[M]+[K]\right)\cos\theta+[D]\sin\theta\right)\{\hat{X}\}=\{\hat{F}\}$$

$$\left(\left(-\omega^2[M]+[K]\right)\sin\theta-[D]\cos\theta\right)\{\hat{X}\}=\{0\}$$

(5.6) also (2.69)

recalling that

$$\{f\}=\{\hat{F}\}e^{i\omega t}$$

$$\{x\}=\{\hat{X}\}e^{i(\omega t-\theta)}$$

(5.7) also (2.68)

If the phase lag, θ, between (all) the forces and (all) the responses is to have a value of exactly 90° (for resonance) then it can be seen that equation (5.6) reduces to:

$$\left(-\omega^2[M]+[K]\right)\{\hat{X}\}=\{0\}$$

(5.8) also (2.70)

which is clearly the equation to be solved to find the undamped system properties, including its undamped (i.e. real) mode shapes. It can also be shown that the following two equations hold:

$$\text{Re}[\alpha(\omega)]\{\hat{F}\}=\{0\}$$

$$\{\psi_u\}=\text{Im}[\alpha(\omega)]\{\hat{F}\}$$

(5.9)

Thus the sequence of steps required to determine this solution is as follows:

(4)   compute $[\alpha(\omega)]$ from the complex modal model;

(2)   determine the undamped system natural frequencies, ω_r^2, by solving the determinantal equation: $\det|\text{Re}[\alpha(\omega)]|=0$;

(4)   calculate the mono-phase vector for each mode of interest using: $\text{Re}[\alpha(\omega)]\{\hat{F}\}=\{0\}$;

(4)   calculate the undamped system mode shapes, $\{\psi_u\}$, using the just-derived force vector, $\{\hat{F}\}$, and $\{\psi_u\}=\text{Im}[\alpha(\omega)]\{\hat{F}\}$.

5.3.2.3 Matrix transformation

The third method introduced here is one which seeks a numerical solution to the expression linking the known damped (complex) modes and the unknown undamped (real) modes. The essential relationship between these two sets of eigenvectors is simple:

$$[\Phi_u] = [\Phi_d][T_1]$$

The method is developed in detail in Reference [52] but in summary can be described as the following steps:

(4) assume that $\text{Re}[T_1]$ is unity and calculate $\text{Im}[T_1]$ from:
$$\text{Im}[T_1] = -[\text{Re}[\Phi_d]]^T [\text{Re}[\Phi_d]]^{-1} [\text{Re}[\Phi_d]]^T \text{Im}[\Phi_d];$$

(2) calculate $[M_1]$ and $[K_1]$ from:
$$[M_1] = [T_1]^T [T_1] \; ; \; [K_1] = [T_1]^T [\lambda^2][T_1]$$

(3) solve the eigenproblem formed by $[M_1]$ and $[K_1]$ leading to:
$[\omega_r^2]$; and $[[T_2]]$

(4) calculate the real modes using:
$$[\Phi_u] = [\Phi_d][T_1][T_2]$$

As previously mentioned, the application of these methods to experimentally-measured data can be difficult to achieve and reference to the detailed description on the methods is strongly recommended.

5.3.3   Expansion of Models

The next most important model refinement that is often required is that of expansion: the addition to the actually-measured modal data of estimates for selected DOFs which were not measured for one reason or another. Prior to conducting each modal test, decisions have to be made as to which of the many DOFs that exist on the structure (and which would be included in a theoretical model) will be measured. These decisions are made for various practical reasons which include: limited test time, inaccessibility of some DOFs; anticipated low importance of motion in certain DOFs; and so on. One major application of expansion is to the acquisition of data relating to RDOFs (rotational DOFs). As mentioned elsewhere, these can be very difficult to measure directly (see Section 3.11) yet they are critical for some applications (see Section 6.4). One means of including these data is to derive them by one of the expansion methods described below.

Later, during the exploitation or application of the test-derived model, the absence of some of the data which have thus far been omitted can present difficulties. For example, if relatively few DOFs have been measured, then it may be difficult to interpret visually the animated displays that are used to illustrate mode shapes. At the same time, these displays show the displacement at DOFs which have not been measured as having zero value and that is not necessarily the case. Further, there may well be several DOFs which are difficult to access or which are difficult to measure (such as rotational DOFs) and for these it can be interesting to seek data from interpolation or expansion based on the data that have been measured.

Three approaches to the expansion of measured modes will be mentioned here:

(1)   Geometric interpolation using spline functions;
(2)   Expansion using the analytical model spatial properties, and
(3)   Expansion using an analytical model's modal properties.

In all three we are in effect seeking a transformation matrix that allows us to construct a long eigenvector, $\{\phi_{COMPLETE}\}_{N \times 1}$, from knowledge of a short (incomplete) one, $\{\phi_{INCOMPLETE}\}_{n \times 1}$, plus some additional information. In effect, that means finding a transformation matrix, $[T]$, that satisfies:

$$\{\phi_{COMPLETE}\}_{N \times 1} = [T]_{N \times n} \{\phi_{INCOMPLETE}\}_{n \times 1}$$

5.3.3.1   Interpolation

The first of these — simple interpolation — has a limited range of application and can only really be used on structures which have large regions of relatively homogeneous structure: those with joints or abrupt changes in section are much less likely to respond to this form of expansion. The method is simply geometric interpolation between the measured points themselves, such as by fitting a polynomial function through the measured points and using that to interpolate for intermediate, unmeasured, DOFs. This provides a very simple approach to the task, but one of limited applicability.

5.3.3.2 Expansion using theoretical spatial model (Kidder's method)

The second approach to the problem of interpolation is provided by using a theoretical model's mass and stiffness matrices in a form of an inverse Guyan reduction procedure. If we partition the eigenvector of interest, $\{\phi_A\}_r$, into $\{_A\phi_1\}_r$ (the DOFs to be included) and $\{_A\phi_2\}_r$ (those which are **not** available from measurements), then we may write:

$$\left(\begin{bmatrix} {}_A K_{11} & {}_A K_{12} \\ {}_A K_{21} & {}_A K_{22} \end{bmatrix} - \omega_r^2 \begin{bmatrix} {}_A M_{11} & {}_A M_{12} \\ {}_A M_{21} & {}_A M_{22} \end{bmatrix}\right) \begin{Bmatrix} {}_A \phi_1 \\ {}_A \phi_2 \end{Bmatrix}_r = \{0\} \qquad (5.10)$$

$$\underset{N \times N}{} \qquad \qquad \underset{N \times 1}{}$$

We can use this relationship between the measured and unmeasured DOFs (exact for the analytical mode shapes) as the basis for an expansion of the incomplete measured mode shapes, as follows:

$$\{{}_X \tilde{\phi}_2\}_r = -\left([{}_A K_{22}] - \omega_r^2 [{}_A M_{22}]\right)^{-1} \left([{}_A K_{21}] - \omega_r^2 [{}_A M_{21}]\right)\{{}_X \phi_1\}_r$$

$$\underset{(N-n) \times 1}{} \quad \underset{(N-n) \times (N-n)}{} \qquad \underset{(N-n) \times n}{} \quad \underset{n \times 1}{}$$

or

$$\{{}_X \tilde{\phi}_2\}_r = [T_{21}]\{{}_X \phi_1\}_r \qquad (5.11)$$

where

$$[T_{21}] = -\left([{}_A K_{22}] - \omega_r^2 [{}_A M_{22}]\right)^{-1} \left([{}_A K_{21}] - \omega_r^2 [{}_A M_{21}]\right)$$

This expression is suitable for use to fill out the unmeasured elements in the mode shape matrix and can be related to the earlier general relationship between the incomplete measured vector and the complete, expanded, vector as follows:

$$\{\tilde{\phi}_X\}_r = [T]\{{}_X \phi_1\}_r = \begin{bmatrix} [I] \\ [T_{21}] \end{bmatrix}\{{}_X \phi_1\}_r \qquad (5.12))$$

As always, attention must be paid to the potential ill-condition of the matrix which is to be inverted and in this case the requirement is for a non-singular matrix: $([K_{22}] - \omega_r^2 [M_{22}])$ and this will generally be satisfied as the natural frequencies for the whole model are very unlikely to be those for the partitioned submatrices found in this expression. In fact, if necessary, alternative expressions can be derived for the expansion, such as:

$$\{{}_X \tilde{\phi}_2\}_r = -\left([{}_A K_{12}] - \omega_r^2 [{}_A M_{12}]\right)^+ \left([{}_A K_{11}] - \omega_r^2 [{}_A M_{11}]\right)\{{}_X \phi_1\}_r \qquad (5.13)$$

$$\underset{(N-n) \times 1}{} \quad \underset{(N-n) \times n}{} \qquad \underset{n \times n}{} \quad \underset{n \times 1}{}$$

for which a suitable generalised inverse computation can be used in place of the direct inverse for equation (5.11).

5.3.3.3 Expansion using analytical model mode shapes

The most popular method of expansion, however, is another one which uses the analytical model for the interpolation but bases that process on the mode shapes derived from the analytical model spatial matrices, rather than on these matrices themselves. There are, in fact, a number of options for this version, using different mixtures of data from the analytical and test-derived models and an extensive body of work has been reported in this area under the general title of SEREP (System Equivalent Reduction and Expansion Processes) methods, [53].

Using the notation introduced above, we may write the following expression which relates the experimental model mode shapes to those of the analytical model:

$$\{\phi_X\}_r = [\Phi_A]\{\gamma\}_r$$

or   (5.14)

$$\begin{Bmatrix} X\phi_1 \\ X\phi_2 \end{Bmatrix} = \begin{bmatrix} [_A\Phi_{11}] & [_A\Phi_{12}] \\ [_A\Phi_{21}] & [_A\Phi_{22}] \end{bmatrix} \begin{Bmatrix} \gamma_1 \\ \gamma_2 \end{Bmatrix}_r$$

The basis of this method is to assume that the measured mode shape submatrix can be represented exactly (although not uniquely) by the simple relationship (which assumes that $\{\gamma_2\}_r$ can be taken to be zero):

$$\{_X\phi_1\}_r = [_A\Phi_{11}]\{\gamma_1\}_r \qquad\qquad (5.15)$$

so that an estimate can be provided for the unmeasured part of the eigenvector from:

$$\{_X\tilde{\phi}_2\} = [_A\Phi_{21}][_A\Phi_{11}]^{-1}\{_X\phi_1\}_r = [T_{21}]\{_X\phi_1\}_r \qquad (5.16)$$

Thus we can write the full transformation as:

$$\{\tilde{\phi}_X\}_r = \begin{Bmatrix} X\phi_1 \\ X\tilde{\phi}_2 \end{Bmatrix} = \begin{bmatrix} [_A\Phi_{11}] \\ [_A\Phi_{21}] \end{bmatrix} [_A\Phi_{11}]^{-1}\{_X\tilde{\phi}_1\}_r \qquad (5.17)$$

This formula can be generalised to a single expression which covers several measured modes, m in our usual notation:

$$\left[\tilde{\Phi}_X\right]_{N \times m_X} = \left[\Phi_A\right]_{N \times m_A} \left[_A\Phi_{11}\right]^+_{m_A \times n} \left[_X\Phi_1\right]_{n \times m_X} \tag{5.18}$$

where m_X, m_A are the numbers of experimental and analytical modes used, respectively. This equation can be further simplified, to the form:

$$\left[\tilde{\Phi}_X\right]_{N \times m_X} = \left[T\right]_{N \times n} \left[_X\Phi_1\right]_{n \times m_X} \tag{5.19}$$

The expression we have defined above for the overall expansion transformation matrix, $[T]$, is one of a family of alternatives. The one already defined bases the expansion purely on the analytical model mode shapes. Other formulations are possible involving various combinations of the available experimental mode shape data and those for the analytical model. Below we give the various possibilities for $[T]$:

$$\left[T_{(1)}\right] = \left[\Phi_A\right]\left[_A\Phi_1\right]^+ \qquad \text{(A model - based)}$$

$$\left[T_{(2)}\right] = \left[\Phi_A\right]\left[_X\Phi_1\right]^+ \qquad \text{(X model - based)}$$

$$\left[T_{(3)}\right] = \begin{bmatrix} _X\Phi_1 \\ _A\Phi_2 \end{bmatrix}\left[_A\Phi_1\right]^+ \qquad \text{(Mixed/A - based)} \tag{5.20}$$

$$\left[T_{(4)}\right] = \begin{bmatrix} _X\Phi_1 \\ _A\Phi_2 \end{bmatrix}\left[_X\Phi_1\right]^+ \qquad \text{(Mixed/X - based)}$$

These various options comprise the basis of the SEREP family of methods. As a final comment, it must be pointed out that all the above expressions are approximate because of the initial assumption that the higher modes are not required to be included in the process (i.e. that $\{\gamma_2\}$ is zero). Whilst this may be a reasonable assumption, it is just that — an assumption — and so the resulting expressions for the expanded mode shapes are simply an interpolation and are not exact. Indeed, it would be somewhat remarkable if there were any other result as that would suggest that it was never necessary to measure any more than a few points on the structure as the rest could be deduced by expansion. Expansion does not change the fact that the basic model is incomplete and, consequently, if the expanded models are to be used to validate an analytical model, it must be remembered that much of the expanded mode shape data has come indirectly from the very model that is under evaluation. Do not be surprised if expanded mode shapes correlate very well with the analytical model!

5.3.4   Reduction of Models

The final process to be considered here is model reduction, the inverse of the expansion process, which is used when it is decided to obtain compatibility between two otherwise disparate models by reducing the size of the larger of the two models — almost always, the analytical model. Strictly speaking, model reduction is not a procedure that is generally applied to test-derived modal models, and so is slightly out of place in this section. However, it is relevant to the context of model expansion and so will be included, albeit briefly.

Although model reduction was an important and widely-used technique in structural analysis just a few years ago, because of the limitations imposed by the available computing power, it has more recently become less necessary, partly because of an extraordinary increase in computing power but also because of the various approximations it introduces. In the present context of comparing test- and analysis-derived dynamic models, reduction is clearly less popular than the alternative expansion approach. Indeed, reduction is not a process that is applied to a test-derived model at all, but rather to the analytical models with which the experimental data are to be compared.

There are basically two different types of model reduction, both of which are applied to the spatial model, as opposed to the modal model as is the case in model expansion. Both approaches achieve the same end result of yielding a smaller-order model, with system matrices which are nxn instead of NxN, but the difference is between (i) a condensed model which seeks to represent the entire structure completely at a smaller number of DOFs (and is thus a 'complete' model, albeit approximate) and (ii) a reduced model which has removed information related to the DOFs that are eliminated from the model, and which is thus an incomplete model. In the former type, the condensation introduces an approximation because it is, in effect, deriving a coarser model than the larger original version. The latter type of model has the advantage that the information related to those DOFs which are retained is as accurate as was that from the original model, but the resulting reduced model is incomplete in the sense that mass and stiffness features pertaining to the eliminated coordinates are not included, nor are they compensated for. A reduced model of this type is relatively useful if it is a modal model, or a response model, but it is very difficult to use if the model is of the spatial type. The real difficulty in this application is that the condensed model is the most appropriate type of reduced analytical model, but the corresponding experimental models which are also of reduced order are not of this type. A small-order test-derived model of a complex structure comprises a selection of data from the full model. The missing data are simply unmeasured and the behaviour of and at the unmeasured DOFs is in no way accounted for in that reduced model.

We shall summarise below the basic features of model reduction by condensation so that this process can be seen in relation to the preceding techniques of expansion. The basic equation of motion for the original model can be expressed as:

$$[M]\{\ddot{x}\} + [K]\{x\} = \{f\} \tag{5.21}$$

and this can be partitioned into the kept DOFs, $\{x_1\}$, and the eliminated DOFs, $\{x_2\}$, (which by definition cannot have any excitation forces applied to them) as follows:

$$\begin{bmatrix} M_{11} & M_{12} \\ M_{21} & M_{22} \end{bmatrix} \begin{Bmatrix} \ddot{x}_1 \\ \ddot{x}_2 \end{Bmatrix} + \begin{bmatrix} K_{11} & K_{12} \\ K_{21} & K_{22} \end{bmatrix} \begin{Bmatrix} x_1 \\ x_2 \end{Bmatrix} = \begin{Bmatrix} f_1 \\ 0 \end{Bmatrix} \tag{5.22}$$

As seen in the earlier section on expansion, a relationship between the kept and eliminated DOFs can be written in the form:

$$\begin{Bmatrix} x_1 \\ x_2 \end{Bmatrix}_{N \times 1} = \begin{bmatrix} [I] \\ [T] \end{bmatrix}_{N \times n} \{x_1\}_{n \times 1} \tag{5.23}$$

where the transformation matrix, $[T]$, can be defined by:

$$[T] = (1 - \beta)\left(-[K_{22}]^{-1}[K_{21}]\right) + \beta\left(-[M_{22}]^{-1}[M_{21}]\right) \tag{5.24}$$

in which β is a reduction coefficient whose limiting values are $\beta = 0$ **for static reduction** and $\beta = 1$ for **dynamic reduction**. The reduced mass and stiffness matrices which are produced by this process are:

$$[M^R]_{n \times n} = [\, [I] \quad [T]^T \,]_{n \times N} \begin{bmatrix} M_{22} & M_{21} \\ M_{12} & M_{22} \end{bmatrix}_{N \times N} \begin{bmatrix} [I] \\ [T] \end{bmatrix}_{N \times n} \tag{5.25a}$$

and

$$[K^R]_{n \times n} = [\, [I] \quad [T]^T \,]_{n \times N} \begin{bmatrix} K_{22} & K_{21} \\ K_{12} & K_{22} \end{bmatrix}_{N \times N} \begin{bmatrix} [I] \\ [T] \end{bmatrix}_{N \times n} \tag{5.25b}$$

The two limiting cases of static and dynamic reduction are of particular interest. In each case, one of the two system matrices is unchanged and the other one is, for static (Guyan) reduction and dynamic reduction, respectively:

$$\beta = 1: \quad \left[M^{Rstatic} \right] = [M_{12}] \left(-[M_{22}]^{-1}[M_{21}] \right) + [M_{11}];$$

$$\left[K^{Rstatic} \right] = [K]$$

$$\beta = 0: \quad \left[M^{Rdynamic} \right] = [M];$$

$$\left[K^{Rdynamic} \right] = [K_{12}] \left(-[K_{22}]^{-1}[K_{21}] \right) + [K_{11}]$$

(5.26)

From a theoretical standpoint, these reduction procedures, when properly used, can provide useful though approximate models of the structures to which they relate. 'Proper' application means an optimum choice of which DOFs to retain and which can be eliminated, and this choice is case-dependent. The resulting models are capable of representing the eigenproperties of the structure with reasonable accuracy. However, as mentioned in the introduction, a reduced theoretical model of this type does **not** correspond to a similarly low-order model which is obtained from experiments since that is formed simply by ignoring the eliminated DOFs. The measured data for the included DOFs are the same no matter how many DOFs are eliminated. The retained properties of a condensed model are different for different degrees of reduction. Thus, there are inherent difficulties involved in using this mixture of condensed (but complete) theoretical models and reduced (but incomplete) experimental models.

5.4 DISPLAY OF MODAL MODELS

One of the attractions of the modal model is the possibility of obtaining a graphic display of its form by plotting the mode shapes, thereby giving some visual insight into the way in which the structure is vibrating. As there are a number of alternatives for this phase, and a number of important features, it is worth discussing them briefly.

Once the modal model has been determined and tabulated according to the description given in Section 5.2, there are basically two choices for its graphical display: a static plot or a dynamic (animated) display. While the former is far less demanding than the latter in respect of the material necessary to produce the display, it does have serious limitations in its ability to illustrate some of the special features of complex modes. In cases where the modes have significant complexity and individual displacements have phase angles which are not simply 0° or 180° from the others, only an animated display is really capable of presenting a realistic image.

5.4.1 Static Displays
5.4.1.1 Deflected shapes

A static display is often adequate for depicting relatively simple mode shapes and, in any case, this is the only format suitable for permanent documentation in reports. The simplest procedure is to draw first an outline of the test structure using a viewpoint which permits visibility of all important points on the structure. Usually, this drawing is formed of a frame linking the various coordinates included in the modal survey, such as that shown in Fig. 5.3(a), and often this datum grid is drawn in faint or broken lines. Then, the grid of measured coordinate points is redrawn on the same plot but this time displaced in each direction (x, y and z) by an amount proportional to the corresponding element in the mode shape vector. The elements in the vector are scaled according to the normalisation process used — and are usually mass-normalised — and their absolute magnitudes have no particular significance in the present process. It is customary to select the largest eigenvector element and to scale the whole vector by an amount that makes that displacement on the plot a viable amount. It is not possible to dictate how large a deflection is 'viable' as this depends on the particular mode shape as well as on the complexity of the structural form itself. It is necessary to be able to see how the whole structure deforms but the displacements drawn on the plot must not be so large that the basic geometry of the structure appears to be violated. Fig. 5.3(b) shows a suitable plot for a mode of the plate previously illustrated in Fig. 5.3(a).

In this process, it will be necessary to assign a positive or negative phase to each element in the mode shape vector. Only phase angles which are effectively 0° or 180° with respect to the norm can be accommodated on this type of plot even though the results from the modal test may indicate marked deviations from such a pattern, as in the case of complex modes. Thus, it is often necessary to perform a 'whitewashing' exercise on the modal data and sometimes this requires making difficult judgements and decisions, such as how to incorporate an eigenvector element whose phase angle is closer to 90° than to 0° or 180° (see the previous section). Fortunately, this dilemma is most often encountered on modal deflections which are relatively small so that they do not influence the overall shape of the plot very much. However, the selection of positive or negative phase for such a point usually has the effect of determining the location of a nodal point or line, and this in itself may be important.

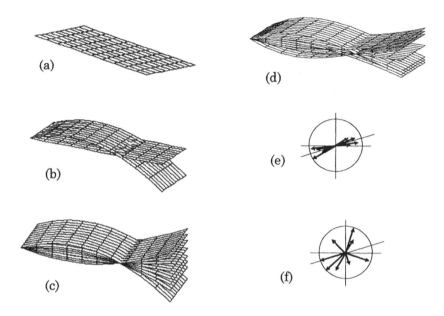

Fig. 5.3   Static displays of mode shapes.
(a) Basic grid; (b) Single-frame deflection pattern; (c) Multiple-frame deflection pattern; (d) Complex mode; (e) Argand diagram 'starburst' plot — 1 Quasi-real mode; (f) Argand diagram 'starburst' plot — 2 Complex mode

5.4.1.2   Multiple frames

A logical sequel to the previous presentation is that in which a series of deflection patterns are superimposed, each one computed for a different instant in time. If all these are plotted together, see Fig. 5.3(c) then some indication of the motion of the structure can be conveyed, and especially, lines or points of zero motion (nodes) can be clearly identified. It is also possible, in this format, to give some indication of the essence of complex modes, as shown in Fig. 5.3(d). As has been explained in earlier chapters, complex modes do not, in general, exhibit fixed nodal points and this is a feature that can be illustrated in the static multi-frame presentations that we have shown here.

5.4.1.3   Argand diagram ('starburst') plots

Another form of presentation which is useful for complex modes has already been introduced earlier in the book but is worth repeating here.

Each eigenvector is, in effect, a vector of complex numbers: even for cases of real modes, these complex eigenvector values simply have arguments of 0° or 180°. Thus it is useful to plot the individual elements of the eigenvector on a polar plot, as shown in the examples of Figs. 5.3 (e) and (f). Although there is no attempt to show the physical deformation of the actual structure in this format, the complexity of the mode shape is graphically displayed, even when that feature is negligible and the mode is, effectively, real. These presentations can be very helpful when seeking to understand or demonstrate modal complexity.

5.4.2   Dynamic Display

The above-mentioned difficulties are avoided by the alternative display, that of the animated mode shape. Using suitable computation facilities, it is possible to display on the screen a plot of the type just described, and to update this picture at regular and frequent intervals of time so that a simulation of the vibration is displayed in slow motion. Usually what is done is that the coordinates for the basic picture are computed and stored, as are a corresponding set for a fraction of a vibration cycle later, and then for a further fraction of a cycle later, and so on. Each of these sets of data constitutes one 'frame' and some 10-20 frames are used for a complete cycle, i.e. at intervals of some 20° to 40° in angular frequency. Once this data set has been constructed, successive frames are displayed with an update rate which is suitable to give a clear picture of the distortion of the structure during vibration; a rate which can be adjusted to accommodate the varying phase angles of complex modes. Indeed, the dynamic character of animation is the only really effective way to view modal complexity. Another advantage of most animated mode shape displays is the additional facility of changing the viewpoint from which the picture is drawn, often necessary as different modes can be best viewed form different orientations.

A static plot can be obtained from a dynamic display simply by 'freezing' the animation (i.e. by requesting a zero update rate).

5.4.3   Interpretation of Mode Shape Displays

There are a number of features associated with mode shape displays that warrant a mention in the context of ensuring that the correct interpretation is made from viewing these displays.

The first two observations concern the consequences of viewing what is very clearly an incomplete model. Where there are no mode shape data for some of the points which comprise the grid which outlines the structure, the indicated result is zero motion of those DOFs and this can be very misleading. The most obvious manifestation of this problem can be seen at grid points where measurements have been made in one direction only — perhaps the x direction — while no data exist for the other two, y and z, directions. Because there are no data for these other directions, the corresponding motion at those points is indicated as being zero. Quite often, when viewing an animated mode shape display, we can see significant x-direction motion of a particular point of interest accompanied by no movement in the other transverse directions and, not surprisingly, we tend to interpret this as a motion which is purely in the x-direction. There are very few displays that will differentiate between a measured response level of zero amplitude and the absence of any data (which by default is displayed as a zero response). Although this may sound trivial, it can easily be the source of serious misinterpretations of the results of a modal test; results that may be of very good quality but which are, as is so often the case, incomplete.

There is a second version of this same problem that is more subtle, and is the basis of several sections in the next chapter. This second problem is one which arises when the grid of measurement points that is chosen to display the mode shapes is too coarse in relation to the complexity (geometric, not mathematical) of the deformation patterns that are to be displayed. The problem can be illustrated graphically using a very simple example: suppose that our test structure is a straight beam, and that we (ill-advisedly) decide to use just three response measurement points to describe the vibrations that are measured — one point at each end and a third point in the middle of the beam (which we shall assume here to be uniform). If we consider the first six modes of the beam, whose mode shapes are sketched in Fig. 5.4, then we shall 'see' the mode 'shapes' indicated by the discrete points in Fig. 5.4. It soon becomes clear that with this few measurement points, modes 1 and 5 look the same as each other, as do 2, 4 and 6, and indeed all the higher modes will be indistinguishable from these first few. This is a well-known problem of 'spatial aliasing' and takes its name from the well-documented phenomenon encountered in signal processing. For our purposes here, however, it is sufficient to note that when viewing mode shapes of complex structures, one must always be on guard for deficiencies of this type: it will not always be as obvious as this illustration suggests to establish whether the density of points is adequate for the proper discrimination of the models of interest. The mode shapes displayed will not show evidence of anything untoward: they will simply be difficult to distinguish from other modes, usually those with higher natural frequencies.

The final comments in this section are a warning to beware of misinterpretation from unusual perspectives. Wire meshes are frequently used as the basis for mode shape displays and without the benefit of hidden line removal, or shading (that are thankfully

396

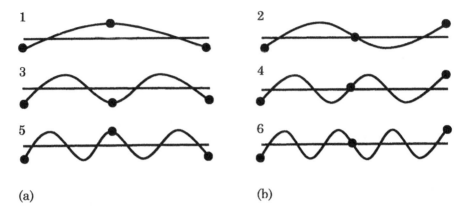

Fig. 5.4   Misinterpretation of mode shapes by spatial aliasing.
(a) True beam mode shapes; (b) Displayed mode shapes

becoming more widely used in this context), these sketches can give problems of distortion of perspective that lead to misinterpretation of how the structure is vibrating — and these interpretations are entirely what the mode shape displays are for.

5.5   RESPONSE MODELS

There are two main requirements demanded of a response model, the first being the capability of regenerating 'theoretical' curves for the frequency response functions actually measured and analysed and the second being that of synthesising the other response functions which were not measured. In general, the form of response model with which we are concerned is an FRF matrix whose order is dictated by the number of coordinates included in the test, n. (Note that this is not necessarily equal to the number of modes studied, N.) Also, as explained in Section 5.1, it is not normal practice to measure and to analyse all the elements in the FRF matrix but rather to select a small fraction of these, usually based on one column or row with a few additional elements included as a backup. Thus, if we are to construct an acceptable response model it will be necessary to synthesise those elements which have not been directly measured. However, in principle this need present no major problem as it is possible to compute the full FRF matrix from a modal model using:

$$[H]_{n \times n} = [\Phi]_{n \times m} \left[\left(\lambda_r^2 - \omega^2 \right) \right]_{m \times m}^{-1} [\Phi]_{m \times n}^T \tag{5.27}$$

5.5.1 Regenerated FRF Curves

It is usual practice to regenerate an FRF curve using the results from the modal analysis as a means of checking the success of that analysis, as described in Chapter 4. However, if the collected results from several FRF curves are subjected to an averaging process, such as that described in Section 5.1, then a new regenerated curve should be produced and, if necessary, a new set of residuals computed (see Section 4.3.6). It should be noted at this stage that in order to construct an acceptable response model, it is essential that all the modes in the frequency range of interest be included, and that suitable residual terms are added to take account of out-of-range modes. In this respect, the demands of the response model are more stringent than those of the modal model.

5.5.2 Synthesis of FRF Curves

One of the implications of equation (5.27) is that it is possible to synthesise the FRF curves which were not measured. In simple terms, this arises because if we measure three individual FRF curves such as $H_{ik}(\omega)$, $H_{jk}(\omega)$ and $H_{kk}(\omega)$, then modal analysis of these yields the modal parameters from which it is possible to generate, or 'synthesise' the FRF $H_{ij}(\omega)$, $H_{jj}(\omega)$, etc.. Indeed, application of this principle has already been suggested as a means of checking the overall performance of a modal analysis exercise.

However, it must be noted that there is an important limitation to this procedure which can sometimes jeopardise the success of the whole exercise. This limitation derives from the fact that only that part of the relevant FRF which is due to the modes whose properties are available can be computed. The remaining part, due to the out-of-range modes — the residual contribution — is **not** available by this method of synthesis and, as a result, a response model thus formed is liable to error unless values for the relevant residual terms are available from other sources.

As an example, the result of applying this synthesis procedure to measurements made on a turbine rotor are shown in Fig. 5.5. FRF data H_{11} and H_{21}, at and between the ends of the rotor, were measured and analysed and the resulting modal parameters used to predict or 'synthesise' the other FRF, H_{22}, initially unmeasured. This predicted curve was then compared with measurements producing the result shown in Fig. 5.5(a). Clearly, the agreement is poor and would tend to indicate that the measurement/analysis process had not been successful. However, the 'predicted' or 'synthesised' curve contained only those terms (from the complete modal series) relating to the modes which had actually been studied from H_{11} and H_{21} and this set of modes (as is often the case) did not include **all** the modes of the structure. Thus our predicted curve, H_{22}, omitted the influence of out-of-range modes or, in other words, lacked the residual terms. The inclusion of these two additional terms (obtained here only after measuring and analysing H_{22} itself) resulted in the greatly improved predicted vs. measured comparison shown in Fig. 5.5(b).

The appropriate expression for a 'correct' response model, derived via a set of modal properties is thus:

$$[H] = [\Phi]\left[\left(\lambda_r^2 - \omega^2\right)\right]^{-1}[\Phi]^T + [Res] \tag{5.28}$$

In order to obtain all the data necessary to form such a model, we must first derive the modal model on which it is based (as described in Section 5.1) and then find some means of determining or estimating the elements in the **residual matrix**, [Res]. This latter task may be most accurately achieved by measuring all (or at least something over half) of the elements in the FRF matrix, but this would constitute a major escalation in the quantity of data to be measured and analysed. A second possibility, and a reasonably practical one, is to extend the frequency range of the modal test beyond that over which the model is eventually required. In this way, much of the content of the residual terms is included in separate modes and their actual magnitudes can be reduced to relatively unimportant dimensions. The main problem with this approach is that one does not generally know when this last condition has been achieved, although a detailed examination of the regenerated curves using all the modes obtained and then again less the highest one(s) will give some indications in this direction.

A third possibility is to try to assess which of the many FRF elements are liable to need large residual terms and to make sure that these are included in the list of those which are measured and analysed. We noted earlier that it is the point mobilities which are expected to have the highest-valued residuals and the remote transfers which will have the smallest. Thus, the significant terms in the [Res] matrix will generally be grouped close to the leading diagonal, and this suggests making measurements of most of the point mobility parameters. Such a procedure would seldom be practical unless analysis indicates that the response model is ineffective without such data, in which case it may be the only option.

5.5.3 Direct Measurement

Finally on this topic, it should be noted that it is quite possible to develop a response model by measuring and analysing all the elements in one half of the FRF matrix (this being symmetric, only one half is essential) and by storing the results of this process without constructing a modal model, or 'modal data base' as this is sometimes called. Such a

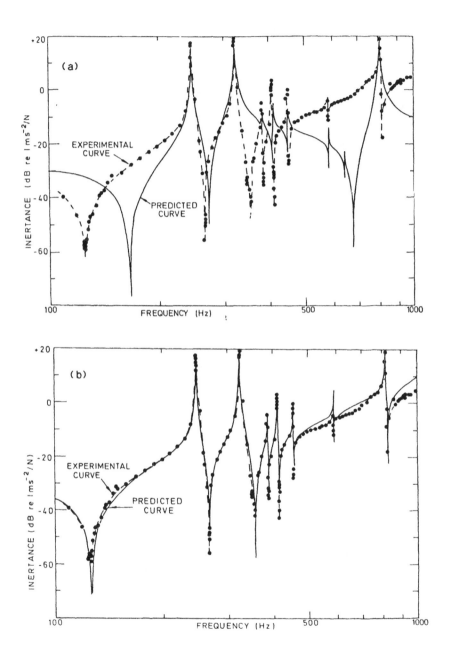

Fig. 5.5 Synthesised FRF plot.
(a) Using measured modal data only; (b) After inclusion of residual terms

procedure clearly solves the residual problem discussed above but it is likely to present another one by introducing inconsistencies into the model. Unless the structural behaviour, the measurements and the analysis are all of a remarkable calibre, the small differences described in Section 5.1 will be locked into the response model thus formed and will undoubtedly cause serious difficulties when that model is put to use. At the very least, the natural frequencies and damping factors of the individual modes should be rationalised throughout the model but even that is insufficient to ensure a satisfactory model.

Many of the same comments apply to the very crude method of obtaining a response model by simply storing the raw measurements made of each of the elements in the FRF matrix, a technique which bypasses the data reduction and smoothing facilities afforded by modal analysis. Although there are some instances where this is a viable procedure, they are rare and rather special.

5.5.4   Transmissibilities

One vibration parameter which has not been mentioned so far in this book is that of transmissibility. This is a quantity which is quite widely used in vibration engineering practice to indicate the relative vibration levels between two points. It is perhaps surprising that transmissibility has not featured in our studies or descriptions thus far.

In general, transmissibility is considered to be a frequency-dependent response function, $T_{jk}(\omega)$, rather like the frequency response functions that we rely on so heavily, which defines the ratio between the response levels at two points (or DOFs), j and k. Simply defined, we can write:

$$T_{jk}(\omega) = \frac{X_j e^{i\omega t}}{X_k e^{i\omega t}} \tag{5.29}$$

but, in fact, we need also to specify the excitation conditions that give rise to the two responses in question and these are missing from the above definition, which is thus not complete, or rigorous. It does not give us enough information to be able to reproduce the conditions which have been used to measure $T_{jk}(\omega)$. If the transmissibility is 'measured' during a modal test which has a single excitation, say at DOF i, then we can define the transmissibility thus obtained more precisely, as $_iT_{jk}(\omega)$:

$$_iT_{jk}(\omega) = \frac{H_{ji}(\omega)}{H_{ki}(\omega)} \tag{5.30}$$

and this $\neq {}_q T_{jk}(\omega)$, where q is a different DOF to i.

Plots of two examples of this transmissibility function are shown in Fig. 5.6, which also displays the natural frequencies of the structure which do not coincide with peaks on the transmissibilities. Thus we can see that there is no such thing as a simple transmissibility, $T_{jk}(\omega)$, and so this may explain why it has not been more extensively used in modal analysis. By studying the detailed expression (5.30), we can see that in general the transmissibility depends significantly on the excitation point (or points): hence the preceding conclusion. However, there are some special circumstances which, when encountered, result in a relaxation of that general result and lead to a situation where the transmissibility becomes almost independent of the excitation conditions. We can understand this result if we consider the expression that is used to compute each of the FRF properties which make up the expression in (5.30):

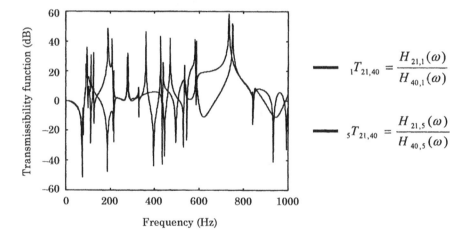

Fig. 5.6 Transmissibility plots

$$_i T_{jk}(\omega) = \frac{H_{ji}(\omega)}{H_{ki}(\omega)} = \frac{\sum_r \frac{\phi_{jr}\,\phi_{ir}}{\left(\omega_r^2 - \omega^2\right)}}{\sum_r \frac{\phi_{kr}\,\phi_{ir}}{\left(\omega_r^2 - \omega^2\right)}} \tag{5.31}$$

At frequencies of vibration close to a natural frequency, ω_r, we find an approximate expression for the transmissibility as:

$$_iT_{jk}(\omega)_{\omega \approx \omega_r} \approx \frac{\dfrac{\phi_{jr}\,\phi_{ir}}{\left(\omega_r^2 - \omega^2\right)}}{\dfrac{\phi_{kr}\,\phi_{ir}}{\left(\omega_r^2 - \omega^2\right)}} = \frac{\phi_{jr}}{\phi_{kr}} \tag{5.32}$$

This interesting result suggests that there are situations in which transmissibility may be useful in modal testing, but only when the appropriate conditions are satisfied: here, a structure which is sufficiently lightly damped that near resonance only one mode dominates the response.

5.5.5 Base Excitation

The one application area where transmissibilities can be used as part of modal testing is in the case of base excitation. Base excitation is a type of test where the input is measured as a *response* at the drive point, $x_0(t)$, instead of as a *force*, $f_1(t)$, as illustrated in Fig. 5.7. A schematic model for each of these configurations is shown in Fig. 5.8 and the analysis which follows refers to this simple system, indicating how a more general analysis can be made for the more realistic configuration of a real structure. For the essential system/structure shown in Fig. 5.7(a), we can construct mass and stiffness matrices for the case shere x_0 is grounded, $[M]$ and $[K]$, and these, in turn, will yield the modal properties of interest, $[\Phi]$ and $[\lambda]$, and the associated FRF characteristics, $[H(\omega)]$. If we turn next to the second model (in Fig. 5.7(b)) which relates to the base excitation configuration, we see that we can construct two different system matrices, $[M']$ and $[K']$, which have their own (different) modal and FRF properties, $[\Phi']$, $[\lambda']$ and $[H'(\omega)]$.

We can now re-write the equations of motion for this second configuration, using new coordinates, $\{y\}$ instead of $\{x\}$, where:

$$\{y\} = \{x\} - \{g\}x_0 \tag{5.33}$$

where $g_i = 1.0$ if x_i and x_0 are in the same direction.

Then, the equations of motion for the base excitation configuration can be expressed as:

$$[M]\{\ddot{y}\} + [K]\{y\} = -\ddot{x}_0\,[M]\{g\} \tag{5.34}$$

If we now describe steady forced vibration, where the excitation is provided by a harmonic notion of the base: $x_0(t) = X_0\,e^{i\omega t}$, then we obtain:

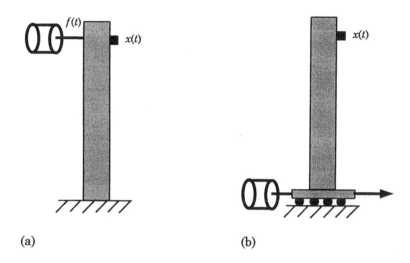

Fig. 5.7 Base excitation configuration.
(a) Conventional modal test setup; (b) Base excitation setup

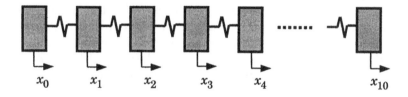

Fig. 5.8 Mathematical model for base excitation test

$$\big([K] - \omega^2\,[M]\big)\{Y\} = \omega^2\,X_0\,[M]\{g\}$$

or

$$[H(\omega)]^{-1}\,\big(\{X\} - X_0\,\{g\}\big) = \omega^2\,X_0\,[M]\{g\}$$

which can be written:

$$[H(\omega)]^{-1}\,\frac{\big(\{X\} - X_0\,\{g\}\big)}{\omega^2 X_0} = [H(\omega)]^{-1}\,\{Q\} = [M]\{g\} = \{\mu\} \tag{5.35}$$

where

$$\{Q\} = [H(\omega)]\{\mu\}$$

A typical element, $Q_i(\omega)$, in the vector $\{Q(\omega)\}$, can be related to the modal properties of the required fixed-base configuration of the test structure as follows:

$$Q_i(\omega) = \sum_j H_{ij}\mu_j = \sum_j \sum_r \frac{(\phi_{ir})(\phi_{jr})}{(\omega_r^2 - \omega^2)}\mu_j = \sum_r \frac{(\phi_{ir})d_r}{(\omega_r^2 - \omega^2)} \qquad (5.36)$$

where

$$d_r = \sum_j (\phi_{jr})\mu_j$$

Thus it is possible to determine $Q_i(\omega)$ from measurements of X_i and X_0, and if we treat $Q_i(\omega)$ as if it were an FRF, then we can extract the modal properties of natural frequency, damping factor and unscaled mode shape for each of the modes that are visible in the frequency range of measurement. It should be emphasised that the modes that are found in this way relate to the structure in its fixed-base configuration and not the moving-base condition in which the measurements are made. Special considerations must be applied for any DOFs which are not in the same direction as the base excitation, but the basis of a method for conducting modal analysis via base excitation tests has been illustrated. The fact that the excitation force is never measured is responsible for the lack of formal scaling of the mode shapes, a limitation that can be corrected by suitable calibration.

5.6   SPATIAL MODELS
It would appear from the basic orthogonality properties of the modal model that there exists a simple means of constructing a spatial model from the modal model, but this is not so. From Section 2.3 we have that:

$$[\Phi]^T [M][\Phi] = [I]$$
$$[\Phi]^T [K][\Phi] = \begin{bmatrix} \lambda_r^2 \end{bmatrix} \qquad (5.37)$$

from which it would appear that we can write:

$$[M] = [\Phi]^{-T}[\Phi]^{-1}$$
$$[K] = [\Phi]^{-T}\begin{bmatrix} \lambda_r^2 \end{bmatrix}[\Phi]^{-1} \qquad (5.38)$$

Indeed, we can, but this latter equation is only applicable when we have available the complete $N \times N$ modal model. This is seldom the case and it is much more usual to have an incomplete model in which the eigenvector matrix is rectangular and, as such, is non-invertible. Even if we constrain the number of modes to be the same as the number of DOFs (an artificial and often impractical restriction) so that the mode shape matrix is square, the mass and stiffness matrices produced by equation (5.34) are mathematical abstractions only and carry very little physical significance, unless n is almost equal to N.

One step which can be made using the incomplete data is the construction of 'pseudo' flexibility and inverse-mass matrices. This is accomplished using the above equations in the form:

$$
\begin{aligned}
[K]_{n\times n}^{-1} &= [\Phi]_{n\times m} \left[\lambda_r^2\right]_{m\times m}^{-1} [\Phi]_{m\times n}^T \\
[M]_{n\times n}^{-1} &= [\Phi]_{n\times m} [\Phi]_{m\times n}^T
\end{aligned}
\qquad (5.39)
$$

It is clear that a pair of pseudo matrices can be computed using the properties of just a single mode. Further, it can be seen that the corresponding matrices are simply the arithmetic sums of those for each mode individually. Because the rank of each pseudo matrix is less than its order, it cannot be inverted and so we are unable to construct stiffness or mass matrices from this approach.

Further discussion on the construction of spatial models may be found in the section concerned with the correlation of theory and experiment, Section 6.2.

5.7 MOBILITY SKELETONS AND SYSTEM MODELS

We have seen earlier how mobility and other FRF plots tend towards mass-like or stiffness-like behaviour at frequencies well away from resonance (and antiresonance). We have also suggested (in Section 4.1), that a 'skeleton' of mass and stiffness lines can be constructed based on the FRF curve and that this can be used to check the overall quality of the measured curve. We shall now examine these skeletons in rather more detail and show how they may be used to construct simple spatial models of a test structure.

We shall establish the basic features of the skeleton using a very simple mass-spring-mass 2DOF system, shown in Fig. 5.9(a) and for which the point mobility Y_{11} has the form shown in Fig. 5.9(b).

Certain basic features of this plot may be predicted from knowledge of the system, **without** necessarily computing the FRF in detail. These features are that:

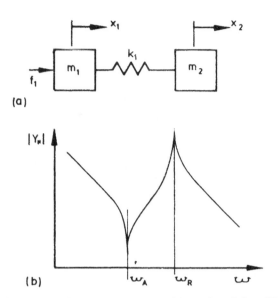

Fig. 5.9   Basic mass-spring-mass system, (a), and mobility FRF plot, (b)

(i)   there will be an antiresonance at $\omega_A^2 = k_1/m_2$ (note that this frequency is independent of the value of m_1 and would still apply if m_1 were doubled, or trebled or replaced by any combination of masses and stiffnesses);

(ii)   there will be a resonance at $\omega_r = k(1/m_1 + 1/m_2)$;

(iii) at very low frequencies ($\omega \ll \omega_A$) the mobility FRF will be dominated by the rigid body motion of the system, since it is ungrounded, and will approximate to

$$Y_{11}\left(\omega \ll \omega_A\right) \cong -\frac{i}{\omega\left(m_1 + m_2\right)} \qquad (5.40)$$

(iv)   similarly, at high frequencies, the mobility FRF will be dominated by the mass of the drive point, so

$$Y_{11}\left(\omega \gg \omega_R\right) \cong -\frac{i}{\omega m_1} \qquad (5.41)$$

Now, it is possible to draw a skeleton of mass and stiffness lines on this FRF plot, changing from stiffness to mass at each resonance and from mass back to stiffness at each antiresonance, as sketched in Fig. 5.10(a).

The question is raised as to whether, when we construct such a skeleton starting from the low frequency asymptote of the FRF curve, the final arm (above ω_r) will also be asymptotic to the mobility curve as in Fig. 5.10(b), or not (as in Fig. 5.10(a)). If it can be shown that the former case applies, then it is likely that a more general rule which, in effect, requires the skeleton to 'follow' the FRF curve. The proof of such a property may be made using the 2DOF system, and referring to Fig. 5.11(a), as follows.

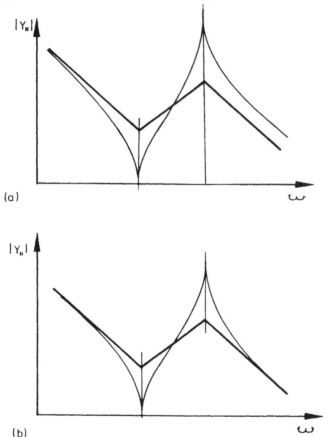

(a)

(b)

Fig. 5.10 Concept of the mobility skeleton

Suppose we define the skeleton as consisting of an initial mass line (m_1') for a mass of $(m_1 + m_2)$ plus a final mass line (m_2') corresponding to the mass m_1, connected by a stiffness line (k_1') which meets the first branch at the known antiresonance, ω_A. This skeleton satisfies the overall requirement that it 'follows' the FRF away from resonance and antiresonance but we have not yet imposed or met the condition that

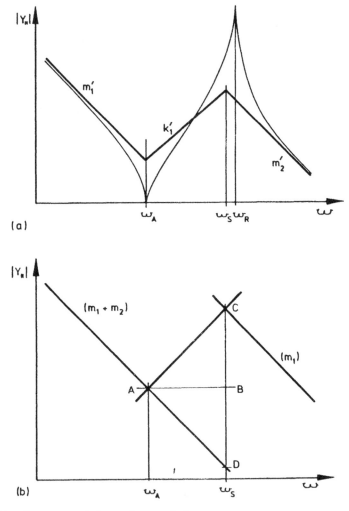

Fig. 5.11 Geometry of the mobility skeleton

the change from stiffness (k_1') to mass (m_2') occurs at the resonant frequency, ω_R. Suppose these two branches meet at ω_S. Using the geometry of the skeleton shown in Fig. 5.11(b), we see that

$$AB = BC = BD = \frac{CD}{2}$$

Now

$$AB = \beta \left(\log \omega_S - \log \omega_A \right) = \beta \log \frac{\omega_S}{\omega_A}$$

Also

$$CD = \beta \left(\log \left| Ym_2'(\omega_S) \right| - \log \left| Ym_1'(\omega_S) \right| \right)$$

$$= \beta \log \frac{\left| Ym_2'(\omega_S) \right|}{\left| Ym_1'(\omega_S) \right|} = \beta \log \frac{m_1 + m_2}{m_1}$$

So

$$2\beta \log \frac{\omega_S}{\omega_A} = \beta \log \frac{m_1 + m_2}{m_1} \tag{5.42}$$

or

$$\omega_S^2 = \omega_A^2 \frac{m_1 + m_2}{m_1} = \frac{k(m_1 + m_2)}{m_1 m_2}$$

$$= \omega_R^2 \tag{5.43}$$

Hence, the skeleton connecting the extreme asymptotic behaviour of the FRF changes from mass-like to stiffness-like and back to mass-like elements at antiresonance and resonance, respectively.

This basic idea can be extended to more complex systems and the general rule for constructing skeletons is that the first will be mass-like (slope = −1 on mobility or alternate values for other FRF forms) or stiffness-like (mobility slope = +1) depending upon whether the structure is freely supported or grounded, respectively. Thereafter, the slope of the skeleton changes by +2 at each antiresonance and by −2 at each resonance. Thus, for a point mobility the skeleton branches are all of slope +1 or −1 but for a transfer mobility the slopes will be +1, −1, −3, −5, ... and so on as the absence of an antiresonance between two resonances will cause a general downward drift of the skeleton (and, of course, of the FRF curve itself). By way of example, two of the sets of FRF curves shown in Fig. 2.32 are repeated in Fig. 5.12 together with their skeletons.

It will be noted from the first example above, Fig. 5.9(a), that the physical system has three components, two masses and a spring (m_1,

410

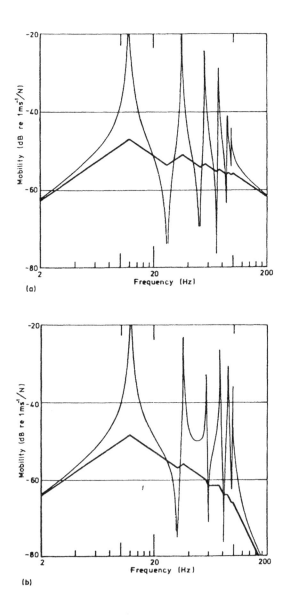

(a)

(b)

Fig. 5.12 Mobility skeletons for 6DOF system

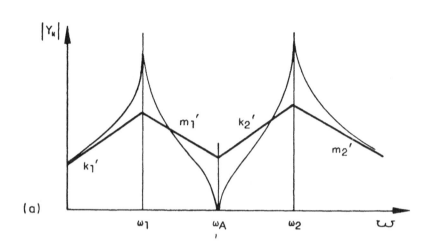

(a)

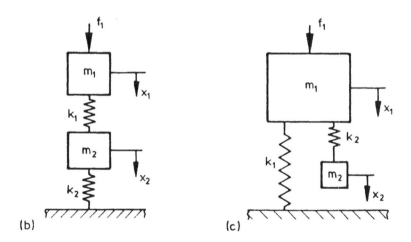

Fig. 5.13 Mobility skeletons for 2DOF system.
(a) Mobility curve and skeleton; (b) Possible system configuration; (c) Possible system configuration

m_2 and k_1) and that the corresponding skeleton has two mass lines and one spring line (m_1', m_2' and k_1'). While it must be acknowledged that there is no direct correspondence between the two similar sets of parameters, they are related and each set may be derived from the other. Indeed, this is a general rule that the skeleton for the point FRF contains just as many mass and stiffness links as there are corresponding elements in the physical system. This fact provides us with a mechanism for deriving spatial models from measured data.

It must be observed at the outset that it is left to the user to decide upon the configuration of the suitable model: analysis of the skeleton will then furnish the values for the model parameters. Consider the FRF indicated in Fig. 5.13(a).

Clearly, this relates to a 2DOF system such as that shown in Fig. 5.13(b)) or the one shown in Fig. 5.13(c).

In the first case, it can be shown that the model parameters may be determined from the skeleton parameters using the following formulae:

$$m_1 = m_2'$$

$$k_1 = m_2'\left(\omega_1^2 + \omega_2^2 - \omega_A^2\right) = m_2'\left(\frac{k_1'}{m_1'} + \frac{k_2'}{m_2'} - \frac{k_2'}{m_1'}\right)$$

$$k_2 = (k_1)\frac{k_1'}{(k_1) - k_1'} \tag{5.44}$$

$$m_2 = (k_1) + \frac{(k_2)}{\omega_A^2}$$

Alternatively, if the second configuration was the correct one, then the model parameters would be

$$k_1 = k_1'$$

$$m_1 = m_2'$$

$$k_2 = (m_2)\left(\omega_1^2 + \omega_2^2 - \omega_A^2\right) - (k_1) \tag{5.45}$$

$$m_2 = \frac{(k_2)}{\omega_A^2}$$

Hence the solution obtained is not unique and additional data would be required in the example above in order to establish which of the two configurations was the more representative. The additional data could be provided by other FRF plots. In his book, Salter [3] develops the skeleton idea in greater detail, presenting a useful additional tool to the modal analyst.

CHAPTER 6

Applications

6.1 INTRODUCTION

In this chapter, we turn our attention to the destination of the final results from conducting a modal test: namely, the problem which we set out to resolve. As mentioned in the first chapter, there are many possible application areas for modal tests and the major ones were described there. We shall now consider these in turn and examine in more detail some of the specific procedures and methods which are available for each. As already mentioned in the introduction to this book, many of the methods and procedures available in modal testing and analysis have become extensive subjects in their own right, and this is particularly true in some of the application areas. As a result, it is not feasible to include in a chapter such as this details or even mention of all the many different methods approaches and variants that have been explored and reported in the past 15 years. Hence, this chapter seeks to guide the reader through some of the more established methods and to guide him or her towards the more advanced ones through a judicious selection of references.

It should be noted at the outset, however, that the application of the techniques described below to practical engineering structures is often found to be more difficult or more onerous than at first expected. This is due to many factors, not least the considerable volume of data usually required for real cases (by comparison with that used in simpler illustrative examples) and the inevitable incompleteness of the data which can be acquired under typical practical testing conditions and constraints. Nevertheless, this 'fact of life' should not be permitted to provide a deterrent to the ambitious or the tentative application of any of the procedures listed below. The techniques can be used with great effect, especially if the user is fully aware of the extent or limitations of his of her data. Furthermore, dramatic advances are still being made in measurement and analysis techniques which will reduce the limitations and enhance both the availability and quality of good data, thus enabling the modal analyst to make even more precise and confident assessments of structural dynamic behaviour.

6.2   COMPARISON AND CORRELATION OF EXPERIMENT AND PREDICTION

6.2.1   Different Methods of Comparison and Correlation

Probably the single most popular application of modal testing is to provide a comparison between predictions for the dynamic behaviour of a structure and those actually observed in practice. Sometimes this process is referred to as 'validating' a theoretical model, although to do this effectively several steps must be taken. The first of these steps is to make a direct and objective **comparison** of specific dynamic properties, measured vs. predicted. The second (or, perhaps, still part of the first) is **correlation**, which means to quantify the extent of the differences (or similarities) between the two sets of data. Then, the third step is to identify or to **locate** the sources of any discrepancies between the two models and the final step is make **adjustments** or modifications to one or other set of results in order to bring them closer into line with each other. These last two steps are often referred to as '**updating**', although '**reconciliation**' is perhaps a better description. When this is achieved, the theoretical model can be said to have been **validated** and is then fit to be used for further analysis. In this section we shall be concerned with the first and second of these steps, dealing with the third and fourth steps in the next section.

In most cases, a great deal of effort and expense goes into the processes which lead to the production of an experimentally-derived model on the one hand (subsequently referred to as the 'experimental' model or data) and a theoretically-derived (or 'analytical' or 'predicted') model on the other. This being so, it is appropriate to make as many different types or levels of comparison between the two sets of data as possible. As discussed much earlier in the work, we have identified three types of dynamic model, loosely called 'Spatial', 'Modal' and 'Response' models. It is now convenient to return to this classification and to try to make comparisons between experiment and prediction for each (or at least more than one) of these types of model. Thus we shall discuss comparisons of response characteristics and of modal properties, as both of these provide many opportunities for useful correlation between experiment and theory. Comparisons of spatial properties are more difficult, however, and we shall leave discussion of this aspect until the next Section (6.3).

Whichever medium is used for comparison and correlation purposes, either one or the other model will have to be developed fairly extensively from its original form and what is the most convenient format for one case will often be the least accessible for the other. This situation derives from the different routes taken by theoretical and experimental approaches to structural vibration analysis, as shown in Figs. 2.1 and 2.2. However, in closing these general remarks, it is appropriate to reinforce the recommendation to make as many different types of comparison as possible and not just to rely on one, usually the first one that comes to hand, or mind.

One further comment, and concept, which should be introduced here (although it will be discussed in more detail in subsequent sections) is that of **verifying** the models to be compared. The concept of a verified model is different to that of a validated model in the following respects: a model can be said to be **verified** if it contains the correct features, most importantly the appropriate number and choice of DOFs, to render it capable of representing the dynamic behaviour of the structure; a model is said to be **valid** if the coefficients in that model are such as to provide an acceptable representation of the actual behaviour; 'acceptable' being to some extent a matter of judgement and certainly to be determined afresh for each individual application. It will be seen that a model can only be validated after it has been verified. This means that we should not embark on lengthy comparison or correlation procedures unless we are first satisfied that the two models to be used are compatible with each other, and their intended roles.

6.2.2   Comparison of Modal Properties
While there is no compelling reason for choosing one rather than the other, we shall start our comparison procedures with those based on modal data and follow with those which use response properties. Although the response data are those most directly available from test for comparison purposes, some theoretical analysis packages are less than convenient when it comes to predicting FRF plots. This is largely because of the requirement that all (or at least a large proportion) of the modes must be included in the calculation of a response characteristic. By contrast, modal properties can be predicted individually and comparisons can be confined to specific frequencies or to specific frequency ranges with much greater facility for the analyst. Nevertheless, such a comparison of modal properties does place additional demands on the experimental route as it requires the measured data to have been subjected to a modal analysis or curve-fitting procedure in order to extract the corresponding modal properties from the test. In spite of this requirement, comparisons of modal properties are perhaps the most common and we shall now describe a number of methods which may be employed to that end.

6.2.2.1   Comparisons of natural frequencies
The most obvious comparison to make is of the measured vs. the predicted natural frequencies. This is often done by a simple tabulation

418

of the two sets of results but a more useful format is by plotting the experimental value against the predicted one for each of the modes included in the comparison, as shown in Fig. 6.1(a). In this way it is possible to see not only the degree of correlation between the two sets of results, but also the nature (and possible cause) of any discrepancies which do exist. It is important to stress, however, that the points plotted in this way must be of the measured and predicted natural frequencies of corresponding (or 'correlated') modes: it is not sufficient simply to plot the first, second, third, ... measured natural frequencies against the first, second, third ... predicted values as there is no guarantee that the first three measured modes correspond one-for-one with their predicted counterparts. Some positive identification of each measured mode with its predicted counterpart is essential, to provide a set of Correlated Mode Pairs (CMPs), and in order to achieve this, recourse must usually be made to the mode shape correlation methods discussed in the next section. An example of such a situation is shown in Figs. 6.1(b) and (c), where the first of these graphs shows the natural frequencies plotted simply in ascending order, while the second shows them correctly paired (using information about their modes shapes — not shown here) with the poorer level of correlation which actually exists in this case.

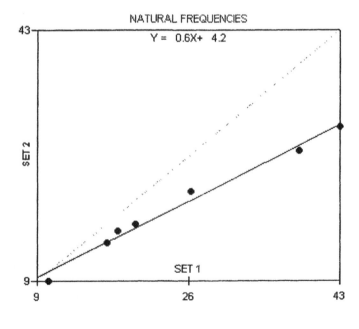

(a)

Fig. 6.1 Measured and predicted natural frequencies.
(a) General

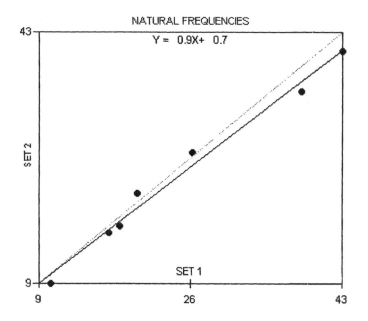

(b)

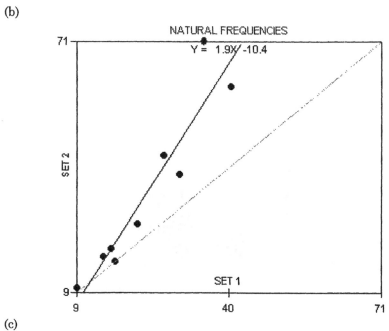

(c)

Fig. 6.1   Measured and predicted natural frequencies.
(b) Ordered frequencies; (c) Correlated mode pairs

Once sorted, the points should lie on or close to a straight line of slope 1. If they lie close to a line of a different slope then almost certainly the cause of the discrepancy is an erroneous material property used in the predictions. If the points lie scattered widely about the 45° straight line then there is probably a serious failure of the model to represent the test structure and a fundamental re-evaluation is called for. If the scatter is small and randomly distributed about a 45° line then this may be expected from a normal modelling and measurement process. However, a case of particular interest is where the points deviate slightly from the ideal line but in a systematic rather than a random fashion as this situation suggests that there is a specific characteristic responsible for the deviation and that this cannot simply be attributed to experimental errors.

There is an inclination to quantify the deviation of the plotted points from the ideal straight line as a means of assessing the quality of the comparison in a single correlation factor. Although this is indeed useful, it cannot replace the benefit gained from the plot itself as (without employing complicated functions) it is generally insensitive to the randomness or otherwise of the deviations and this is an important feature.

Another possible form of plotting these same data is provided by the Natural Frequency Difference (NFD) diagram — as shown in Fig. 6.2. This is a table which plots simply the natural frequency difference between all possible combinations of experimental and analytical model modes. This can be used in the automatic selection of correlated mode pairs which is a feature in more advanced correlation packages.

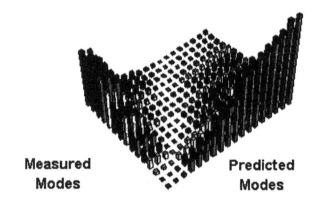

Fig. 6.2   Natural Frequency Difference (NFD) diagram

6.2.2.2 Comparisons of mode shapes — graphical

When the above procedure is applied in practical cases, it will often be found more difficult than first anticipated because of the problems of matching the experimental modes with their analytical counterparts. Whereas on simple structures with well-separated modes this pairing often presents no difficulty, on more complex structures — especially ones with closely-spaced natural frequencies — ensuring that the correlated mode pairs are correctly identified becomes more difficult and requires the additional information in each case of the mode shape as well as the natural frequency. Hence it is appropriate to make comparisons of mode shapes at the same time as those of natural frequencies.

In this case, we have rather more data to handle for each mode and one possible way of performing the comparison is by plotting the deformed shape for each model — experimental and predicted — along the lines described in Chapter 5, and overlaying one plot on the other. The disadvantage of this approach is that although differences are shown up, they are difficult to interpret and often the resulting plots become very confusing because there is so much information included. An alternative plot comprising a single picture which is a display of the mode shape difference can also be difficult to interpret. A more convenient approach is available by making an x-y plot along similar lines to that used for the natural frequencies in which each element in the mode shape vector is plotted, experimental vs. predicted, on an x-y plot such as is shown in Fig. 6.3. The individual points on this plot relate to specific DOFs on the model and it is to be hoped that they should lie close to a straight line passing through the origin. If, as is often the case, both sets of mode shape data consist of mass-normalised eigenvectors, then the straight line to which the points should be close will have a slope of ± 1. Once again, the pattern of any deviation from this requirement can indicate quite clearly the cause of the discrepancy: if the points lie close to a straight line of slope other than ± 1, then either one or other mode shape is not mass-normalised or there is some other form of scaling error in the data. If the points are widely scattered about a line, then there is considerable inaccuracy in one or other set and if the scatter is excessive, then it may be the case that the two eigenvectors whose elements are being compared do not relate to the same mode.

This form of presentation has particular value when the deviations of the points from the expected line are systematic in some way, such as is the case in Fig. 6.3(b). In this event it can be useful to superimpose the plots for several modes so that the basis of the comparison is broadened, and this has been done in Fig. 6.3(b) for the first three modes of the structure. We now see that three of the points on the structure (4, 5 and 6) systematically produce a poor correlation between experiment and prediction although we are not yet in a position to identify which set are in error. From Fig. 6.3(b) it is clear that most of the points for DOFs 4, 5 and 6 do in fact lie close to a straight line but one with a slope considerably different from 45°. If the discrepancy is due to poor analytical modelling (the natural assumption of the experimentalist!), then it might reasonably be expected to differ in extent from one mode to the next. However, this is not the case here and it can be seen that the deviations are consistent with the result which would follow from an incorrect scaling factor on the measured FRF plots pertaining to points 4, 5 and 6 (since all modes would be equally affected by such an error). A repeat of the measurement (and modal analysis) phase in this case, together with the inclusion of some additional coordinates, resulted in the revised plot shown in Fig. 6.3(c): clearly, a much more satisfactory comparison and one achieved using the original analytical model.

At this juncture, it should be observed that the above method assumes implicitly that the mode shapes in both cases are real (as opposed to complex) and while it is highly likely that the results from a theoretical analysis will indeed comply with this assumption, those from an experimental source will, in general, not be so simple. Although it is possible to envisage a complex version of the type of plot discussed above, by using a third axis to display the imaginary part of the complete eigenvector elements, this is not recommended as it tends to disguise the essential conflict which is inherent in comparing complex (experimental) data with real (predicted) values. It is necessary to make a conscious decision on how to handle this particular problem and that usually adopted is to 'whitewash' the measured data by taking the magnitude of each eigenvector element together with a + or − sign, depending on the proximity of the phase angle to 0° or 180°. In many cases this is adequate but it is not satisfactory for highly complex modes: no form of direct graphical comparison between these modes and the real data produced by a typically undamped theoretical model is likely to be effective. In such cases, it becomes necessary to employ one of the complex-to-real transformation or 'realisation' procedures which were discussed in Section 5.2, or to rely more on the numerical correlations that are described next.

6.2.2.3 Comparison of mode shapes — numerical correlation (MAC)

Several workers have developed techniques for quantifying the comparison between measured and predicted mode shapes (in fact, these methods are useful for all sorts of comparisons — not just experiment vs. theory — and can be used for comparing any pair of

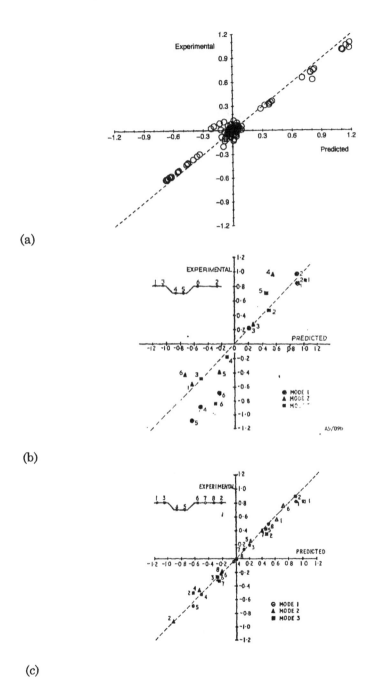

(a)

(b)

(c)

Fig. 6.3 Measured and predicted mode shape vectors.
(a) General case; (b) Systematic errors; (c) Corrected data

mode shape estimates). As an alternative to the above graphical approach, we can compute some simple statistical properties for a pair of modes under scrutiny. The formulae given below assume that the mode shape data may be complex, and are based on a comparison between an experimentally-measured mode shape, $\{\psi_X\}$, and a theoretically-predicted or analytical one, $\{\psi_A\}$.

The first formula is for a quantity sometimes referred to as the 'Modal Scale Factor' (MSF) and it represents the 'slope' of the best straight line through the points as plotted in Fig. 6.3. This quantity is defined as:

$$MSF(X,A) = \frac{\displaystyle\sum_{j=1}^{n}(\psi_X)_j(\psi_A)_j^*}{\displaystyle\sum_{j=1}^{n}(\psi_A)_j(\psi_A)_j^*} \tag{6.1a}$$

where n is the number of DOFs for which both A and X data are available, and there are two possible expressions relating the two mode shapes, depending upon which is taken as the reference one:

$$MSF(A,X) = \frac{\displaystyle\sum_{j=1}^{n}(\psi_A)_j(\psi_X)_j^*}{\displaystyle\sum_{j=1}^{n}(\psi_X)_j(\psi_X)_j^*} \tag{6.1b}$$

It should be noted that this parameter gives no indication as to the quality of the fit of the points to the straight line; simply its slope.

The second parameter is referred to as the 'Mode Shape Correlation Coefficient' (MSCC) or, more-popularly, as the 'Modal Assurance Criterion' (MAC) and this provides a measure of the least-squares deviation or 'scatter' of the points from the straight line correlation. This parameter is defined by:

$$MAC(A,X) = \frac{\left|\displaystyle\sum_{j=1}^{n}(\psi_X)_j(\psi_A)_j^*\right|^2}{\left(\displaystyle\sum_{j=1}^{n}(\psi_X)_j(\psi_X)_j^*\right)\left(\displaystyle\sum_{j=1}^{n}(\psi_A)_j(\psi_A)_j^*\right)} \tag{6.2a}$$

or

$$MAC(A,X) = \frac{\left|\{\psi_X\}^T\{\psi_A\}\right|^2}{\left(\{\psi_X\}^T\{\psi_X\}\right)\left(\{\psi_A\}^T\{\psi_A\}\right)}$$

(6.2b)

and is clearly a scalar quantity, even if the mode shape data are complex. In the same way that the Modal Scale Factor does not indicate the degree of correlation, neither does the Modal Assurance Criterion discriminate between random scatter being responsible for the deviations or systematic deviations, as described earlier. Thus, whereas these parameters are useful means of quantifying the degree of correlation between two sets of mode shape data, they do not present the whole picture and should preferably be considered in conjunction with the plots of the form shown in Fig. 6.3. (The close similarity between the MAC and the coherence function used in signal processing, Chapter 3, should be noted here.)

It is worth considering two special cases: (i) that where the two mode shapes are identical and (ii) where they differ by a simple scalar multiplier. Thus in case (i), we have:

$$\{\psi_X\} \equiv \{\psi_A\}$$

for which it can be seen that

$$MSF(X,A) = MSF(A,X) = 1$$

and also that:

$$MAC(X,A) = 1$$

In the second case, (ii), we have $\{\psi_X\} = \gamma\{\psi_A\}$ and we find that

$$MSF(X,A) = \gamma \quad \text{while} \quad MSF(A,X) = \frac{1}{\gamma}$$

although, since the two modes are still perfectly correlated, we still have:

$$MAC(X,A) = 1$$

426

In practice, typical data will be less ideal than this and what is expected is that if the experimental and theoretical mode shapes used are in fact from the same mode, then a value of the Assurance Criterion of close to 1.0 is expected, whereas if they actually relate to two different modes, then a value close to 0.0 should be obtained. Given a set of m_X experimental modes and a set of m_A predicted modes, we can compute a set of $m_X \times m_A$ Modal Assurance Criteria and present these in a matrix which should indicate clearly which experimental mode relates to which predicted one. Such a table is shown in Fig. 6.4(a), together with some of the common graphical presentations used to display these data, in Figs. 6.4(b) and (c). It is difficult to prescribe precise values which the Assurance Criterion should take in order to guarantee good results. Generally, it is found that a value in excess of 0.9 (or 90%) should be attained for well-correlated modes and a value of less than 0.1 (or 10%) for uncorrelated modes. In some situations, the boundaries for 'acceptable' and 'non' correlation are quoted as above 80 per cent and less then 20 per cent, respectively. However, the significance of these quantities depends considerably on the specific data points used in the correlation (see below) and on the subsequent use planned for the model — some are much more demanding than others — and so considerable caution should be used in attaching quantitative significance to the absolute values of MAC obtained in practical cases. The greatest value of these coefficients lies in their use for comparison purposes.

Analytical mode number	Experimental mode number									
	1	2	3	4	5	6	7	8	9	10
1	100	0	1	0	0	0	0	0	0	0
2	0	100	1	1	0	0	0	0	0	0
3	0	1	94	3	2	0	0	0	0	0
4	0	0	2	92	5	3	0	0	0	0
5	0	0	0	4	86	7	4	0	0	0
6	0	0	0	0	7	81	9	5	0	0
7	0	0	0	0	0	10	75	10	5	0
8	0	0	0	0	0	0	12	71	11	5
9	0	0	0	0	0	0	0	14	68	11
10	0	0	0	0	0	0	0	0	16	65

MODAL ASSURANCE CRITERION (MAC) %

(a)

Fig. 6.4 Common presentations of MAC properties.
(a) Tabular

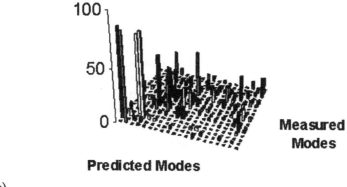

Predicted Modes

(b)

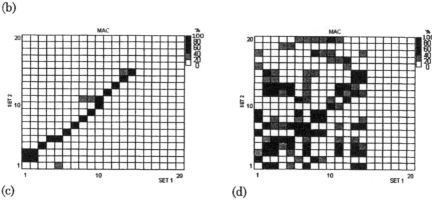

(c)   (d)

Fig. 6.4   Common presentations of MAC properties.
(b) Chart; (c) Isometric; (d) Reduced data

It is worth noting some of the causes of less-than-perfect results
from these calculations. Besides the obvious reason — that the model is
incorrect — values of the MAC of less than the expected value of unity
(or other elements being noticeably greater than the expected zero) can
be caused by:

(i)   non-linearities in the test structure;
(ii)   noise on the measured data;
(iii)   poor modal analysis of the measured data; and
(iv)   inappropriate choice of DOFs included in the correlation.

6.2.2.4 Features of the MAC
AutoMAC

In view of the widespread use of the MAC, and of the risk of its misinterpretation, it is appropriate to include here a fuller discussion of some of the features of this approach.

The first consideration concerns the choice of DOFs which are included in the calculation. It is clear that if all the degrees of freedom in the model are included then a very accurate measure of the correlation between the two vectors will result. However, it is also clear that a different result will be obtained for the MAC if only a small fraction of the full set of DOFs are included. A second set of results from the case shown above are now presented in Fig. 6.4(d), this time using only about 30 per cent of the DOFs which were included in the earlier result. From this new calculation it is immediately clear that a much more confusing result has been obtained, especially in respect of the purpose of the table in identifying the correlated mode pairs: there are several cases where a given analytical mode appears to correlate equally well with several of the experimental modes. The problem we have encountered here is a spatial version of the time-signal-processing problem of aliasing which was discussed in Chapter 3. Simply put, the problem here is that there are insufficient data points (measured DOFs) for us to be able to discriminate between the different modes. The solution is equally simple: it is necessary to include more DOFs in the correlation, and that is what was done in the first version of this example, shown in Fig. 6.4(c).

In practical terms, there is some difficulty in deciding how many, or more precisely, which DOFs need to be included in order to avoid the spatial aliasing problem. A detailed procedure for answering this question is given later, in Section 6.6, but here it will suffice to show how the MAC can be used to check whether a given selection of DOFs is adequate, or not. This is done using a version of the MAC called the AutoMAC in which a set of mode shape vectors are correlated *with themselves*. If, for example, we take the mode shape vectors for the analytical model but defined only at the DOFs which are to be used in the correlation with the experimental model (i.e. those DOFs which are included in the modal test) and compute the MAC table with themselves, we produce results such as those shown in Figs. 6.5(a) and (b). In the first of these, (a), a 'full' set of DOFs is included while in the second and third ones, (b) and (c), only a reduced subset that were used in the above example are included. From these AutoMAC plots we observe a number of features: (i) all the diagonal values are identically unity — they must be 100 per cent, by definition, because each mode shape must correlate perfectly with itself; (ii) the AutoMAC matrix is symmetric and (iii) there are a number of non-zero off-diagonal terms,

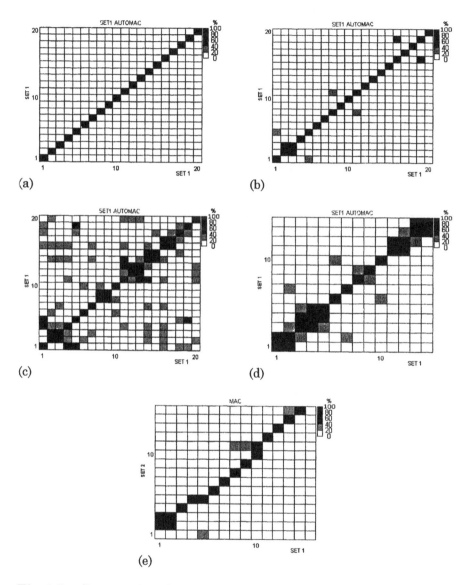

Fig. 6.5 Presentation of MAC and AutoMAC data.
(a) AutoMAC$_A$ for full set of 102 DOFs; (b) AutoMAC$_A$ for reduced set of 72 DOFs; (c) AutoMAC$_A$ for reduced set of 30 selected DOFs; (d) AutoMAC$_X$ for reduced set of 30 selected DOFs; (e) MAC for reduced set of 30 DOFs

which means that some of the modes appear to exhibit a degree of correlation with others, a result which is not immediately expected, since the modes are supposed to be 'orthogonal' to each other (see Chapter 2). However, there are two reasons why this orthogonality property does not translate to a perfectly diagonal AutoMAC matrix. First, is because the orthogonality property is only strictly applicable when the mass matrix is used (see equation (2.32)), and second, because the orthogonality condition is only applicable when all the DOFs are included in the calculation. We shall return to the question of the mass matrix later, but for the moment our interest is focused on the matter of the selection of the DOFs that are or should be included. In the limit, it can be seen that if we only define each mode shape by the amplitudes at just two DOFs, then most modes would look very similar to each other and, indeed, if we did a formal correlation, would be found to be highly correlated with roughly half of all the modes included in the process. It is clearly necessary to include sufficient DOFs to ensure the effective discrimination between the various modes. In fact, it is necessary to include in the selection of measured DOFs only those which are required to ensure that the eigenvector submatrix $[\Psi_1]_{n \times m}$, which is formed from:

$$
\begin{bmatrix}
[\Psi_1]_{n \times m} & [\Psi_3]_{n \times (N-m)} \\
[\Psi_2]_{(N-n) \times m} & [\Psi_4]_{(N-n) \times (N-m)}
\end{bmatrix}
$$

where n represents the included DOFs, and m the measured modes, is non-singular. As shown in Section 6.6.2, achieving this condition is not a trivial matter, but the degree to which a given choice of DOFs satisfies it can be readily demonstrated by using the AutoMAC. Figs. 6.5(c), (d) and (e) show the suitability of the selected DOFs for an industrial structure, first on the FE model results and then on the experimental data themselves, both results confirming that the relatively small number of DOFs included in the correlation are suitable for the task of matching the correlated mode pairs.

Normalised MAC

Reference was made in the previous section to the absence of the mass matrix in the MAC calculation, an absence which means that the MAC is not a true orthogonality check. It is possible to remedy this limitation by including information which may be available on the mass of the system or, equally, of its stiffness, but to do so is relatively expensive, and constitutes a significant extension to the effort required to perform these checks. Bearing in mind the essentially comparative nature of the MAC coefficients, this extra effort is seldom warranted but in more advanced cases, including those where an automatic correlation procedure is sought, and where the numerical values of the correlation coefficients are likely to be used in subsequent stages of validation, the extension of the concept to the mass-normalised version may be considered. The formula for this version of the MAC, sometimes referred to as the Normalised Cross Orthogonality (NCO), is given by:

$$NCO(A,X) = \frac{\left|\{\psi_X\}^T [W]\{\psi_A\}\right|^2}{\left(\{\psi_X\}^T [W]\{\psi_X\}\right)\left(\{\psi_A\}^T [W]\{\psi_A\}\right)} \tag{6.3}$$

where the weighting matrix, $[W]$, can be provided either by the mass or stiffness matrices of the system.

The main difficulty to be overcome, even in those cases where a full mass and/or stiffness matrix is available from the analytical model, is that of reducing or condensing this mass matrix to the order of the specific DOFs for which data are available. A Guyan-type or equivalent reduction must be made if the mass matrix is to be used explicitly. One of the more practical approaches uses the SEREP-based reduction process discussed in Section 5.3. In this approach, a pseudo-mass matrix of the correct size is computed from the simple formula:

$$\left[M^R\right] = [\Psi]^{+T} [\Psi]^{+} \tag{6.4}$$

using either the limited measured eigenvectors or the corresponding analytical ones (preferred because of their greater accuracy). This pseudo-mass matrix can then be used in the NCO calculation as a weighting matrix and a readily-accessible version of a Normalised MAC — sometimes referred to as the SEREP-Cross-Orthogonality (SCO) coefficient — is thereby computed:

$$SCO(A,X) = \frac{\left|\{\psi_X\}^T [\Psi]^{+T} [\Psi]^{+} \{\psi_A\}\right|^2}{\left(\{\psi_X\}^T [\Psi]^{+T} [\Psi]^{+} \{\psi_X\}\right)\left(\{\psi_A\}^T [\Psi]^{+T} [\Psi]^{+} \{\psi_A\}\right)} \tag{6.5}$$

Examples of both AutoMAC and of AutoSCO are shown in Figs. 6.6(a) and (b).

Improved MAC (IMAC)
Sometimes, the MAC values computed according to the above formulae belie the level of correlation which is evident from a visual inspection of the actual mode shapes, particularly when viewed in animation. In these situations, the reason is usually to be found in the specific selection of DOFs which have been included in the correlation

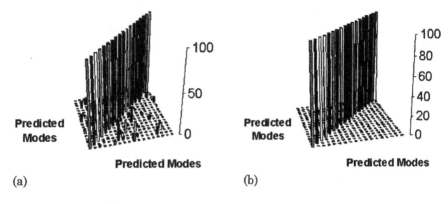

(a)   (b)

Fig. 6.6   Weighted MAC function.
(a) Unweighted AutoMAC; (b) Mass-weighted AutoMAC

calculation. For a variety of reasons, important individual DOFs may be omitted, perhaps because they were difficult to measure or, simply, were unmeasured. If, amongst these omitted DOFs, were some with the largest amplitudes of the true mode shape, then the remaining and included DOFs might well represent only the lesser motions of the mode, and may well constitute the measurement 'noise' on that mode shape, yielding apparently poor correlations as a result. For this and other related reasons, there have been proposals for 'enhancements' to the standard MAC calculation by selectively eliminating those DOFs from the set used in the calculation which contribute most to a reduction in the MAC value. Such a procedure can be dangerous as it can be used to serve the purpose of confirming a pre-judged result. However, if it is used to eliminate inaccurate data from the correlation process, then it can serve a valid and useful role. Fig. 6.7 presents an illustration of the application of this concept to an industrial structure.

There are one or two other related issues which concern the question of choice of DOFs. It must be remembered that the accuracy of amplitude measurements made with attached accelerometers can be subject to considerable errors in cases where the motion in directions perpendicular to the axis of actual measurement is considerably greater than that being measured. In these circumstances, errors of 100 per cent or more in the (small) amplitudes being recorded are common, and can easily contaminate correlation calculations as a result. Also, the difference between the units used in translational and rotational DOFs means that if both types of response are included in such a calculation, then one or other of these two sets of DOFs will be weighted quite differently to the other. Thirdly, there is the important matter of the exactness with which the two sets of DOFs match each other. The

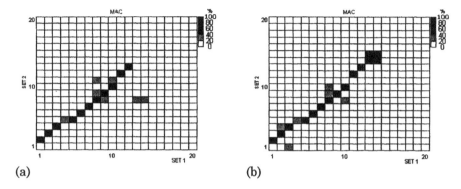

Fig. 6.7 Improved MAC function.
(a) Unimproved original (140 DOFs); (b) Improved (110 DOFs —
30 DOFs deleted)

precise location and/or orientation of the measurement sites in relation
to the FE model node points can also be a critical feature. It will be seen
in the next section just how relatively slight discrepancies in the
location of a DOF can influence the numerical value of the recorded
mode shape amplitude, and of the resulting correlation coefficients.

These, and other, considerations show how sensitive the correlation
calculations can be to the choice and accuracy of the mode shape
amplitude data which are used in the process.

Frequency-scaled MAC (FMAC)
It is often found that it is necessary to examine several comparison
plots in order to construct a comprehensive picture of the full extent of
the correlation between two sets of modal properties. Certainly, it is
necessary to examine the natural frequency comparison plot as well as
the simple MAC table, and usually it is helpful if the AutoMAC plot is
taken into consideration as well. It is possible to combine all three of
these presentations into a single plot, referred to here as the FMAC and
illustrated by the example shown in Fig. 6.8 (from [54]). In this
diagram, the mode number scales of the standard MAC have been
replaced by natural frequency scales so that proximity or distance of
natural frequencies of adjacent modes is immediately apparent.
Further, the diameter of each circle indicates the extent of the
correlation for that pair of modes. The background (grey) plot is for the
AutoMAC based on the reference model and using the same degrees of
freedom as used in the 'active' comparison. The bolder black points
comprise the *X-A* comparisons, showing the frequency correlation and,
by the circle diameters, the shape correlation simultaneously while the

434

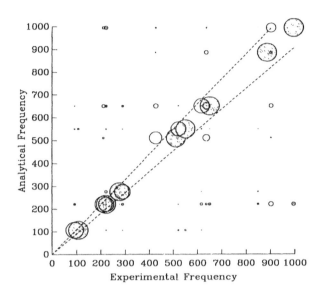

Fig. 6.8   Frequency-scaled MAC (FMAC)

lighter, grey, points relate to the AutoMAC plots for the analytical data
set. With this type of diagram it is possible to draw together all the
disparate features that are necessary to make a judgement about the
degree of correlation between the two models. It is immediately clear,
for example, whether or not the choice of measured DOFs is such that
spatial aliasing might be encountered. This provides valuable and
immediate guidance to the interpretation of significant off-diagonal
MAC values.

6.2.2.5   COMAC
The preceding paragraphs have all been concerned with the influence
on the correlation process of the various DOFs which are included in
the calculations. While these DOFs do not appear explicitly in any of
the MAC coefficients, their importance is evident by comparison of the
values produced using different selections of DOFs. Clearly, there is a
spatial dependence of the correlation parameters and our goal in the
present section is to seek a way of expressing that dependence directly,
so that a measure of the degree of correlation is presented as a function
of the individual DOFs. This goal can be realised by rearranging the
order in which the correlation calculations are performed and by
defining a quantity called the 'Coordinate MAC' or COMAC.

    In the calculation of the MAC between two vectors, a summation is
made over all the DOFs included, resulting in a single coefficient for

that pair of modes. The first step in the calculation of the COMAC is to preserve the individual elements in that summation, noting that each refers to one particular DOF, as illustrated schematically in Fig. 6.9(a). If we then take another pair of vectors, or modes, from the same two sets and repeat this step, we arrive at a second set of individual terms, relating to the same set of DOFs, and so on for as many mode pairs as we choose to include (again, see Fig. 6.9(a)). If we restrict the pairs of modes thus included to the already-identified correlated mode pairs, then the data we have gathered in this way contain information about the quality of the correlation between properly-matched vectors and so we can use them to define this correlation in more detail. In effect, the MAC value for each of the selected mode pairs is obtained by summing the contributions along one row in the table in Fig. 6.9(a), while a summation down each individual column yields information about the degree of correlation observed for that individual DOF. Suitably normalised to present a value between 0 and 1, the COMAC parameter for an individual DOF, i, is expressed as:

$$COMAC(i) = \frac{\sum_{l=1}^{L} |(\psi_X)_{il} (\psi_A)_{il}|^2}{\sum_{l=1}^{L} (\psi_X)_{il}^2 \cdot \sum_{l=1}^{L} (\psi_A)_{il}^2} \qquad (6.6)$$

Here, l represents an individual correlated mode pair, of which a total of L are available, where L may well be less than the total number of modes in both sets, m_A for the analytical model or an m_X being the total number of experimentally-determined modes.

The COMAC can be displayed in different ways, the most obvious being simply a diagram of its value against the DOF number, as shown in Fig. 6.9(b). Alternatively, it is possible to use a display of the actual structure as the basis for a diagram such as that shown in Fig. 6.9(c), which is a much more graphic illustration of the result.

As with many of these parameters, the correct interpretation can be difficult to make. One is tempted to conclude that regions of the structure which show up as having (relatively) low values of COMAC are those regions which harbour the discrepancies that are responsible for the differences observed between the two models. This is seldom the case, for the simple reason that regions of low COMAC correspond to regions where the *consequences* of any discrepancies between the two models are felt, rather than where they are actually located. Thus, large reductions in COMAC are often observed at regions of large amplitude, such as at the free ends of beams, where the effects of inaccurate

CMP No.	i (Co-ordinate Number)				MAC Value
L	1	2	3	n	
1	MAC $(1, 1)_1$ + MAC $(1, 1)_2$ + MAC $(1, 1)_3$ + ... MAC $(1, 1)_m$				= MAC (1, 1)
2	MAC $(2, 3)_1$ + MAC $(2, 3)_2$ + ...				= MAC (2, 3)
3	MAC $(3, 4)_1$ + MAC $(3, 4)_2$ + ...				= MAC (3, 4)
4	MAC $(4, 5)_1$ + ...				= -
5	+				= -
.	+				= -
L	MAC $(p, q)_1$ +				= MAC (p, q)
	COMAC (1) .... COMAC (n)				

(a)

(b)

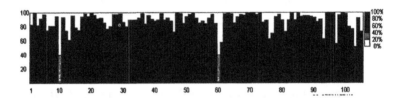

(c)

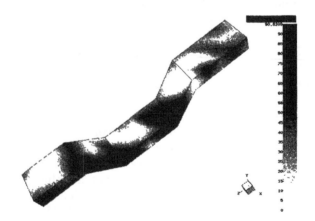

Fig. 6.9   Coordinate MAC (CoMAC).
(a) Constructional MAC and CoMAC properties; (b) CoMAC diagram; (c) CoMAC contour plot

flexibility data in other parts of the structure are most dramatically felt. Care and ingenuity must be exercised in making such interpretations but the fact remains that the existence of systematic patterns of COMAC values almost always indicate systematic sources of discrepancy between the two models and, even if these are not immediately located, this constitutes valuable information.

6.2.3   Comparison and Correlation of Response Properties
6.2.3.1   Comparison of individual response functions
If we start with the experimental model, we find that the raw data available in this case are those describing the time histories of the excitation and response properties of the test structure during the measurement. Although it is true that these time histories constitute the most direct measurement of the structure's actual dynamic behaviour, it is difficult to make comparisons between these data and the corresponding quantities computed from the analytical model. This is so for several reasons, the most important of which is that the actual time histories are very sensitive to certain properties in the analytical model which are very difficult to estimate: most critically, the damping, but also a range of other features which, in themselves, are not critical but which combine to make useful comparisons of time histories difficult to achieve. As a result, attempts to extract useful information from comparisons between prediction and observation of these raw response data are not usually made.

The next level of proximity to the actual measurements (a condition which is important to achieve if an honest comparison of observation and prediction is to be made) is in the form of the response functions which are derived from spectral analysis and further processing of the original time records. These response functions are generally presented as FRFs (or IRFs), or sometimes as ODSs, and it is on these formats that we shall focus our efforts for further comparison formats. In its simplest form, this level of comparison is made with an individual response function, and is shown by overlaying the measured curve on its analytically-predicted counterpart, although it must be borne in mind right at the outset that two important estimates will have to be made in order to be able to compute the theoretical curve: the nature and level of damping, which is not usually strictly part of the modelling process, and the number of modes which will be included in the summation which is made to compute the response functions. The first of these two estimated parameters affects predicted FRFs only in the immediate vicinity of resonances or antiresonances by limiting the sharpness of their peaks and troughs but the second parameter can have more significant effects on the general shape of the curves in all regions away from the resonances if sufficient modes are not included.

438

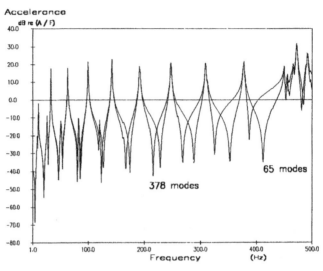

Acceleronce

(a)

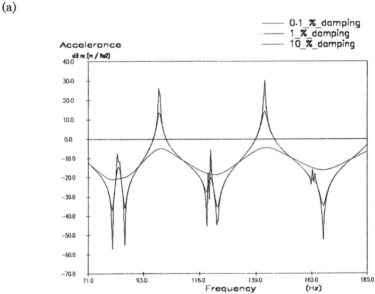

0.1_%_damping
1_%_damping
10_%_damping

Acceleronce

(b)

Fig. 6.10 Generation of FRFs from modal data.
(a) Effect of number of included modes; (b) Effect of increased damping level

Strictly, a check that a sufficient number have been included should be made before any response function comparisons are attempted.

Fig. 6.10(a) shows the result of such a check where a given FRF is re-computed several times based on the same finite element model but including progressively more of the modes from the model in the calculation of each curve. It will be seen that once sufficient modes have been included — and that number may be very difficult to predict without performing this type of check — then adding yet more modes in the series serves little added benefit. At the same stage, it is worth noting that in many cases, the theoretical model is effectively undamped and any damping which is added in order to make comparisons of response possible at all will be crudely estimated. However, in cases of very complex structures which have many modes, some of which are 'significant', global, ones while others are 'secondary', or local, ones, the re-computation of the FRFs using different levels of damping may help in identifying which modes are global and which are local, a feature which is illustrated in Fig. 6.10(b).

In Fig. 6.11(a) we show a comparison between direct measurement and prediction (via a finite element model) of a point FRF for a simple beam-like structure. The plot clearly shows a systematic discrepancy between the two sets of data (resulting in a steady frequency shift between the two curves) while at the same time indicating a high degree of correlation in the amplitude axis. Also of interest are the relative values of the frequencies of resonance and of antiresonance, close examination of which can indicate whether the discrepancies are due to localised errors (loss of stiffness at joints, etc.) or to more general factors (such as incorrect values of elastic modulus or material density, etc.). Nevertheless, it is a frequent requirement for some quantitative measure of the difference between two such response functions, and several proposals have been made for a parameter to do this: the area between the two curves (or under the ΔFRF curve) is often suggested but this is very sensitive to relatively small errors (often of secondary importance) between the measured and predicted natural frequencies. Indeed, this is a common difficulty in making comparisons of response quantities and we shall see in the following paragraphs just how much influence secondary effects can have. It should also be noted that most of the FRF plots shown here use a logarithmic scale for the magnitude (i.e. the modulus is plotted as dB) and this has the effect of showing percentage discrepancies rather than absolute ones, resulting in as much attention being drawn to what are minor discrepancies (in absolute terms) near antiresonances as to major differences at the resonances. There are arguments for using linear scales instead, but these can be countered by those which led us to use logarithmic units in the first place.

A second example is shown in Fig. 6.11(b) where a transfer mobility for a different structure is illustrated, again for both experimental and

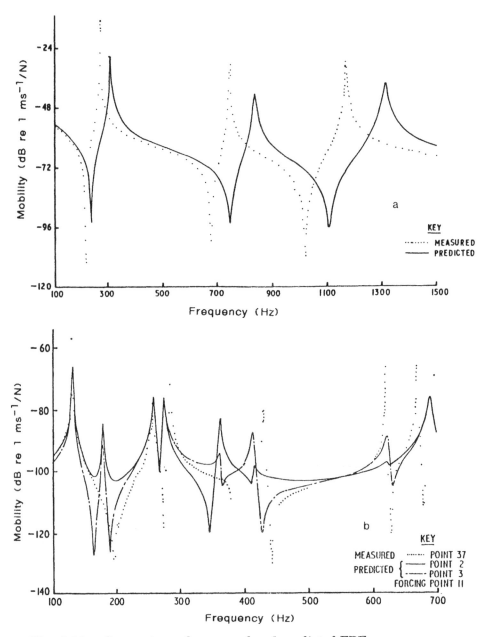

Fig. 6.11 Comparison of measured and predicted FRFs.
(a) Simple structure — point FRF; (b) Complex structure — transfer FRFs

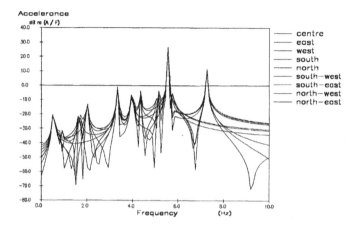

Fig. 6.11 Comparison of measured and predicted FRFs.
(c) Effects on FRFs of uncertainty of DOF location

predicted data. However, in this case the location of the response point used in the modal test does not coincide exactly with any of the mesh of grid points used in the analytical model, thereby making a direct comparison impossible. In order to proceed, the predicted curves relating the two grid points closest to the test position (and these were only a few mm away, on a plate-like structure of some $1m \times 2m$) are used and are displayed in Fig. 6.11(b). In this example, it is clear that not only are there marked differences between the two models (albeit of a different type to the previous case), but also there are striking differences between the two predicted curves which relate to two points very close to each other on the structure. This last observation is very important when we consider how to assess the degree of correlation between the experimental and predicted models. Because the particular parameter being measured (an FRF) can be very sensitive to the exact location of the response point (and, possibly, to the excitation point, although that does not suffer from the same difficulty as does the response in the example cited), major differences may be apparent at the comparison stage which do not directly reflect on the quality of the model, but on something much more basic — namely, the coordinate geometry used in both instances. A third example is included in Fig. 6.11(c), which demonstrates the uncertainty of the exact theoretical FRF curve by plotting a group of nine point FRFs corresponding to the target DOF and the eight immediately-adjacent DOFs.

6.2.3.2 Correlation of complete set of FRFs

It can be noted that in a typical modal test a set of FRFs are measured consisting of at least one column (i.e. one vector) in the FRF matrix based on the measured DOFs, and sometimes including data from several such columns (vectors). It is thus possible to envisage the curve-to-curve comparison described above having to be applied to a large number of such data in order to gain an overall impression of the degree of correlation between measurement and prediction. This is a daunting task, and difficult to perform effectively because there is simply so much information for the analyst to retain and sort. However, it is not unlike the problem faced by the need to compare several mode shapes simultaneously except that in this case there are of the order of 400 or 800 such vectors because there is one for each excitation frequency used in defining the FRFs. This observation leads to the idea of applying the MAC approach to the correlation of two vectors, one from the measured data and the other from a corresponding analytical model prediction. Thus we can define a frequency domain assurance criterion, or FDAC, as follows:

$$FDAC(A(\omega_j), X(\omega_i))_k$$

$$= \frac{\left| \{H_X(\omega_i)\}_k^T \{H_A(\omega_j)\}_k \right|^2}{\left(\{H_X(\omega_i)\}_k^T \{H_X(\omega_i)\}_k \right) \left(\{H_A(\omega_j)\}_k^T \{H_A(\omega_j)\}_k \right)} \tag{6.7}$$

Clearly, a diagram of the type previously used for the MAC can be used to display this function, although it will have a much denser form as a result of the large number of frequencies (typically, several hundreds) by comparison with the usual number of modes (typically, tens). Examples are shown in Figs. 6.12(a) and (b) for the AutoFDACs of both test and analysis results followed by the test-analysis correlation for the same structure in Fig. 6.12(c). Likewise, it is possible to define an FRF equivalent to the COMAC and this has been proposed as the Frequency Response Assurance Criterion, or FRAC, with the following definition:

$$FRAC(j)_k = \frac{\sum_{i=1}^{L} \left| \left(_X H_{jk}(\omega_i) \right) \left(_A H_{jk}^*(\omega_i) \right) \right|^2}{\sum_{i=1}^{L} \left| _X H_{jk}(\omega_i) \right|^2 . \sum_{i=1}^{L} \left| _A H_{jk}(\omega_i) \right|^2} \tag{6.8}$$

where $i = 1, L$ represents individual frequencies for which both test and

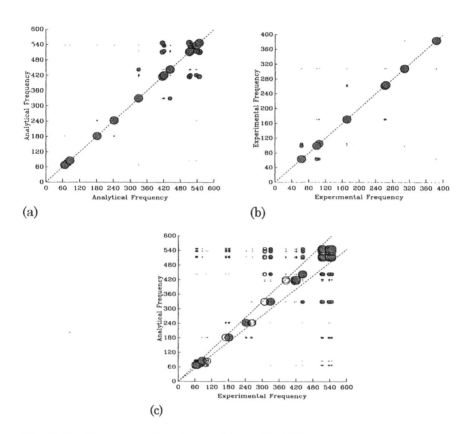

Fig. 6.12 Frequency-domain correlation (FDAC)
(a) AutoFDAC — predicted FRFs; (b) AutoFDAC — measured
FRFs; (c) Correlation of measured and predicted FRFs

analysis data exist and the suffix $_k$ indicates that the relevant FRFs
have DOF k as their reference (k^{th} row or column in the FRF matrix).
One of the difficulties encountered in seeking to make direct
comparisons of FRF curves is the problem which arises when there is a
shift in the location of resonance frequencies between the two sets of
data. Such not infrequently-encountered features give rise to
indications of very poor correlation and adjustments have been
proposed to compensate for this effect.

6.2.3.3 Operating deflection shapes and hybrid modal-response correlation (MFAC)

Finally, at the end of this section, we arrive at a proposal for perhaps
the most realistic comparison between an analytical model and its

experimental counterpart; namely, a correlation between the analytical model *modal* properties and the experimental measured *response* functions. Such a process avoids the complications involved in (a) computing accurate FRF data for the theoretical model (the problem associated with residual terms) and (b) performing a modal analysis on the measured response data, with the inherent approximation that this incurs.

What we have computed in the above-mentioned FDAC parameter is a correlation of operating deflection shapes (ODSs) for the system under study with the excitation being that which is applied to perform the modal test and we can consider the FDAC and FRAC as just that: the correlation between predicted and measured ODSs. If we confine our interest to frequencies in the immediate vicinity of resonances, then in many cases the FDAC will approximate to the MAC calculations because at frequencies which are very close to a natural frequency, most ODS vectors will reflect closely the shape of the mode whose natural frequency is adjacent. If we assume that distortions will exist (because more than just that one mode contributes to the ODS), but that they will be similar in both measured and predicted cases, then the FDAC correlations near natural frequencies may well serve as a very useful approximation to the MAC correlations themselves, without the need for a modal analysis to be performed. An example of these features is illustrated in Fig. 6.13. In the first figure, 6.13(a), is shown a MAC-type of correlation between the ODS vectors derived from the analytical model at each of the (known) natural frequencies of the structure with the corresponding analytical mode shape vectors derived from the same model. This shows the degree of correlation which might be expected between ODS, $\{H_X(\omega_r)\}$, and mode shape, $\{\psi_X\}_r$, for a typically complex practical engineering structure. Also shown, in Fig. 6.13(b), is the result of performing an FDAC correlation between the same ODS vectors derived from the analytical model and the corresponding ODS characteristics for the experimental model deduced from the measurements by a simple peak-picking process at each of the resonance frequencies observed in the measured data set. The next comparison, in Fig. 6.13(c), shows the traditional MAC results, using the results of modal analysis of the measured data and, finally, in Fig. 6.13(d), the proposed hybrid correlation between the experimental ODS properties at the measured resonance frequencies and the analytical model mode shapes, called the MFAC, and defined by the following equation:

445

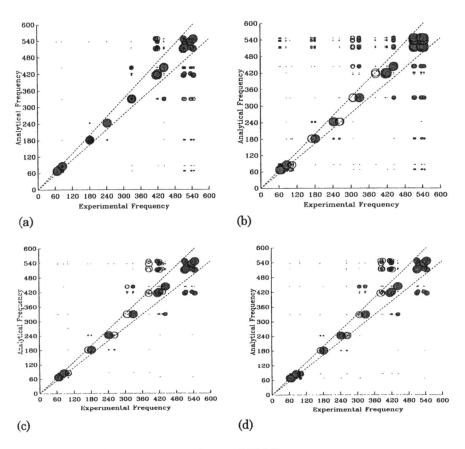

Fig. 6.13 Mode-response correlation (MFAC).
(a) Correlation of predicted FRF vectors with predicted mode shapes; (b) Correlation of measured and predicted FRF vectors; (c) Correlation of measured and predicted mode shapes; (d) Correlation of measured FRF vectors with predicted mode shapes

$$MFAC(A_r, X(\omega)) = \frac{\left|\{H_X(\omega_q)\}^T\{\phi_A\}_r\right|^2}{\left(\{H_X(\omega_q)\}^T\{H_X(\omega_q)\}\right)\left(\{\phi_A\}_r^T\{\phi_A\}_r\right)} \qquad (6.9)$$

Close inspection of these results reveals that Figs. 6.13(b), (c) and (d) all tell much the same story, but the hybrid plot (d) can be produced with much less effort than the others (see Reference [54]).

6.2.4 Concluding Discussion

It will be appropriate to conclude this, and each of the major application areas with a few summarising remarks, in order to help put the preceding section into perspective. The main activity which has been discussed in this section has been that of providing a measure of the distance between two models — usually (but not always) an experimentally-derived model and a theoretically-developed one. There are two main requirements for this correlation capability. The first is the need to be able to identify the pairing between predicted and measured modes (to identify the Correlated Mode Pairs) so that subsequent reconciliation procedures of error location and model updating can proceed on the basis of the initial discrepancies between the key parameters. The second requirement is to provide a comprehensive but simple measure of the proximity of the two models in order to check the progress or success of any attempts to reconcile the initial discrepancies between them. It can be said that the tools required for this task of comparison and correlation are now quite well developed and that efficient and sophisticated measures of the differences between two nominally-similar models are available. As is often the case with such complex procedures, considerable care is required in making the proper interpretations of the indicators that can be computed and the reader is warned of the dangers of oversimplification, or over-hasty conclusions being drawn from incomplete data. As will be seen on many occasions in this work, the problems of incompleteness in the data which can be supplied from tests pose a greater hazard to the successful implementation of the results of modal testing than do the inaccuracies that we expect to find in our data.

6.3   ADJUSTMENT OR UPDATING OF MODELS
6.3.1   Rationale and Ground Rules for Model Adjustment

The subject of model updating, in which an initial theoretical model constructed for analysing the dynamics of a structure can be refined, corrected or updated using test data measured on the actual structure, has become one of the most demanding and demanded applications for modal testing. The stakes are high: if the process can be performed successfully, then the approximations and limitations which are inherent on current analytical (i.e. finite element) modelling can be identified and corrected on a case-by-case basis, and the correct ways of overcoming them in future designs learned in the process. On the other side, the costs of both modelling and of testing are high and if the updating processes are not successful then much time effort and credibility may be lost. It is important, therefore, that the fundamentals of the subject are well understood, as also are its scope and limitations and the demands that are placed on those who wish to pursue such an application.

The subject has become a very extensive one, with already one text book [55] and several hundred papers devoted to its details. It is not realistic to attempt to condense all this material into a single section of this text, or even into a single chapter, but what we shall seek to do is to present the fundamentals of the subject, as these must be mastered before any attempt is made to use the detailed methods of updating themselves. We shall offer some definitions and propose some ground rules which are considered necessary as the basis for the development or implementation of model updating in practical situations. We shall then, in the following sections, present a very concise summary of each of the major algorithms for the updating problem together with some discussion of how and when each of these might be considered for use in practice.

First, it is appropriate to make some definitions for use throughout the rest of this text, noting the precision and subtlety of distinction between various of these terms.

Analytical model — a model comprising $N \times N$ mass and stiffness matrices, usually based on finite element modelling methods, occasionally including an associated damping matrix, or modal damping factors. While the model is defined in terms of these spatial parameters, it is understood that any of the corresponding modal or response parameters can be obtained by suitable analysis of the given matrices.

Experimental model — a model consisting originally of a set of FRF data (or equivalent) from which a limited modal model can be obtained by modal analysis of the measured response functions. It is generally assumed that the experimental model comprises an $m \times m$ eigenvalue matrix and an $n \times m$ eigenvector matrix, both of which are usually complex.

Valid model — a model which predicts the required dynamic behaviour of the subject structure with an acceptable degree of accuracy, or 'correctness'. This criterion is itself subject to qualification, not only in terms of the precision of the various parameters — natural frequencies, response amplitudes etc. — but also in their extent. A series of different levels of 'correctness' have been proposed in this respect, as follows:

- Level 1: prediction (by the model and accurate to a within specified tolerance) of the various modal properties extracted from modal test (i.e. measured natural frequencies, and mode shapes defined at measured DOFs) in the frequency range covered by the test;

- Level 2: accurate prediction of measured response functions in the frequency range covered by the test;
- Level 3: accurate prediction of measured modal properties, including at unmeasured DOFs, in the frequency range covered by the test;
- Level 4: accurate prediction of response properties over the measured frequency range but including unmeasured DOFs;
- Level 5: accurate prediction of response properties over the full frequency range and at all DOFs (this implies an 'exactly-correct' model).

Data — quantitative items which describe specific features of a model or a structure or its response or other dynamic behaviour.

Information — in effect, *independent* items of data that can be used together to solve problems of model adjustment, correction and updating. The amount of information available must generally be equal to or greater than the number of unknowns or variables for a unique solution to be available.

Comparison — the process of setting two sets of data side-by-side so that a direct comparison can be made of corresponding properties. This means presenting the two sets in the same format so that the comparison can be made objectively, although it is used in a passive, and qualitative, rather than quantitative way.

Correlation — is the process of quantifying the degree of similarity and dissimilarity between two models and is the numerical conclusion of the comparison process. Different indices and weighting factors can be involved to provide the most useful measures of correlation for the subsequent use of the model.

Localisation (or location) — the process of locating the whereabouts of differences between two models. Equivalent to the specification of the parameters to be updated: a necessary step before updating can take place.

Optimisation — the process of determining a set of values for given parameters such that a pre-defined objective (penalty) function is minimised. In the context of model updating, it is to determine the values for a predefined set of model parameters such that the discrepancy between measured and predicted dynamic behaviour is minimised. Only in cases where all the erroneous parameters have been located and included in the optimisation process can an 'exact' updating be performed. In all other cases, the solution is a compromise of unknown quality.

Reconciliation — the process of explaining the origins of differences between two models which result in there being discrepancies observed in their respective dynamic behaviour (usually measured by comparison with predicted, but may be applied to two different analytical models).

Updating — the process of correcting the numerical values of individual parameters in a mathematical model using data obtained from an associated experimental model such that the updated model more correctly describes the dynamic properties of the subject structure (see above for a discussion of 'correctness').

Verification (of a model) — the process of determining whether a given model is capable of describing the behaviour of the subject structure, if all the individual model parameters are assigned the correct values. A model may not be verified if it lacks certain features or freedoms which are present in the actual structure since, in this case, no amount of parameter correction can compensate for the errors embedded in the basic model.

Validation (of a model) — the process of demonstrating or attaining the condition that the coefficients in a model are sufficiently accurate to enable that model to provide an acceptably correct description of the subject structure's dynamic behaviour. It is clear that a model which is not verified cannot be validated, and, indeed, that such a validation procedure should not be attempted on an unverified model.

A set of ground rules can be constructed from the above definitions. It is first necessary to decide upon the level of accuracy, or correctness, which is sought from the adjustment of the initial model, and this will be heavily influenced by the eventual application of the refined model. Then, it is necessary to determine whether or not the initial model can be updated, a difficult task which, in effect, calls for it to be verified. This means ensuring that all the important features of the actual structure are included in the model, even if only approximately at the outset, and that there are no actual features which have been omitted from the model. Of particular concern at this stage is the inclusion of sufficient flexibility at the joints of an assembled structure and, most importantly, of sufficient fineness of mesh such that the model has converged. Model updating cannot be used to improve a model which is too coarse; only to refine one which is basically correct, but inaccurate in some of its components.

Next, it is necessary to determine the order of the problem: and by that is meant to establish how many of the model's coefficients need to be corrected. This is, in effect, the same task as locating the regions which contain all the errors to be corrected, but it is necessary to identify all of these so that the full scale of the problem — the number of unknowns to be identified in the updating process — is established. Sometimes referred to as 'locating the errors' or as 'specifying the updating parameters', this step is one of the most difficult yet most critical in the whole updating process.

The last stage before actual computation of the updating corrections themselves involves the specification of the data which need to be obtained in the validation tests so that the updating procedure can succeed. This is not simply a question of quantity of measured data (as is often thought to be the case) but, more importantly, of its selection so that the maximum amount of *information* about the experimental model is made available from the measured *data*.

Once all these preparatory stages have been undertaken, it is possible to embark on the updating calculation itself, with the equation to be solved defined in as satisfactory state a possible. There are several numerical difficulties encountered in the process of performing an updating computation and most of these derive from ill-conditioned matrices which are themselves the result of poorly-defined equations together with a general insufficiency of useful information.

6.3.2   Basic Concepts of Model Updating

It will be helpful to review the basic concept and objectives of the model updating process prior to a description of the major algorithms that have been developed for its implementation. These concepts can be presented concisely as follows:

- The starting point is the existence of an initial analytical model that is to be updated, so that we have available mass and stiffness matrices, $[M_A]_{N \times N}$ and $[K_A]_{N \times N}$ and, by direct analysis of these, the corresponding modal matrices, $[\omega_A]_{N \times N}$ and $[\Phi_A]_{N \times N}$ and the FRF matrix, $[H_A(\omega)]_{N \times N}$;
- Also, we have access to a limited number of response and modal properties for the experimental model in the form of $[H_X(\omega)]_{n \times p}$, $[\omega_X]_{m \times m}$ and $[\Phi_X]_{n \times m}$. However, it must be noted at this stage that the sizes of these experimental model matrices are smaller,

even much smaller, than those of the analytical model;

• The task of the model updating process is to determine the changes which must be made to the initial analytical model ($[\Delta M]$ and $[\Delta K]$) so that the modal and response properties of the thus-corrected analytical model match those of the experimental data.

The practical implication of this situation is that the problem we face is seriously under-determined because in general there will be many more elements in the initial model that may need to be updated (or corrected) than there are data items for the experimental measurements. Thus there will be more unknowns than there are equations and so the problem is not soluble in a mathematical sense. However, if it can be established that, in fact, the number of parameters in the initial model that actually do need to be corrected is considerably smaller than the total number of parameters (or that most of the initial model is already acceptable), and if we can identify which are the parameters that need to be corrected, then it is possible that the problem may be converted to an over-determined one that is therefore amenable to solution. There are several 'ifs' in this statement but they represent the conditions that must be met if model updating is likely to be applicable in a practical situation. In fact, the first 'if' is a way of saying that the initial model must be close enough to the correct one so that just a manageable number of parameters need to be changed. More precisely, this means that the errors in the initial model need to be relatively few in number, perhaps localised in certain regions of the structure, and not distributed throughout all the elements. It is less critical that the errors be small than that they be few in number.

The second 'if' is perhaps the most difficult aspect of model updating: it is the requirement that we can successfully identify the locations of the parameters that need to be corrected. This is often the most demanding step in the application of model updating procedures in practice and will be further discussed later in this section. However, if that question is set on one side for the moment, we can now examine the various algorithms that are available for carrying out the updating computations which will lead to a revised analytical model that approaches the objective stated above by exhibiting the same dynamic properties as measured on the experimental test structure.

6.3.3 General Methods of Model Updating

Before offering a synopsis of the different methods of updating themselves, it is appropriate to describe the general classes of method which are available. It is convenient to group them first into two major types:

- direct matrix methods: those methods in which the individual elements in the system matrices are adjusted directly from comparison between test data and initial analytical model predictions; and
- indirect, physical property adjustment methods: in which changes are made to specific physical or elemental properties in the model in a search for an adjustment which brings measured and predicted properties closer together.

The first of these two groups are generally non-iterative methods which share the feature that the changes they introduce may not be physically-realisable changes: they are simply new values for individual elements in the system $[M]$ and $[K]$ matrices, some of which may be applied to elements which are initially (and which, for reasons of the model configuration or connectivity should remain) zero. They generally require complete mode shape vectors as input but are, nevertheless, computationally very efficient. They have as their target the ability to reproduce the measured modal properties of m natural frequencies and mode shapes.

The second group of methods are in many ways more acceptable in that the parameters which they adjust are, or are much closer to, physically-realisable quantities. In the simpler versions, a single correction factor might be applied to the entire elemental stiffness submatrix for a particular (finite) element, but this is much more supportable than a correction factor which introduces a finite value into an off-diagonal stiffness matrix element which links two physically-disconnected DOFs. Methods in this second group are generally iterative and, as such, more expensive of computer effort. To offset this disadvantage, they will generally work with incomplete mode shape vectors (mode shapes defined at the n DOFs of a typical modal test in place of the complete N DOFs of the full analytical model which are required by most of the first group of methods).

Developments are continuing in methods of both groups, although it is the second group that have emerged as the most widely-used in general practical application. As a preface to the following sections, which summarise each of the main approaches, it should be stressed that the subject is still relatively immature and that success in its application is not assured. One important feature of this state of development is that while successes are often reported, these are quite case-dependent. This is taken as a sign that the essential technology is there for the developing, but that the necessary developments are not yet complete. Potential users of the methods should therefore be forewarned but encouraged!

In the following sections we shall summarise the essential features of the following methods, while referring the interested reader to the more detailed literature for a full explanation and exposition of each. Two surveys that provide a useful entree to this subject are reported in References [56] and [57]). The methods summarised include, from the first group:

- Direct Matrix Updating (DMU) Method
- Error Matrix Method (EMM)

and from the second group:

- Eigendynamic Constraint Methods (ECM)
- Inverse Eigensensitivity (IES) Methods
- Response Function Methods (RFM)

6.3.4   Direct Matrix Updating
The Direct Matrix Updating method was one of the earliest to be developed for industrial application, dating back to the 1970s. In its basic form, applied to an undamped system, the method performs an adjustment first on the system mass matrix, and then uses the result to perform a similar adjustment on the stiffness matrix. As previously mentioned, the method requires the model and the input mode shape data for the full set of DOFs. This normally presents a problem for the experimental model description and so one of the expansion methods presented in Chapter 5 would normally need to be applied to the incomplete measured mode shape vectors to prepare them for application in this method.

The formula for the updating is simply defined, as follows:

$$[\Delta M] = [M_A][\Phi_X][m_A]^{-1}\left([I]-[m_A]\right)[m_A]^{-1}[\Phi_X]^T[M_A]$$

where

$$[m_A]_{m\times m} = [\Phi_X]^T_{m\times N}[M_A]_{N\times N}[\Phi_X]_{N\times m}$$

followed by:

$$[\Delta K] = [M_A][\Phi_X][\Phi_X]^T[K_A][\Phi_X][\Phi_X]^T[M_A]$$
$$+[M_A][\Phi_X][\omega_X^2][\Phi_X]^T[M_A]$$
$$-[K_A][\Phi_X][\Phi_X]^T[M_A]-[M_A][\Phi_X][\Phi_X]^T[K_A] \qquad (6.10)$$

Additional constraints can be added to restrict the adjustments to the mass and stiffness matrices which these formula introduce so that they are more consistent with the physics of the system. It can easily be seen that the solution obtained is not a unique one, by any means. This can be shown mathematically (by the fact that there are a large number of $[\Delta M]$ matrices which will satisfy the basic requirement of equation (6.10)) as well as heuristically by the fact that the model is of order $N \times N$ and no requirement has been stipulated for the properties of the modes after mode number m. This means that there exist a large number of models which satisfy the specified constraints but which all have different properties for the unspecified modes. It is clear that a solution obtained in this way is a numerical solution and not a physical one. Further details of this method can be found in Reference [58]

6.3.5   Error Matrix Method (EMM)

In this method, an alternative approach is suggested for the task of determining the adjustments to the elements in the mass and stiffness matrices. The method uses the concept of 'error matrices', which have already been introduced in the previous section, which are the matrix differences between the experimental (X) and analytical (A) system matrices, as follows:

$$[\Delta M] = [M_X]-[M_A] \quad ; \quad [\Delta K] = [K_X]-[K_A] \qquad (6.11)$$

We are, in effect, interested in determining these two error matrices. It can be seen that direct computation of the mass or stiffness error matrix from equation (6.11) is hindered by our inability to specify $[M_X]$ or $[K_X]$. Although there exist direct transformations between the modal properties and the spatial properties, in the form of:

$$[M] = [\Phi]^{-T}[\Phi]^{-1}$$
$$[K] = [\Phi]^{-T}[\omega_r^2][\Phi]^{-1} \qquad (6.12)$$

these are only valid in cases where the modal matrices are complete — i.e. contain all modes and all N DOFs — and so are not applicable in the general practical case where only m modes are identified and these defined at only n DOFs. The EMM seeks to obtain an estimate for the error matrices using the following approach which is based on the contribution of the known modes to the system flexibility and inverse mass properties. This we may write as:

$$[\Delta K] = [K_X] - [K_A]$$

$$[K_X]^{-1} = \left([I] + [K_A]^{-1}[\Delta K]\right)^{-1}[K_A]^{-1} \qquad (6.13)$$

$$\approx [K_A]^{-1} - [K_A]^{-1}[\Delta K][K_A]^{-1} + \left([K_A]^{-1}[\Delta K]\right)^2 [K_A]^{-1} + \dots$$

and similarly for the mass properties, leading to the EMM formulae:

$$[\Delta K] \approx [K_A]\left([K_A]^{-1} - [K_X]^{-1}\right)[K_A]$$

$$\approx [K_A]\left([\Phi_A][\omega_{Ar}^2]^{-1}[\Phi_A]^T - [\Phi_X][\omega_{Xr}^2]^{-1}[\Phi_X]^T\right)[K_A] \qquad (6.14)$$

$$[\Delta M] \approx [M_A]\left([\Phi_A][\Phi_A]^T - [\Phi_X][\Phi_X]^T\right)[M_A]$$

in the application of which the data for as many or as few modes can be included, although special precautions must be taken if these data are incomplete in the sense of DOFs included.

An example of the application of this method to a practical aerospace structure is shown in Fig. 6.14, and further details can be found in Reference [59].

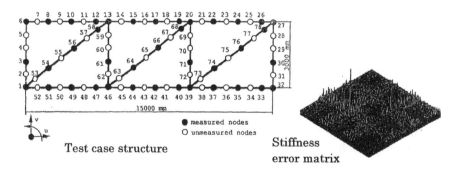

Fig. 6.14 Example of application of EMM updating

6.3.6 Eigendynamic Constraint Methods (ECM)

One of the earliest of the so-called indirect updating methods is the Eigendynamic Constraint Method (ECM) which has a number of different versions, depending upon which of the various eigendynamic equations are employed. These relationships are:

$$\left[[K_X] - (\omega_X)_r^2 [M_X]\right]\{\phi_X\}_r = \{0\} \tag{6.15a}$$

$$\{\phi_X\}_r^T [M_X]\{\phi_X\}_r = 1 \tag{6.15b}$$

$$\{\phi_X\}_r^T [K_X]\{\phi_X\}_r = (\omega_X)_r^2 \tag{6.15c}$$

The ECM is typical of all the indirect methods in that the unknowns of the problem are no longer the individual elements in $[\Delta M]$ and $[\Delta K]$ (as is the case for the DMU and EMM methods just described), but, instead, are a set of individual element correction factors, a_s and b_s for element (or element group) s. This transformation can be summarised by the expressions:

$$[\Delta M] = \sum_{s=1}^{L} a_s [m_s] \quad ; \quad [\Delta K] = \sum_{s=1}^{L} b_s [k_s] \tag{6.16}$$

where matrices $[m_s]$, $[k_s]$ are the mass and stiffness matrices for the sth element (or element group, if several elements are grouped together for the purposes of updating, as may be done if these elements are assumed to carry the same error, or correction factor); L is the total number of elements or element groups to be updated; and a_s and b_s are the correction factors which are applied to every element in the respective elemental mass and stiffness matrices (and are sometimes referred to as the 'p' factors).

The ECM updating algorithm is based on application of the eigendynamic conditions above and the determination of a set of correction or 'p' factors which allow the modified model to satisfy these conditions. The most general version of this family of methods is the ECM itself, which can be described as follows:

$$(\lambda_X)_r [\Delta M]\{\phi_X\}_r - [\Delta K]\{\phi_X\}_r = (\omega_X)_r^2 [M_A]\{\phi_X\}_r - [K_A]\{\phi_X\}_r$$

and $\hspace{9cm}$ (6.17)

$$\{\phi_X\}_r^T [\Delta M]\{\phi_X\}_r = \{\phi_X\}_r^T [M_A]\{\phi_X\}_r - 1$$

These equations can be rearranged into a standard set of linear equations which will lead to a solution for the unknown correction factors:

$$[A_i]_{(N+1)\times 2L} \{p\}_{2L\times 1} = \{\delta_i\}_{(N+1)\times 1} \tag{6.18}$$

where N = the number of DOFs in the analytical model; $2L$ = the number of independent design variables (or element (group)s to be corrected); $[A]$ and $\{\delta\}$ are a coefficient matrix and vector, respectively, which are formed using the (known) elements in the initial analytical model $([M_A]$ and $[K_A])$ and the experimental properties of the rth mode $((\omega_X)_r^2; \{\phi_X\}_r)$; and $\{p\}$ is a vector which contains the unknown (and sought) correction factors.

If data are available for a number (m) of modes, then m equations of the form of (6.18) can be written and combined to form:

$$\begin{bmatrix} [A_1] \\ [A_2] \\ \dots \\ [A_m] \end{bmatrix}_{m(N+1)\times 2L} \{p\}_{2L\times 1} = \begin{Bmatrix} \{\delta_1\} \\ \{\delta_2\} \\ \dots \\ \{\delta_m\} \end{Bmatrix}_{m(N+1)\times 1}$$

or

$$[A]\{p\} = \{\delta\} \tag{6.19}$$

When equation (6.19) is over-determined (i.e. when $m(N+1) > 2L$), then a least-squares solution for $\{p\}$ can be found using a generalised inverse of $[A]$, preferably computing this by means of the SVD or an equivalent approach that enables the condition of the solution to be assessed:

$$\{p\} = [A]^+ \{\delta\} \tag{6.20}$$

Two other versions of the ECM have been used in the past — one known as the Force Balance Method which makes use only of equation 6.15(a) resulting in a slightly incomplete set of equations:

$$[A']_{mN\times 2L} \{p\}_{2L\times 1} = \{\delta'\}_{mN\times 1} \tag{6.21a}$$

and the so-called Orthogonality Constraint Method which uses equations 6.15(b) and (c), but not (a), resulting in an even smaller subset of the full eigendynamic equations:

$$[A'']_{m(m+1)\times 2L} \{p\}_{2L\times 1} = \{\delta''\}_{m(m+1)\times 1} \tag{6.21b}$$

Both these derivatives make use of less than the full set of data and equations and so are less effective than the primary version, the ECM.

However, it should be noted that there is a major drawback to the use of this group of methods in that while not all modes must be included (m can be less than N), the methods do require that all DOFs are included in the mode shapes of those modes that are used in the analysis. In practice, this means that a mode expansion procedure (see Chapter 5) must be applied to the measured data, and this is a task that is not only troublesome but also of dubious validity for this application since the expansion is likely to be based on the very analytical model which is the subject of the updating, or correction process. Perhaps for this reason, these methods are less frequently used than the others in this class, such as the Inverse Eigensensitivity Method, which is described next.

6.3.7 Inverse Eigensensitivity (IES) Methods

The second of the indirect methods to be summarised is the one which has found the greatest application in practice, with a number of commercially-available codes already available, and is referred to as the 'Inverse Eigensensitivity (IES) Method'. The method is based on an equation of exactly the same general form as the ECM methods (equations 6.19 and 6.21) with the difference that the system matrix and vector, $[A]$ and $\{\delta\}$, are composed of properties which derive from the analytical model modal sensitivities and the discrepancies between predicted and measured modal properties, respectively. The essential theory upon which the method is based is summarised below.

The basis of the method is the assumption that the differences between measured and predicted modal properties (of natural frequencies, $\Delta\omega_r$, and mode shapes, $\{\Delta\phi\}_r$), can be described in terms of the relevant modal sensitivities (rates of change of natural frequencies ($\delta\lambda_{Ar}/\delta m_s$, ...) and mode shapes ($\delta\{\phi_A\}_r/\delta m_s$, ...) with respect to changes in individual mass and stiffness terms, (m_s, k_s, ...)) and small adjustments to selected mass and stiffness elements in the model (a_s, b_s). This can be expressed as follows:

$$\Delta\omega_r^2 \approx \sum_{s=1}^{L} \frac{\partial\omega_{Ar}^2}{\partial m_s} a_s + \sum_{s=1}^{L} \frac{\partial\omega_{Ar}^2}{\partial k_s} b_s = \sum_{s=1}^{2L} \frac{\partial\omega_{Ar}^2}{\partial p_s} p_s$$

$$\{\Delta\phi\}_r \approx \sum_{s=1}^{L} \frac{\partial\{\phi_A\}_r}{\partial m_s} a_s + \sum_{s=1}^{L} \frac{\partial\{\phi_A\}_r}{\partial k_s} b_s = \sum_{s=1}^{2L} \frac{\partial\{\phi_A\}_r}{\partial p_s} p_s \qquad (6.22)$$

where

$$[\Delta M] = \sum_{s=1}^{L} a_s [m_s] \quad ; \quad [\Delta K] = \sum_{s=1}^{L} b_s [k_s] \qquad (6.23)$$

using the same notation as for the ECM method described in the previous section. These expressions implicitly assume a small discrepancy between the two models — experimental and analytical — and this is indicated by the approximation rather than equality in equation (6.22). If data are available relating to measured and predicted values of several (m) modes of the structure, then m sets of equations of the type shown in equation (6.22) can be derived and assembled together into the single equation below which forms the basis of this method:

$$
\begin{bmatrix}
\frac{\partial\omega_{A1}^2}{\lambda_1 \partial m_1} & \cdots & \frac{\partial\omega_{A1}^2}{\lambda_1 \partial m_L} & \frac{\partial\omega_{A1}^2}{\lambda_1 \partial k_1} & \cdots & \frac{\partial\omega_{A1}^2}{\lambda_1 \partial k_L} \\
\frac{\partial\{\phi_A\}_1}{\partial m_1} & \cdots & \frac{\partial\{\phi_A\}_1}{\partial m_L} & \frac{\partial\{\phi_A\}_1}{\partial k_1} & \cdots & \frac{\partial\{\phi_A\}_1}{\partial k_L} \\
\cdots & \cdots & \cdots & \cdots & \cdots & \cdots \\
\cdots & \cdots & \cdots & \cdots & \cdots & \cdots \\
\frac{\partial\omega_{Am}^2}{\lambda_m \partial m_1} & \cdots & \frac{\partial\omega_{Am}^2}{\lambda_m \partial m_L} & \frac{\partial\omega_{Am}^2}{\lambda_m \partial k_1} & \cdots & \frac{\partial\omega_{Am}^2}{\lambda_m \partial k_L} \\
\frac{\partial\{\phi_A\}_m}{\partial m_1} & \cdots & \frac{\partial\{\phi_A\}_m}{\partial m_L} & \frac{\partial\{\phi_A\}_m}{\partial k_1} & \cdots & \frac{\partial\{\phi_A\}_m}{\partial k_L}
\end{bmatrix}
\begin{Bmatrix}
a_1 \\ \cdots \\ a_L \\ b_1 \\ \cdots \\ b_L
\end{Bmatrix}
\approx
\begin{Bmatrix}
\Delta\omega_1^2/\omega_1^2 \\ \{\Delta\phi\}_1 \\ \cdots \\ \cdots \\ \Delta\omega_m^2/\omega_m^2 \\ \{\Delta\phi\}_m
\end{Bmatrix}
$$

or

$$[A]_{m(n+1)\times 2L} \{p\}_{2L\times 1} = \{\delta\}_{m(n+1)\times 1} \qquad (6.24)$$

which is an expression of exactly the same form as that derived earlier for the ECM approach with the only difference being in the contents of the system matrix and vector. Here these are comprised of sensitivity properties of the initial analytical model (which are available from the eigensolution of that model) together with the observed discrepancies between prediction and measurement of the various modal properties, and these are available once a modal test has been conducted on the test structure. It must be noted here, however, that the computational cost of deriving the eignevector sensitivity terms is much higher than for the eigenvalue sensitivities. As a result, some early implementations of the method eschewed the mode shape data and applied the method based only on the eignevalue sensitivities. As will be seen below, this approach severely limits the applicability of the method because of the very small number of experimental data that can be input to the method if only natural frequencies are used.

A solution to the problem presented by this expression can be obtained provided that the equation in (6.24) is over-determined: i.e. if there are more difference data available than there are unknown correction parameters to be found. These two critical parameters are determined by (i) the number (m) of correlated mode pairs that are available (it is important to note that only those modes for which measured and predicted properties have been established can be included) (ii) the number (n) of measured response points for each mode shape and (iii) the number ($2L$) of model parameters than are considered to need correction, or updating. It is clear that unless this over-determination is obtained, then a meaningful solution cannot be found, even if a numerical solution can be computed. Even when the essential condition for over-determination (that $m(n+1) > 2L$) is attained, there is still no guarantee that a viable solution can be found because the matrix $[A]$ will still be singular unless all the data supplied from the modal test are independent (it should be recalled that the choice of the measured response DOFs is just as important as their number in order to achieve the required independence). It is normal to seek a situation where there are two or three times as many equations as unknowns before undertaking a solution to equation (6.24), which can be done by computing a generalised inverse for $[A]$, preferably using the SVD or equivalent approach that can ensure the condition of the matrix. Then we can write:

$$\{p\} = [A]^+ \{\delta\} \tag{6.25}$$

in which the solution obtained for the required correction factors $\{p\}$ is a minimum least-squares solution.

It will be seen that this method, unlike its predecessor, the ECM, does not require complete eigenvector data and can function even when only a limited number of DOFs are available. This gives this approach a

461

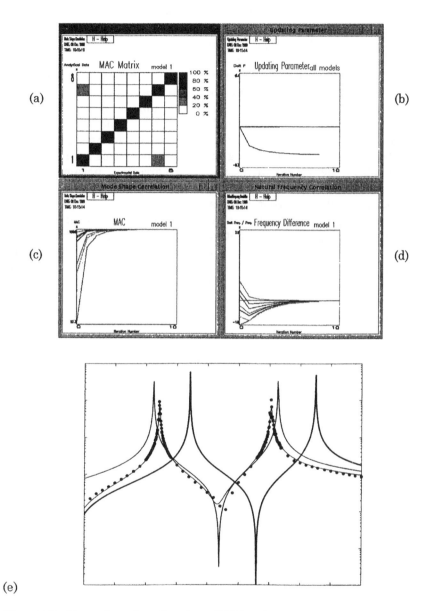

Fig. 6.15 Examples of application of IES updating.
(a) Current MAC chart; (b) Evolution of correction factors;
(c) Evolution of MAC values of correlated mode pairs;
(d) Evolution of natural frequency errors for correlated mode
pairs; (e) FRF characteristics of updated rotor system:
············· Measured; ——— Initial model; ——— Updated model

significant advantage over the ECM since the need to expand incomplete measured mode shapes is removed, and there is no unrealistic expectation placed on the selection of points of measurement. However, because of the inherent approximation involved at the outset of this formulation, and of the inevitable incompleteness of the modal data, the numerical solution (6.25) is not an exact one to the physical problem being addressed and should be repeated iteratively to seek a stable solution of practical value. An illustration of such a solution procedure is given in Fig. 6.15 which shows the evolution of an updating calculation using this approach. The figure displays the values of (a) updating correction parameters, (b) natural frequency differences and (c) MAC values as the solution progresses through some 10 iterations.

It will be appreciated that there are many choices that can be made by the analyst in applying this method: (i) how many modes to include, (ii) how many measurement points to include (ii) whether to include mode shape data as well as natural frequency data (recall that the eigenvector sensitivities are computationally more expensive than those for the eigenvalues). Should a convergent solution not be found, or if the results of a converged solution are deemed to be physically unrealistic or unacceptable (in that they demand such major changes to the original model parameters that these are not acceptable to the modeller), then it is likely that the problem has been ill-posed, probably by specifying an incorrect selection of the parameters which need to be corrected. In such cases, it will be necessary to restart the updating process employing a different set of parameters, or additional data.

Further discussion of these aspects of how to use the updating algorithms are presented in a later section of this chapter.

6.3.8   Response Function Methods (RFM)
One of the difficulties in application of any of the aforementioned updating methods is how to obtain enough measured data to ensure that the updating equation can be made over-determined. In most applications, there are a great many more model parameters that might need to be updated than the number of modal properties that can be measured in a conventional modal test. The incompleteness of any measured modal data is twofold — (i) it is difficult (although not impossible) to measure the mode shapes at anything like the full number of DOFs which are present in the corresponding analytical model, and (ii) it is almost impossible to measure the properties of more than a small number of low-frequency modes for the simple reason that in most structures, a combination of modal density and damping level means that the successful extraction of reliable modal properties becomes infeasible above 30 to 50 modes. For structures which possess

tens of thousands of DOFs, this restriction to working with only tens of modes presents a severe constraint to the whole updating process.

One solution to this problem of incompleteness was proposed in the mid 1980s, [60], by seeking to use directly the response function data which are obtained for a modal test. The philosophy behind this approach is to exploit the facts that (i) there are many more 'items' of response data than of modal data (since it is the objective of modal analysis to reduce the amount of data required to describe a structure's dynamic behaviour) so that the under-determination problem might be positively addressed by this route, (ii) the response data are more faithful to the actual behaviour of the test structure than are the modal properties which are extracted from them, necessarily being affected by the approximations and assumptions embedded in the parameter extraction procedures that are used and (iii) the difficult question of inclusion or exclusion of damping from the analysis might be better treated if the sweeping assumptions of damping type and distribution can be delayed until later in the analysis than is possible in any modal approach. Based on these ideas, another updating algorithm has been devised and offers an additional, if not alternative, tool and perspective for the analyst.

The method presents another version of the same type of equation as we have seen in the two preceding methods (as shown in equations 6.20 and 6.25), but in this case the system matrix $[A]$ and vector $\{\delta\}$ are populated by response properties rather than modal properties. The essential features of the analysis (which are detailed in References [60] and [61]) are as follows:

$$[H_A(\omega_i)] = [H_X(\omega_i)] = [H_A(\omega_i)][\Delta Z(\omega_i)][H_X(\omega_i)] \qquad (6.26)$$

or, using only the jth column of each FRF matrix:

$$\{H_A(\omega_i)\}_j - \{H_X(\omega_i)\}_j = [H_A(\omega_i)][\Delta Z(\omega_i)]\{H_X(\omega_i)\}_j$$

where

$$[\Delta Z(\omega_i)] = [\Delta K] - \omega_i^2 [\Delta M] + i\omega_i [C] \dots \text{etc} \qquad (6.27)$$

These expressions can be combined and transformed into a more convenient format:

$$[R(\omega_i)]_{n \times 2L} \{p\}_{2L \times 1} = \{\Delta H(\omega_i)\}_j \qquad (6.28)$$

where the system matrix, $[R(\omega_i)]$, is constructed from the FRF properties of the analytical model, and the system vector, $\{\Delta H(\omega_i)\}_j$, represents the differences between all the measured and corresponding predicted FRFs at the chosen response DOFs for frequency of vibration, ω_i. If we have response function data for several different frequencies, ω_i, $i = 1, S$, then we can construct a single equation of the same form as (6.28) which can have many more components than is generally possible with modal data since, in practice, S is likely to be much greater than m, the number of modes that can be defined and included. Thus we can write:

$$
\begin{bmatrix} [R(\omega_1)] \\ [R(\omega_2)] \\ \dots \\ [R(\omega_S)] \end{bmatrix}_{Sn \times 2L} \{p\}_{2L \times 1} = \begin{Bmatrix} \Delta H_j(\omega_1) \\ \Delta H_j(\omega_2) \\ \dots \\ \Delta H_j(\omega_S) \end{Bmatrix}_{Sn \times 1} \tag{6.29}
$$

This is clearly another version of the same type of equation we have derived in the previous indirect updating methods and so can be written summarily, with the usual form of solution, as follows:

$$
[R]\{p\} = \{\delta\}
$$

so that

$$
\{p\} = [R]^+ \{\delta\} \tag{6.30}
$$

Much the same comments apply to this approach, in terms of the need to over-determine the set of equations before seeking a solution using (6.30), but the difference in this case is the relatively rich data source in the FRF data, usually one or two orders of magnitude greater than that available in modal properties. Nevertheless, care must be taken to ensure the condition of the $[R]$ matrix which is constructed because sheer quantity of measured data points does not ensure a matrix which is non-singular and the selection of an optimum set of measured frequencies as well as of measured response locations remains an important task for the analyst, although this approach offers a wider range of possibilities than do the methods based purely n modal properties.

A practical example of the application of this method is illustrated in Fig. 6.16 which is taken from Rference [61] where further details of the method can be found.

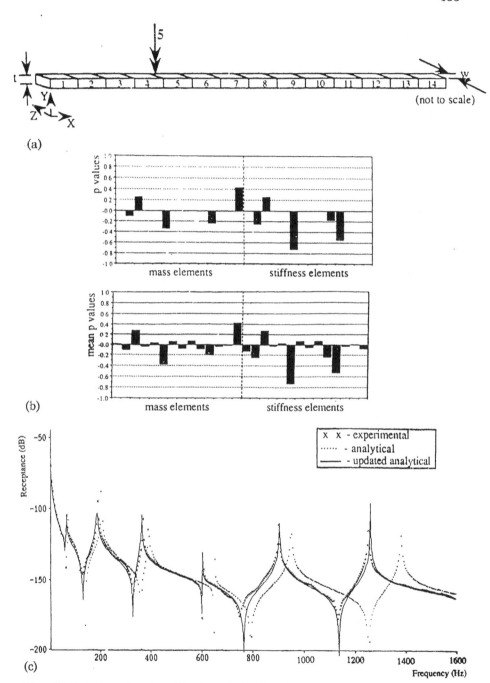

Fig. 6.16 Example of application of RFM updating.
(a) Test structure; (b) Correction factors; (c) FRF comparisons before and after updating

In practice, a combination of this method and the inverse eigensensitivity method is thought likely to emerge as offering the best of both approaches although, as previously mentioned, this technology remains at a relatively early stage of development and there are several improvements to be wrung out of the foregoing analytical considerations.

For further details of these methods, the interested reader is directed to one of the many specialised treatises on the subject, as indicated at the start of this section.

6.3.9 Application of Model Updating to Practical Problems

It is often found that application of these model updating procedures to practical problems meets with mixed results and frequently the whole process is very difficult to conduct. It is true that most methods or algorithms are validated by benchmark case studies, but these are usually based on simulations of experimental data, rather than real test data. This type of validation using simulated data is routinely made for the good reason that the correct answer is 'known' and so the degree of success of each application can be accurately assessed. However, such procedures have the drawback that they risk excluding certain features of real life; features that can have a major influence on the very procedures that are being assessed. This is certainly the case in respect of model updating and considerable care must be exercised in using these methods on real-world practical problems.

It can be seen that there exist more than one type of model error, and only one of these is amenable to correction by model updating procedures. The errors which can exist in an analytical model constructed to describe the dynamics of a structure:

- first, there are **parameter errors**, by which we refer to the numerical values of the individual parameters of mass and stiffness which comprise the model;
- second there are **discretisation errors**, which can result from the use of an over-coarse representation of the actual continuous structure by a finite number of discrete components; and
- third, there are **structural errors**, which result when features in the actual structure are omitted from the idealisation that is created for the mathematical model. These errors include, clearly, the omission of certain structural features but more subtly, they include the omission of flexibilities that can exist at joints or interfaces between components. In all such cases, the essential problem is that the parameters which need to be adjusted to make the analytical model match the experimental observations simply do not exist in the initial model which is thus not capable of being

updated.

It has been found that only errors of the first type — parameter errors — can be corrected by model updating procedures. The other types of error — discretisation and structure errors — cannot be corrected by these methods and so it is important that models to be subjected to updating have been verified as being capable of such correction before embarking on an updating exercise. However, this verification is not so easily achieved, especially in respect of structural errors, and so we must learn how to recognise when an updating process is failing to attain a satisfactory solution

It is appropriate to consider here the consequences of applying model updating to situations which do not satisfy these requirements, and we can do so by using a numerical simulation which is seeded with errors of all three types (rather than just the first type, which is common practice in the development of the updating algorithms themselves). A model of a simple plate shown in Fig. 6.17(a) — the analytical model — is used for this purpose (Reference [62]). A second version of the original model is taken and given different properties for several of the elements — the so-called 'experimental' model, (b). A small number of modal data for the second model are computed, contaminated with representative random noise, and a model updating performed. It is found that a satisfactory result is obtained in this case with the actual discrepancies effectively identified by the updating process. An alternative pseudo-experimental or 'reference' model is then constructed (c), which comprises a model with a much finer mesh, but one which possesses exactly the same material and dimensional properties as the original analytical model (i.e. there are no parameter errors). However, it is found that the fine-mesh model and the coarse-mesh model possess different modal properties. In the second example, modal data from the fine-mesh model is used to update the initial analytical model and it is found that parameter changes ('corrections') can be defined, which bring these two models into closer agreement. These corrections are, of course, compensation for the discretisation errors that result from the coarseness of the original analytical model and do not truly represent errors in the physical parameters that had been used. It will be seen that the 'errors' found in this case are of the same order of magnitude as those which were found in the first example, where the parameters used really were in error. In the third case study, the fine mesh model is then corrupted by parameter changes which reflect those used in the first experimental model and the updating is carried out again. Once more, a set of modified parameters are obtained which satisfy the criterion of minimum difference between the modal properties of the target modes. In this

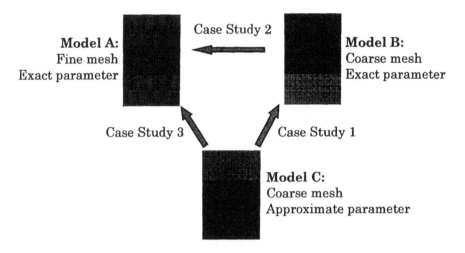

Fig. 6.17 Test case to explore different types of model error.
(a) Model A — Reference model; (b) Model B — Experimental
model; (c) Model C — Analytical model

case, the modifications required combine the adjustments necessary to
correct the true parameter errors and those which are introduced to
compensate for the coarse discretisation of the initial analytical model.
Close inspection of these adjustments in the final case show them to be
sensibly the sum of the corrections found in cases 1 and 2, respectively.

6.3.10   Concluding Discussion

To some extent, many of the summarising comments on this topic have
already been made in the introduction to this section. It will be seen
that there exist a considerable variety of methods and algorithms for
the task of updating a theoretical model to match measured data and,
at first, this may seem to be surprising. Why is there not just a single
approach to the task? It is suspected that the answer lies in the
enormity (some say impossibility!) of the task which seeks to adjust the
values of thousands of coefficients or elements in a theoretical model
using just hundreds of measured data. The problem is further
complicated by the fact that the model being updated may not possess
the correct configuration (so that the coefficients are not strictly valid
ones, anyway) and by the possibility that the experimental data are not
all independent of each other so that there is even less information
available from the tests than initially believed. Against this background
we can present a realistic perspective of the task for the updating
analyst.

Having said that, it is appropriate to view the various methods and algorithms as a set of tools that the analyst (and it remains the role and task of the modeller to do this) can use to seek a better model than the one that was created initially. He or she should learn to use these various numerical indicators provided by the updating routines to guide the improvement of the model, first to one that can be verified as possessing the correct configuration (or order, or connectivity), and then to one which has acceptable numerical values so that it can be declared valid for its intended purpose. It is considered important that the second phase should not be undertaken until and unless the first stage, verification, has been accomplished, since much time and effort can be wasted seeking compensation for an inadequate model by updating its parameters. In the majority of cases, a model which cannot be verified cannot be updated. Hence, great care should be taken in interpreting the results of an updating process to check that progress is really being made towards finding the correct values for the model coefficients and that the algorithm is not seeking to compensate for an inadequate form of model.

Concerning the multiplicity of methods and algorithms, as one of the major difficulties is the under-determination of the problem, there is probably some benefit in trying more than one method, or certainly in making more than one try with a given method, in the hope of eliciting more information from the given data. As with other applications, the absence of valuable information regarding the out-of-range high-frequency modes is a drawback, especially with the modal-based methods which expressly exclude any information which is available in the original measured data. Refined methods which can access this information, such as those which make direct use of response data, or which include antiresonances as well as resonant frequencies, may well have an important role to play in the future evolution of the updating technology as it is known that such data are rich in this additional information.

6.4   COUPLED AND MODIFIED STRUCTURE ANALYSIS
6.4.1   Basic Rationale and Concepts of Coupled/Modified Structure Analysis

Preceding sections in this chapter were concerned with the task of validating, or obtaining and refining, a mathematical model of a given structure. Once this task is complete, we are in a position to use that model for some further purpose (this is, after all, why it was desired to produce the model in the first place) and the following sections will address a number of applications to which such models may usefully be put. One of the most powerful applications for the models which can be developed by modal testing is to the general group of analysis methods known as 'structural modification', 'coupled structures', 'substructuring' or 'structural assembly' applications. Although these may appear to be quite disparate applications, they do in fact all share exactly the same analysis procedures and requirements in terms of the models supplied from the modal test. As a result, it is convenient to present them as a group, rather than as different methods, as is often done.

6.4.1.1 Rationale for structural modification

The first application we shall consider follows on directly from the process of obtaining and verifying the basic structure's model in that it seeks to predict the effect of making modifications to the structure. These modifications may be imposed by external factors — e.g. design alterations for operational reasons — and in this case it will generally be necessary to determine what changes in dynamic properties will ensue from introducing the modifications as these might be detrimental, for example by moving closer to a resonance condition than applied before the changes. Another possibility is that it may be required to change the dynamic properties themselves, perhaps to avoid a resonance or to add more damping, and then it is important to know how best to go about modifying the structure so as to bring about the desired changes in dynamic behaviour without at the same time introducing some new unwanted effects. (All too often in the past has a modification been made to move a natural frequency 'by trial and error' only to find a different mode moving into prominence at a different frequency.)

In both these cases, a technique which permits the prediction of all the changes in dynamic properties resulting from a given structural modification will be of considerable value. As with many of these application areas, basic concepts such as the one outlined below have been extensively developed and refined and the reader is referred to the specialist literature for discussion of these more advanced aspects. However, we present here the basis of the techniques generally referred to as 'structural dynamic modification' methods. We have chosen to present the principles using an undamped system as it is usually the location of the resonance frequencies which is of greatest importance and the inclusion of damping makes little difference in this respect. However, the same approach can be made using the general damped case by appropriate extension of the theory.

6.4.1.2 Rationale for structural coupling

There are many instances in which it is convenient to be able to consider a complex engineering structure as an assembly of simpler components, or substructures. For example, the theoretical analysis of a large structure can often be made much more efficiently if this is broken down into its component parts, each of these then analysed separately and the whole assembly reconstituted in terms of the models of each individual component. The increase in efficiency thus gained derives from the facility of describing a (sub)structure's essential dynamic characteristics very compactly in terms of its modal properties. Another way of viewing the process, in matrix terms, is to note that the spatial model of a typical large-scale structure would tend to have extensive regions of its mass and stiffness matrices populated by zeroes. In effect, a substructure type of analysis concentrates on submatrices centred on the leading diagonals of the complete system matrices, not involving the remote, null, regions.

One particularly powerful application of the substructure approach is to a range of problems where it is required to combine subsystem or component models derived from different sources — perhaps from quite disparate analyses but often from a mixture of analytical and experimental studies. Thus we may seek to combine component models from theoretical analysis with others from modal tests.

6.4.1.3 Concepts and notation of coupled/modified structure analysis

The general principles of this class of application can be demonstrated by the simple system shown in Fig. 6.18(a), which consists of two separate and distinct subsystems, or structural elements, A and B, which are to be combined to form the coupled or connected system, C. In this very simple example, we see one system with N_A DOFs being coupled to a second system with N_B DOFs, with the resulting coupled system having N_C DOFs, where $N_C \neq N_A + N_B$. In this particular example, we see that the correct expression for the number of DOFs in the coupled system is $N_C = N_A + N_B - 2$, and this is the case because the two components are connected or coupled at two of the DOFs in each subsystem with the result that the total number of DOFs after the coupling is two less than the sum of the numbers of DOFs on the separate subsystems. Thus we have introduced the concept of the coupling DOFs, or coupling 'coordinates', as being those DOFs which are essential to the analysis of the coupled system and which determine the number of DOFs in the final assembly of the two (or more) subsystems. The procedure described here can, of course, be extended to include as many subsystems as required.

At this early stage we can introduce the difference between a structural assembly and structural modification. If one of the subsystems, say element B, comprised a rather simpler structure with only two DOFs — the 2DOFs which are involved in the coupling — then the assembled structure would have just the same number of DOFs as the original subsystem, A. In this case we can talk of structure A being

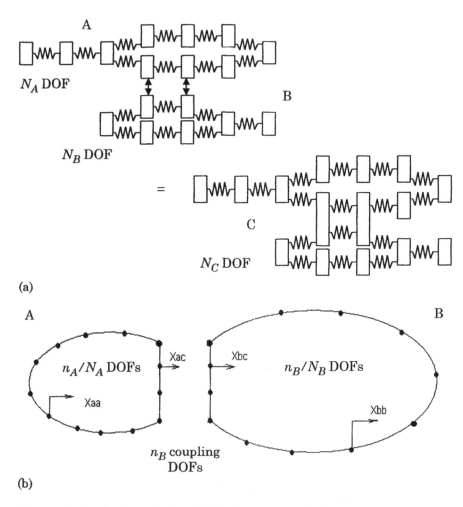

(a)

(b)

Fig. 6.18 Basis of coupled/modified structure analysis.
(a) Simple system; (b) General system

modified by the addition of subsystem B, although clearly the combined
system, C, is still a coupled or assembled structure in the strict sense of
that term. Thus we see that coupled and modified structures really are
one and the same thing at a theoretical level.

An extension of this simple example of the concept of coupled
structures is illustrated in the second Fig. 6.18(b) which again shows
two substructures. Here again we see the separate components, A and
B, which are to be coupled to form the complete system, C. Each
component is described in terms of a number of active DOFs, n_A and

n_B, respectively, so that the assembled or coupled structure is described by n_C DOFs. If the number of common coupling DOFs is n_c, then we can see that $n_C = n_A + n_B - n_c$. The term 'active' DOFs is used in this context because these numbers (n_A and n_B) are not necessarily the numbers of DOFs that are included in the full description of the components, which would be expressed as N_A and N_B. It is perfectly acceptable, for the purpose of a coupled or modified structure analysis such as we are contemplating here, to confine our interest to a limited number of DOFs, without in any way simplifying the description of the model of the dynamic behaviour of the structures. We do this by restricting our access to information about the response and/or excitation at certain DOFs. This is a standard procedure in many types of analysis for complex structures — and many applications of coupled structure analysis are undertaken in this substructuring way in order to make a large and complicated model more efficient - and represents a very practical approach as, in the great majority of situations, there are many DOFs at which there is neither excitation applied nor interest in the response, but there is a compelling need to ensure that the responses which are computed are accurate and not approximated by a reduction or condensation process which has been applied to reduce the size of the model and its computation.

So, to summarise these concepts, it may be useful to show how the vectors containing the DOFs for each component in a coupled or modified structure analysis are subdivided:

$$\{x_A\}_{N_A \times 1} = \left\{ \begin{array}{c} x_a \\ x_{A-a} \end{array} \right\} = \left\{ \begin{array}{c} \text{'active'} \\ \text{'inactive'} \end{array} \right\}$$

$$\{x_a\}_{n_A \times 1} = \left\{ \begin{array}{c} x_{ac} \\ x_{aa} \end{array} \right\} = \left\{ \begin{array}{c} \text{'coupling' or 'master'} \\ \text{'slave' or 'passenger'} \end{array} \right\} \; ; \; \{x_{ac}\}_{n_c \times 1} \; ; \; \{x_{aa}\}_{n_a \times 1}$$

As mentioned earlier, it is a straightforward matter to extend these concepts beyond this simple two-component assembly to include the more general case where the structure is assembled from many components.

6.4.2   Approaches to the Analysis of Coupled Structures

There are two different approaches to the analysis of coupled or modified structures (in addition to the obvious one of constructing a spatial model that contains all the components explicitly — which is not considered here as it offers no advantage or economy, which is the drive behind the whole idea of structure assembly analysis). These are methods based on (i) modal models and (ii) response models of the various structures and substructures involved. Simply put, the former methods use the modal properties of the individual components to yield the modal properties of the assembled or modified structure, while the latter group derive the assembled structure's response characteristics based on the corresponding response properties for each component individually.

Each of these two approaches has its advantages and disadvantages so that there is no clear 'correct' way to proceed. Here we shall seek to explain the principles and the major features of each approach, prior to presenting the essential methodology of each. It should be borne in mind at this stage, however, that we are here concerned with applications of these ideas to experimentally-derived models of the substructures and so we shall be heavily influenced by the ease or difficulty of providing the necessary information about each substructure from that perspective.

6.4.2.1 Modal methods
The modal model methods are more elegant and more concise in presentation, demanding fewer calculations than the response methods. However, it is necessary to define the properties of each substructure as a modal model and that means that a full modal test and analysis must have been successfully completed before the coupling can commence. While this is no more than the expectation of a standard modal test, there can be complications because the basic theory for coupled structure analysis using modes is intended primarily for undamped systems with real modes and although that theory can be extended to damped systems with complex modes, the successful implementation of such an extension is not universally achieved. The other problem with the modal approach is that it is very sensitive to the number of modes that are and need to be included in the modal model. Strictly, all the modes are required for each substructure for the analysis to be correct and this is clearly not feasible in the context of an experimentally-derived model. The issues of just how many modes can be neglected and the model still yield useful results, or how to compensate for the missing modes (usually those at higher frequencies), remain the main preoccupations of the users of this type of analysis.

It should be noted at this stage that 'modes' of a substructure generally refer in this work to the modes of that component for the conditions where the intended coupling boundary is completely free — the sometimes called 'free-interface modes'. The origins of some of the modal coupling analysis procedures can be traced back to the early days of developments of the finite element method when computer capacity continually presented a barrier to the effective application of these methods to increasingly complex structures. A solution to this problem of modal incompleteness, or modal truncation, was often found in those early theoretical applications by the use of a second set of component modes — the so-called 'constraint' modes — which relate to the modes of the substructures with different boundary conditions imposed, specifically ones where the intended coupling boundary was assumed to be grounded. Such a technique is very effective at resolving the problems which arise when only a limited number of the free-boundary modes are available but, unfortunately for our applications here, these constraint modes are not at all easily obtained in the experimental context that we are considering here. It is simply not practical to reconfigure the test set-up so as to provide grounded boundary conditions at each of the coupling DOFs. Even if these DOFs were readily accessible, the feasibility of providing a sufficiently rigid fixture to approximate the required grounded condition is extremely remote.

Some attempts to provide a degree of the compensation afforded by these theoretical constraint modes can be provided by using what we have previously called perturbed boundary conditions, but that complicates the analysis considerably and should not be entered into lightly as a remedy for an inability to measure enough modes to yield a representative modal model of (one of) our substructures.

One final comment regarding the modal approach: it is often the case that the final output of a coupled structure analysis will be of its response properties, and it is also the case that the origins of the modal model of each substructure that is based on a modal test will be response measurements. It should thus be recognised that any coupled structure analysis which uses modes as the basis of the coupling will be required to make a double transformation between the response and the modal models — once, in order to convert measured responses into the modal model and a second time to synthesise the required responses of the final assembly from its constituent modes. The first of these processes introduces errors from the modal analysis process and the second introduces errors from the truncation implicit in the existence of a limited number of modes. It is thus not surprising to find that there are an alternative set of coupled structure analysis methods which are based on the response properties directly, and these are introduced next.

6.4.2.2 Response methods

The alternative approach to analysing the dynamics of an assembly of two or more substructures is by combining the response properties of each component separately. Most of the methods currently in use do this by means of the FRF data: what is required from each component is the FRF matrix relating all the active DOFs for each substructure as a separate component, with its boundaries at the intended coupling DOFs free. In our simple example with two substructures, A and B, the 'input' to this coupling process will comprise two square FRF matrices, one $n_A \times n_A$ and the other $n_B \times n_B$ which will then be combined to yield a corresponding FRF matrix for the coupled structure, C, which is $n_C \times n_C$. The most common method of analysis derives an expression for the coupled structure FRF as a simple combination of the two component structure FRF matrices by applying the conditions of equilibrium and compatibility.

The analysis using this approach is extremely simple and has the advantage of making direct use of the quantities which are actually measured in the modal test. It is particularly important to note that as the analysis is performed directly at the frequencies of interest, there is no inherent limitation due to the truncation problem encountered in the modal approach and so the number of modes encompassed by the modal test is not an issue, so long as the test covers the entire frequency range that is of interest to the final model. However, this approach has as a disadvantage, the requirement that the full matrix of FRFs must be supplied, at every frequency of interest to the final assembled structure. This means, first of all, processing a large amount of data, much more than in the case of the modal model, and specifically of a large number of response functions. In a conventional modal test, not all the FRFs in the matrix are measured: usually, only those in one row or a few, are measured directly and so the need to provide the full matrix either requires the measurement of much more data than would normally be called for, or some means of deriving (synthesising) the unmeasured functions for the purpose of the coupling analysis. This feature is a particularly onerous requirement in the case of systems with non-symmetric matrices, such as the cases discussed earlier in respect of structures with rotating components, although in some of those cases, it may well be possible to exploit the skew-symmetry features which are the primary source of non-symmetry so that only one half of the full FRF matrix needs to be defined in full.

6.4.3   FRF Methods of Coupled Structure Analysis — General

For reasons that will become clear as we progress through this section, it is convenient to describe the response methods of analysis before those of the modal approach. We shall deal first with the method which works in terms of the frequency response properties directly, and which provides as output the frequency response characteristics of the coupled structure. Having said that, however, it is quite possible that the required FRF data may be obtained from a modal model of the components concerned as that is sometimes the most effective way of storing information on a substructure's dynamic characteristics.

This method is often referred to as the 'impedance coupling method'

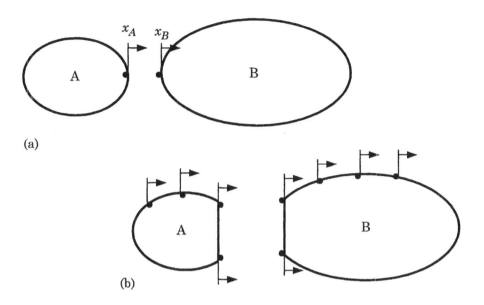

Fig. 6.19 Basis of FRF coupling analysis.
(a) Simple case; (b) General case

or the 'dynamic stiffness method'. The basic principle is demonstrated by the simple example shown in Fig. 6.19(a) in which the two components A and B are to be connected by the single coordinate x to form C. It should be recalled that the number of DOFs used in the coupling process (here, one) does not restrict the number of degrees of freedom which may be possessed by each component except that this latter number should be at least as great as the former. Thus, in this case, components A and B may both possess several degrees of freedom each, even though only one DOF is included for the purpose of coupling the components together. The implication of the notation and analysis which follows is that the system behaviour is not fully described in the spatial sense. However, those coordinates which are included will exhibit the full range of resonances possessed by the system.

If we consider the dynamics of each component quite independently, we can write the following equation for subsystem A when a harmonic force $f_A(t) = F_A e^{i\omega t}$ is applied at the connection DOF, then we can write:

$$X_A e^{i\omega t} = H_A(\omega) F_A e^{i\omega t} \quad \text{or} \quad X_A = H_A F_A \qquad (6.31)$$

and similarly for subsystem B:

$$X_B = H_B \, F_B \tag{6.32}$$

Now, if we consider the two components to be connected to form the coupled system C, and we apply the conditions of compatibility and equilibrium which must exist at the connection point, we find:

$$X_C = X_A = X_B$$
$$F_C = F_A + F_B \tag{6.33}$$

so that:

$$\frac{1}{H_C} = \frac{1}{H_A} + \frac{1}{H_B} \tag{6.34}$$

which can also be written as:

$$H_C^{-1} = H_A^{-1} + H_B^{-1}$$

or

$$H_C^{-1} = Z_C = Z_A + Z_B \tag{6.35}$$

Clearly, these expressions can readily be extended to the case where there are several DOFs involved in the coupling process, although no other DOFs are included in the analysis:

$$[H_C]^{-1} = [H_A]^{-1} + [H_B]^{-1} \tag{6.36}$$

Thus we obtain the FRF properties of the combined system directly in terms of those of the two components as independent or free subsystems. The equation can be expressed either in terms of the receptance (or mobility) type of FRF, as shown, or rather more conveniently in terms of the inverse version, dynamic stiffness (or impedance). Although it makes little difference in this instance, when we extend the analysis to the more general case where the coupling takes place at several DOFs, and there are also DOFs included which are not involved in the coupling, the impedance formulation is much more straightforward algebraically.

It will be noted that equation (6.36) has a form which renders it inconvenient for numerical application because it involves no less than three matrix inverse operations. In order to improve this situation, the following analytical development can be made:

$$[H_C]^{-1} = [H_A]^{-1} + [H_B]^{-1}$$
$$= [H_A]^{-1} \left([I] + [H_A][H_B]^{-1}\right)$$
$$= [H_A]^{-1} \left([H_B] + [H_A]\right)[H_B]^{-1}$$

so that

$$[H_C] = [H_B]\left([H_B] + [H_A]\right)^{-1}[H_A]$$ (6.37a)

which is a much more efficient formula from the numerical viewpoint. Yet another version can be readily derived from (6.37a) in a format which is more appropriate for modification applications, as:

$$[H_C] = \left([H_A] + [H_B] - [H_A]\right)\left([H_B] + [H_A]\right)^{-1}[H_A]$$
$$= [H_A] - [H_A]\left([H_B] + [H_A]\right)^{-1}[H_A]$$ (6.37b)

The above simple analysis can be extended to the more general case illustrated in Fig. 6.19(b). Here it is convenient to note that the receptance FRF properties for component A are contained in a matrix which can be partitioned as shown below, separating those elements which relate to the coupling DOFs from those which do not:

$$[H_A] = \begin{bmatrix} H_{aa}^A & H_{ac}^A \\ H_{ca}^A & H_{cc}^A \end{bmatrix}$$ (6.38)

This receptance FRF matrix can be used to determine the corresponding impedance FRF matrix as follows:

$$[Z_A] = [H_A]^{-1} = \begin{bmatrix} Z_{aa}^A & Z_{ac}^A \\ Z_{ca}^A & Z_{cc}^A \end{bmatrix}_{n_A \times n_A}$$ (6.39a)

Similarly, we can write a corresponding impedance FRF matrix for the other component, B, as:

$$[Z_B] = [H_B]^{-1} = \begin{bmatrix} Z_{bb}^B & Z_{bc}^B \\ Z_{cb}^B & Z_{cc}^B \end{bmatrix}_{n_B \times n_B} \tag{6.39b}$$

By an application of the same equilibrium and compatibility conditions as used before, we can derive both a receptance an impedance version of the FRF matrix for the coupled structure of the form:

$$[H_C]_{n_C \times n_C}^{-1} = [H_A]_{n_A \times n_A}^{-1} \oplus [H_B]_{n_B \times n_B}^{-1}$$

where

$$[Z_C]_{n_C \times n_C} = [Z_A]_{n_A \times n_A} \oplus [Z_B]_{n_B \times n_B}$$

$$= \begin{bmatrix} Z_{aa}^A & 0 & Z_{ac}^A \\ 0 & Z_{bb}^B & Z_{bc}^B \\ Z_{ca}^A & Z_{cb}^B & \left(Z_{cc}^A + Z_{cc}^B \right) \end{bmatrix}_{n_C \times n_C} \tag{6.40}$$

This expression clearly provides a rather inefficient means of deriving the required FRF matrix for the coupled structure, $[H_C]$, because it requires the inverses of three matrices, of orders, n_A, n_B and n_C, respectively. As these numbers are likely to be relatively large, and for the corresponding number of coupling DOFs (n_c) to be relatively small, this is a calculation of a much greater order than the essential one in which only the coupling DOFs are included (equation 6.36). Fortunately, it is possible to make an extension of the analysis shown in equation (6.37) so that we can write:

$$[H_C] = \begin{bmatrix} H_{aa}^C & H_{ac}^C & H_{ab}^C \\ H_{ca}^C & H_{cc}^C & H_{cb}^C \\ H_{ba}^C & H_{bc}^C & H_{bb}^C \end{bmatrix}$$

$$= \begin{bmatrix} H_{aa}^A & H_{ac}^A & 0 \\ H_{ca}^A & H_{cc}^A & 0 \\ 0 & 0 & H_{bb}^B \end{bmatrix} - \begin{bmatrix} H_{ac}^A \\ H_{cc}^A \\ -H_{bc}^{AB} \end{bmatrix}_{n_C \times n_c}$$

$$\left[H_{cc}^A + H_{cc}^B \right]_{n_c \times n_c}^{-1} \left[H_{ca}^A \quad H_{cc}^A \quad - H_{cb}^B \right]_{n_c \times n_C} \tag{6.41}$$

This equation constitutes the most general formula for this approach to coupled/modified structure analysis and can be seen to be dramatically more efficient computationally than the earlier expression in equation (6.40). This characteristic is a direct result of the single (rather than triple) inversion that must be computed at each frequency, ω, and the fact that the order of that single inverse is just n_c, in place of the three matrices of orders n_A, n_B and n_C, respectively. (This is particularly significant because typical practical cases might be expected to possess values for these indices of the order of, say 300, 400 and 650 for n_A, n_B and n_C, and of 50 for n_c, resulting in an enormous difference in the computation time to obtain each point on the frequency scale — of the order of 1000 times, if the time to compute a matrix inverse is taken to be proportional to the (order)^3.

Case study

An example of the application of this method to a practical structure is shown in Figs. 6.20 and 6.21. The problem called for a complete system model of the helicopter/carrier/store assembly shown in Fig. 6.20(a). The essential components or subsystems are illustrated in Fig. 6.20(b) and it was decided to study each by the method felt to be most appropriate in each case: theoretical models for the store and struts plus a modal model from modal tests for the airframe (although a model which is confined to the points of interest on the side of the fuselage) and likewise for the platform. Because of the construction of the system, it was necessary to include several rotation coordinates at the connection points in order to create a truly representative model and where these were required in the experimentally-derived models, the appropriate FRF data were measured using the method described in Section 3.11. An example of a typical measured airframe mobility, together with its regenerated curve based on the SDOF curve-fit method of Section 4.3, is shown in Fig. 6.21(a). Also, a corresponding result for the platform and store substructure, this time analysed using the lightly-damped structures MDOF modal analysis method of Section 4.4.4, is shown in Fig. 6.21(b). One of the set of final results for the complete model, constructed by impedance coupling, is shown in Fig. 6.22 together with measurements made on the complete assembled structure — a result not untypical of what can be expected for this type of complex engineering structure, [63].

If it were required to determine the modal properties of the combined assembly, rather than the FRF properties shown here, then it would be necessary to subject these last FRFs to some form of modal analysis or to use a different approach to the whole analysis, such as described in the next sections.

One final comment which should be made is to note or to recall that

(a)

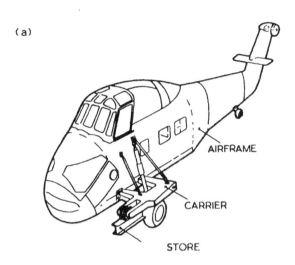

AIRFRAME

CARRIER

STORE

(b)

HELICOPTER
FUSELAGE

ATTACHMENTS WITH
SELF-ALIGNING PINNED JOINTS

STRUTS

CARRIER
PLATFORM

ATTACHMENTS WITH
SIMPLE PINNED JOINTS

STORE

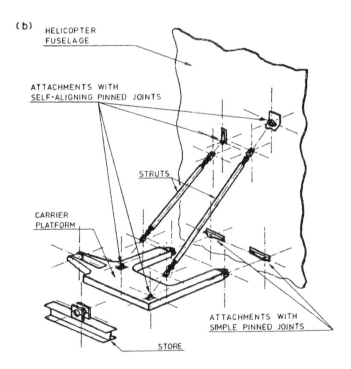

Fig. 6.20 Components and assembly for coupled structure analysis.
(a) Complete system; (b) Components and connections

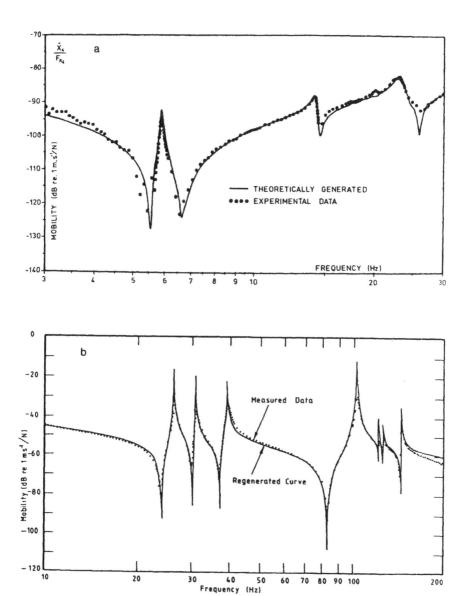

Fig. 6.21 Measured and regenerated FRF properties.
(a) Airframe (using SDOF analysis); (b) Carrier (using MDOF analysis)

484

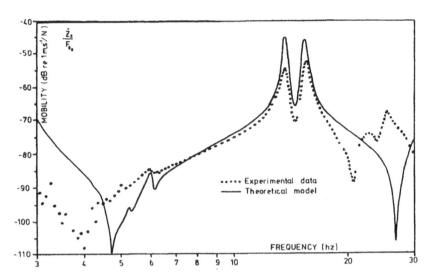

Fig. 6.22 Structural assembly FRF data: measured and predicted from coupled structure analysis

the FRF properties used in this approach must be as accurate as possible. This, in turn, requires that as many of the components' modes as possible should be included in any modal model or data base that is used to regenerate the receptances used in the analysis. Where it is unfeasible to include a near-complete set of modes, then it is necessary to ensure that those excluded are represented by some form of residual terms, as described in Section 4.3.6, in order to ensure that the regenerated FRF data are accurate away from the component resonances as well as close to them. The reader is referred back to Section 5.5 for a discussion on the importance of this aspect.

6.4.4 FRF Methods of Coupled Structure Analysis — Simplified Expressions for SDOF Connections

While the preceding section provides the most general form of the response function method of analysis for coupled and modified structures, there are a number of situations which make use of only the simplest versions of the expressions developed there. These situations tend to be concerned with applications of relatively simple modifications and also those where a study is being made to identify the best places to introduce modifications in order to bring about desired changes to the original structure's performance. The direct modification application poses the question: 'what will be the changes to the structure's dynamic properties if a specific modification is applied at a given point?'. This question can readily be answered using the structural modification/coupling analysis already presented. However, the more

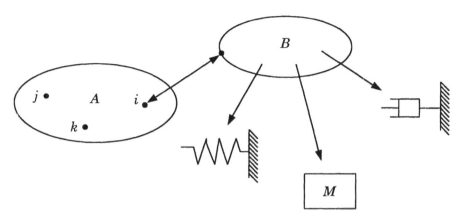

Fig. 6.23 Simple SDOF modification

subtle version of the same question, and the version which has more practical value, is: 'what modification should be made to the original structure in order to bring about a prescribed change to its dynamic properties?'. This second version is much more difficult to answer because there are so many possible solutions (so many alternative modifications that could be applied) but we can devise a systematic approach to answering it by using a development of the analysis presented above.

In this section we shall be concerned with simple, single-DOF, modifications or additions to the original structure and so we shall use the scalar, rather than the matrix, version of the equations given in equations (6.37) and (6.41).

Consider, as shown in Fig. 6.23, the original system and the possibility of adding at DOF i an SDOF modification, B, which consists of a mass, or a stiffness or a damper — or, indeed, an SDOF damped-spring-mass absorber. The FRF of the added component, H_B, can be expressed as:

$$H_B = \frac{1}{-\omega^2 m_B} \qquad \text{for a mass,}$$

$$H_B = \frac{1}{k_B} \qquad \text{for a stiffness, or}$$

$$H_B = \frac{1}{i\omega c_B} \qquad \text{for a damper,}$$

or

$$H_B = \frac{-\left(k + i\omega c - \omega^2 m\right)}{\omega^2 m\left(k + i\omega c\right)}$$ for the mass - spring - damper absorber subsystem

and we can use equation (6.41) to derive a general expression for the FRF of the system which is modified by the addition of B as:

$$H_{jk}^{A'} = H_{jk}^A - H_{ij}^A \, H_{ik}^A \left(H_{ii}^A + H_B\right)^{-1} \tag{6.42}$$

where either j or k might or might not be i. An example of using these formulae is given in Fig. 6.24 which shows the result of using equation (6.42) to compute the effect on the point FRF at a chosen point along a free-free beam of adding a mass to the measurement point. Fig. 6.24(a) shows the measured curve (as a series of points) plus the predictions made of the expected modified curve after a mass of 1 kg has been added to the measurement point. Then, in Fig. 6.24(b), the same predicted FRF is shown again, but this time with a second experimental curve, measured after the proposed modification had been added to the original test structure. The effectiveness of the prediction is clear, even though the quality of the prediction falls away towards the top end of the frequency range. (In fact, this growing discrepancy can be seen to be due to the limitations of assuming a single-DOF coupling: an assumption which becomes blatantly less and less realistic as the frequency of vibration increases, and thus the distortion of the structure, becomes increasingly complex and multi-directional.)

One interesting (and, often, useful) result that can be deduced from these equations is the fact that if we take a structure and ground it at a particular DOF, then the natural frequencies of the thus-grounded structure will be identical to the antiresonance frequencies of the structure at the chosen DOF in its original free configuration. (This result can be obtained from equation (6.42) by letting H_B tend to infinity.)

A second series of simple modifications can be constructed by the addition of a simple spring, or damped spring, which is attached between two DOFs on the primary structure, A, or which connects a single DOF in each of two structures, A and B. These situations are shown schematically in Fig. 6.25, and analysis similar to that used to derive equation (6.42) can also be made to derive the following expressions for the FRF data of the modified/coupled structures, A and B:

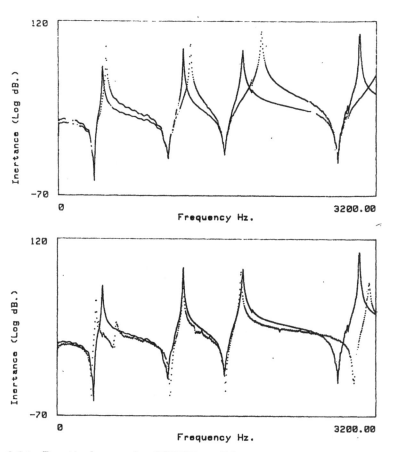

Fig. 6.24 Practical example of SDOF modification.
     (a) —— Original measured FRF; —— predicted modified FRF;
     (b) —— Measured modified FRF; —— predicted modified FRF

$$H_{ii}^{A'} = H_{ii}^{A} - \left(H_{ii}^{A} - H_{ij}^{A}\right)^2 \left(H_k + H_{ii}^{A} + H_{jj}^{A} - 2H_{ij}^{A}\right)^{-1}$$

and   (6.43)

$$H_{ij}^{A'} = H_{ij}^{A} + \left(H_{ii}^{A} - H_{ij}^{A}\right)\left(H_{jj}^{A} - H_{ij}^{A}\right)\left(H_k + H_{ii}^{A} + H_{jj}^{A} - 2H_{ij}^{A}\right)^{-1}$$

Note: if i and j are initially on separate components, then $H_{ij} = 0$.

A second version can also be derived in the case where the added connection is rigid (i.e. $H_k = 0$), in which case we find:

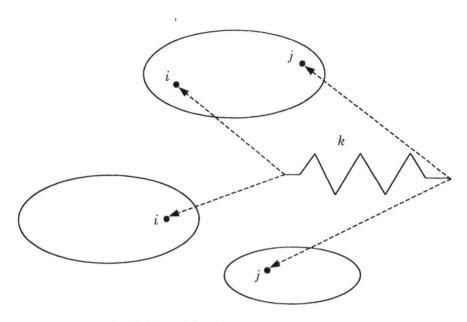

Fig. 6.25 Simple 2DOF modification

$$H_{ii}^{A'} = H_{ii}^{A} - \left(H_{ii}^{A} - H_{ij}^{A}\right)^{2}\left(H_{ii}^{A} + H_{jj}^{A} - 2H_{ij}^{A}\right)^{-1}$$

$$= H_{ij}^{A'} = H_{jj}^{A'} \tag{6.44}$$

because i and j are the same DOF. Also, for two DOFs, p and q, which are independent of the connection DOFs:

$$H_{pq}^{A'} = H_{pq}^{A} + \left(H_{pj}^{A} - H_{pi}^{A}\right)\left(H_{jq}^{A} - H_{iq}^{A}\right)\left(H_{ii}^{A} + H_{jj}^{A} - 2H_{ij}^{A}\right)^{-1} \tag{6.45}$$

In each of these simple cases, the reduced SDOF coupling condition is of restricted validity but is useful nonetheless in identifying trends and comparative modification effects.

6.4.5 Sensitivity to Simple Modifications

One additional feature that is worth noting from this study is the one whereby the sensitivity of the structure to these simple modifications is evaluated and then used to assess the effectiveness of introducing a given modification at a number of alternative sites. Fig. 6.26(a) shows the effect on a particular FRF from a simple beam of adding successively larger masses to the measurement point (calculations

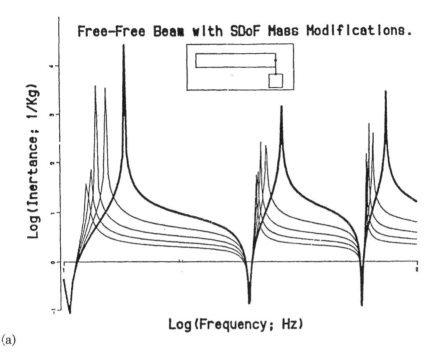

(a)

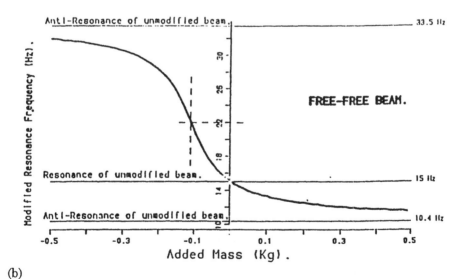

(b)

Fig. 6.26 Natural frequency sensitivity to structural modification.
(a) FRFs for different mass modifications at driving point;
(b) Variation of natural frequencies with added driving point
mass

made using the formula in equation 6.42a). From this plot is clear that adding progressively more and more mass at the drive point results in a progressive lowering all the resonance frequencies (as is expected) but that these drop asymptotically towards fixed values which coincide with the preceding antiresonances. This result can also be inspected using the plot on Fig. 6.26(b) which shows the changes in resonance and antiresonances explicitly. This latter plot also illustrates a property which is of direct use to us in our attempts to optimise structural modifications (by bringing about the desired changes with the minimum added material and with a minimum of unwanted secondary effects). That property is the sensitivity of each resonance frequency to adding a mass (or a spring, for the corresponding plot which could be drawn for added stiffnesses) at the measurement point for the reference FRF. This sensitivity is, simply, the slope of the corresponding natural frequency vs. added mass curve at the reference position of zero added mass — see Fig. 6.26(b). In Reference [64] a derivation is shown of an algebraic expression for this sensitivity which is as follows:

$$
\frac{\partial \omega_r}{\partial m_j} = \frac{C_{jj}\, \omega_r \displaystyle\prod_{s=1}^{m}\left(\omega_s^2\right) \displaystyle\prod_{i=1}^{(m-1)}\left(\left(\Omega_i^{jj}\right)^2 - \omega_r^2\right)}{\displaystyle\prod_{i=1}^{m-1}\left(\Omega_i^{jj}\right)^2 \displaystyle\prod_{s=1;\neq r}^{m}\left(\omega_s^2 - \omega_r^2\right)}
$$

$$
= -\frac{1}{\omega_r^2}\frac{\partial \omega_r}{\partial k_j} \tag{6.46}
$$

where r is the resonance in question, j is the DOF to which the FRF and modification apply, and m is the number of modes visible in the FRF. It will be seen that these sensitivity coefficients can be computed for a selection of potential modification sites by making point FRF measurements at the DOFs in question. Then, based on an inspection of the resulting values, an optimum location can be selected which has the desired effect in terms of changing certain natural frequencies without necessarily changing others. A practical example of the application of this technique to an aerospace structure is shown in Fig. 6.27, where (a) the original measured, (b) the predicted modified and (c) measured modified FRFs are shown, confirming the effectiveness of the simple analysis method in cases such as this.

6.4.6   Derivation of the Component FRFs
There are several ways of providing the FRF data which are required by the various processes described above to analyse coupled structures,

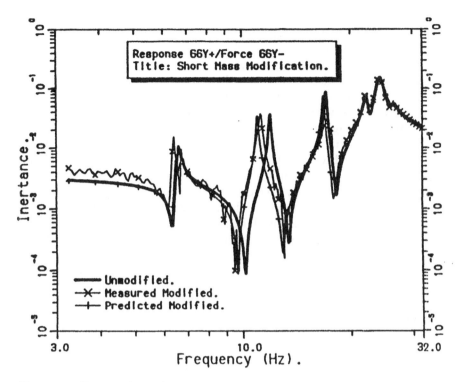

Fig. 6.27 Practical application of structural modification to aerospace structure

and these include:

(i)   direct computation from mass and stiffness matrices: note that
        $[Z(\omega)] = [H(\omega)]^{-1} = ([K] - \omega^2 [M])$;
(ii)   direct analysis of beam-like components for which analytical
        expressions exist for the required receptance or impedance
        properties;
(iii)   direct measurement of a receptance or mobility FRF matrix;
(iv)   extension of (iii) by smoothing the measured FRFs before further
        computation so as to minimise adverse effects of ill-conditioning
        caused by noisy data (this smoothing can be achieved using the
        samecurve-fitting procedures used in the modal analysis process);
        or
(v)   computation of receptance FRF data from a modal model, using
        the formula:

$$[H(\omega)] = [\Phi] \left[\cdot \cdot \left(\lambda_r^2 - \omega^2 \right) \cdot \cdot \right]^{-1} [\Phi]^T \qquad (6.47)$$

In the last three of these options, it may be necessary to invert the receptance FRF matrix originally supplied in order to obtain the impedances which are necessary for the some parts of the coupling process and care must be taken to prevent this inversion from becoming ill-conditioned. In case (iii), this will be a possibility because of the inevitable small errors contained in the measured data. This can be a particular problem with lightly-damped structures when there is likely to be a 'breakthrough' of component resonances in the coupled structure response characteristics: sometimes there appear spurious spikes on the coupled structure FRF plots in the immediate vicinity of the natural frequencies of one or other of the separate components. This is less of a problem with less strongly resonant components, such as those with some damping and a relatively high modal density. In this latter case, the direct use of raw measured FRF data may be a more attractive prospect than the alternative of deriving a modal model by curve-fitting what might well be very complex FRF curves.

In the fifth of the above approaches, there is a requirement that the receptance matrix is not rank-deficient so that its inverse **does** exist. In order for this condition to be satisfied, it is necessary that the number of modes included in the modal model is at least as great as the number of DOFs used to describe that particular component. At the same time, consideration must be given to the possibility that a matrix of residual terms may be required in order to account for the effect of those out-of-range modes which are not included in the modal model, as discussed in Section 5.5. In many cases, the only way of avoiding the need for obtaining such a residual matrix is by including many more modes in the components' modal models than are contained within the frequency range of interest; e.g. where the final coupled structure properties are required between, say, 40 and 400 Hz, it may be necessary to include all the components' modes in the wider frequency range 0 to 1000 Hz (or even higher). **It should be emphasised here that this problem (of incompleteness in the model derived from tests) constitutes one of the most serious limitations to the successful application of coupled and modified structure analysis.** At the same time as this warning is issued, it is also timely to recall that **all** the DOFs which are involved in the coupling must be included in the substructure models. The difficulty that this often represents is the need to include rotational DOFs and these, as has been discussed in earlier chapters, can be much more difficult to obtain than their translational counterparts. It is for this application that methods to derive these RDOF data indirectly using a mode expansion approach were discussed in Section 5.3.3.

6.4.7 Modal Analysis of Coupled and Modified Structures

As mentioned earlier, there are also methods for analysing coupled and modified structures which are based on modal models, and we shall summarise the more important of these next. Also as mentioned earlier, many of these modal methods are derived from the substructuring methods developed for finite element modelling and are somewhat less readily adapted for our current application — where the structural models are to be derived from test data.

6.4.7.1 Simple modification analysis

The first method to be outlined here comprises the basis of many of the first-level structural modification routines that became available as part of the modal analysis packages of the 1980s and early 1990s and is aimed specifically at this application.

For the general NDOF system, we can write the equations of motion for free vibration in the form:

$$[M]_{N \times N} \{\ddot{x}\} + [K]\{x\} = \{0\} \tag{6.48}$$

or, using the standard transformation to modal coordinates, as:

$$[I]_{N \times N} \{\ddot{p}\} + [\omega_r^2]\{p\} = \{0\} \tag{6.49}$$

where

$$\{x\} = [\Phi]\{p\}$$

This full set of equations can be reduced in the event that only m of the modes are known, to:

$$[I]_{m \times m} \{\ddot{p}\} + [\omega_r^2]\{p\} = \{0\}$$

Now suppose that we wish to analyse a modified system, whose differences from the original system are contained in the matrices $[\Delta M]$ and $[\Delta K]$. (Note: it is assumed that the same DOFs are used in both cases: methods for analysing rather more substantial modifications such as adding components will be discussed in the next section.) The equations of motion for the modified system may be written as:

$$([M] + [\Delta M])\{\ddot{x}\} + ([K] + [\Delta K])\{x\} = \{0\} \tag{6.50a}$$

or, using the original system modal transformation (noting that the eigenvectors and eigenvalues are NOT those of the new system but of the unmodified structure), as:

$$\left([I]+[\Phi]^T [\Delta M][\Phi]\right)\{\ddot{p}\} + \left([\omega_r^2]+[\Phi]^T [\Delta K][\Phi]\right)\{p\} = \{0\} \tag{6.50b}$$

In general, we shall have information about only m ($< N$) modes and these will be described at only n ($< N$) DOFs. However, using this equation we have established a new equation of motion with a new mass matrix $[M']$ and stiffness matrix $[K']$, both of which can be defined using the modal data available on the original system (such as might be provided by a modal test) together with a description of the changes in mass and stiffness which are to constitute the structural modification:

$$[M']_{m \times m} = [I]_{m \times m} + [\Phi]^T_{m \times n} [\Delta M]_{n \times n} [\Phi]_{n \times m}$$
$$[K'] = [\omega_r^2] + [\Phi]^T [\Delta K][\Phi] \tag{6.51}$$

The eigenvalues and eigenvectors of these new mass and stiffness matrices can be determined in the usual way, thereby providing the natural frequencies and mode shapes of the modified structure.

It is worth adding one or two comments concerning the implications of using this technique on real engineering structures. These all stem from the fact that it is much easier to specify changing individual elements in a mass or stiffness matrix than it is to realise such changes in practice. For example, if we wish to add a mass at some point on a structure, it is inevitable that this will change the elements in the mass matrix which relate to the x, y and z directions at the point in question and will also have an effect on the rotational motions as well since any real mass is likely to have rotary inertia as well. This means that it is seldom possible or realistic to consider changing elements individually, and also that it may be necessary to include rotational DOFs in the original modal model. This last consideration is seldom made, thanks to the difficulty of measuring rotations, but should be if reliable modification predictions are to be made. Similar comments apply to the stiffness matrix: the attachment of any stiffener, such as a beam or strut, will influence the stiffness in several directions simultaneously, including rotational ones. Lastly, it must be noted that this method, in common with all which rely on a modal data base that may not include all the structure's modes, is vulnerable to errors incurred if the effects of the modes omitted from the modal model — typically, the high-frequency modes — are (a) not negligible and (b) ignored. This point will be discussed further in the next section.

6.4.7.2   General coupled structure modal analysis

The second method to be outlined here is a more general approach to the analysis of coupled structures using the modal models of each component and deriving the model of the assembled structure directly in a modal format. As with the other methods which have been described in this section, we present here just the basic method and the reader is referred to more specialist works for a detailed discussion of the methods.

The basic method applies to the two-substructure system shown in Fig. 6.28 with a connector element, *CC*, which comprises just stiffness

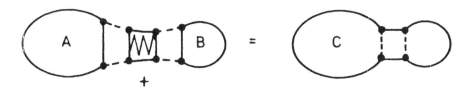

Fig. 6.28   Assembled structure with connector component

and damping elements and represents the existence in real structures of an interface between any two components. As in the previous section, we can write the equations of motion for each of the two components separately as:

$$[M_A]_{N_A \times N_A} \{\ddot{x}_A\} + [K_A]\{x_A\} = \{f_A\}$$

and   (6.52)

$$[M_B]_{N_B \times N_B} \{\ddot{x}_B\} + [K_B]\{x_B\} = \{f_B\}$$

where the only forces present are those at the connection DOFs. These equations can be transformed into the modal coordinates as follows:

$$[I]_{m_A \times m_A} \{\ddot{p}_A\} + [\omega_A^2]\{p_A\} = [\Phi_A]_{m_A \times m_A}^T \{f_A\}$$

and

$$[I]_{m_B \times m_B} \{\ddot{p}_B\} + [\omega_B^2]\{p_B\} = [\Phi_B]_{m_B \times m_B}^T \{f_B\}$$

and combined to yield:

$$
\begin{bmatrix} I & 0 \\ 0 & I \end{bmatrix} \begin{Bmatrix} \ddot{p}_A \\ \ddot{p}_B \end{Bmatrix} + \begin{bmatrix} \omega_A^2 & 0 \\ 0 & \omega_B^2 \end{bmatrix} \begin{Bmatrix} p_A \\ p_B \end{Bmatrix} = \begin{bmatrix} \Phi_A^T & 0 \\ 0 & \Phi_B^T \end{bmatrix} \begin{Bmatrix} f_A \\ f_B \end{Bmatrix} \tag{6.53}
$$

where m_A, n_A, etc. represent the numbers of included modes and component DOFs, respectively, and are both $< N_A$, etc.. These two sets of equations are in fact linked through the connecting element, CC, by the relationship:

$$
\begin{Bmatrix} f_A \\ f_B \end{Bmatrix}_{(n_A + n_B) \times 1} = [K_{CC}] \begin{Bmatrix} x_A \\ x_B \end{Bmatrix} = [K_{CC}] \begin{bmatrix} [\Phi_A] & 0 \\ 0 & [\Phi_B] \end{bmatrix} \begin{Bmatrix} p_A \\ p_B \end{Bmatrix} \tag{6.54}
$$

If there are no external forces applied to the coupled structure, then we may write:

$$
[I] \begin{Bmatrix} \ddot{p}_A \\ \ddot{p}_B \end{Bmatrix} + \left(\begin{bmatrix} \Phi_A^T & 0 \\ 0 & \Phi_B^T \end{bmatrix} [K_{CC}] \begin{bmatrix} \Phi_A & 0 \\ 0 & \Phi_B \end{bmatrix} + \begin{bmatrix} \omega_A^2 & 0 \\ 0 & \omega_B^2 \end{bmatrix} \right) \begin{Bmatrix} p_A \\ p_B \end{Bmatrix} = \{0\} \tag{6.55}
$$

and this equation can be solved for the eigenvalues and eigenvectors of the coupled system. The number of modes of this coupled system will be seen to be determined by the number of modes included in the coupling process (m_A, m_B), rather than by the numbers of DOFs in the subsystems, (N_A, N_B). The main drawbacks of this approach are the exclusion from the analysis of (a) many of the DOFs which are required to describe the dynamic behaviour of the structures and (b) the higher modes whose contribution to the analysis can be very important, even though their own natural frequencies may be much higher than the frequency range of interest (the modal truncation problem, referred to previously). Both of these limitations can be addressed by including more information in the analysis, although that is done at the expense of a loss of the relative simplicity of the formulation. Full details of these refinements of this approach can be found in the relevant literature but we shall here show an extension to deal with the latter problem as that is a major obstacle to the successful use of coupled structure analysis.

It has been shown in earlier chapters that the effect of out-of-range high-frequency modes can be approximated by residual terms which are essentially damped springs and the high-frequency residual flexibility matrix for a structure such as subsystem, A, referred to the DOFs at the coupling points, can be represented approximately by:

$$[R_A] = \left[\Phi_{ce}^A\right]\left[\left(\omega_{re}^A\right)^2\right]\left[\Phi_{ce}^A\right]^T \tag{6.56}$$

where the ω^2 terms are eliminated because they are much smaller than the ω_r^2 terms. There will be a similar residual flexibility for substructure B and these two matrices can be combined to constitute the stiffness matrix of the connecting element, $[K_{CC}]$, as follows:

$$\left[R_{CC}^*\right] = \left([R_A] + [R_B]\right)$$

$$[K_{CC}] = \begin{bmatrix} \left[R_{CC}^*\right]^{-1} & -\left[R_{CC}^*\right]^{-1} \\ -\left[R_{CC}^*\right]^{-1} & \left[R_{CC}^*\right]^{-1} \end{bmatrix} \tag{6.57}$$

This equation provides a useful means of predicting reliable estimates of the coupled structure's behaviour, as illustrated by the example shown in Fig. 6.29, which relates to a test structure for studying vibration isolation in ships. The analysis performed coupled the beam to the hull section at the junction point by coupling these two components in the three in-plane DOFs (x, y, θ_z). The results shown in the figure refer to the tip FRF at the free end of the beam after coupling and the two curves show the predicted curves with and without the correction for the out-of-range high-frequency modes, and are compared with the actual measurements on the assembly. It is clear that in this example the residual effects play a major role in determining the vibration properties of assemblies such as this, but also that it is possible to account for these effects and to make a useful analysis of the coupled structure.

6.4.8 Concluding Discussion

As with the earlier application of model validation, this section has shown a multiplicity of methods and approaches to the task of predicting the dynamics of an assembled or modified structure from a knowledge of the corresponding dynamics of its component parts. Once again we find that a simple and well-defined requirement is more difficult to attain than first expected. Here, as in the other applications, the main obstacles to be overcome are, of course, inaccuracy in the data which are supplied from measurement and, although less immediately obvious, the incompleteness of information which is an inescapable fact of life when dealing with practical testing.

The major problem is that of modal truncation — the difficulty of compensating for the absence of explicit information concerning the out-of-range modes of the substructures — and many of the special

498

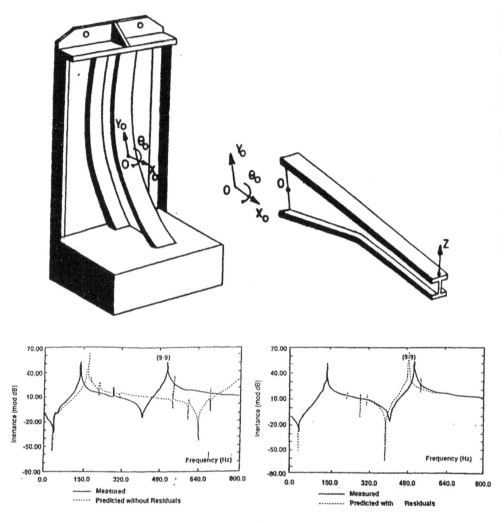

Fig. 6.29 Practical application of modal coupling analysis

techniques are aimed at minimising the consequences of this problem. In addition, there are related problems of incompleteness which arise if the coupling procedures demand information for DOFs which are difficult to obtain experimentally, such as is the case for many rotational motions (the RDOFs). Careful consideration must always be given to the completeness of the data obtained for the substructures as well as to its quality.

The other problem often encountered in application of the various coupling analysis methods described above is ill-conditioning in some of the matrix operations, especially those of inversion. This can happen if the elements of the matrix contain noise, as is likely in the case of matrices populated with raw measured data, but it also occurs if the essential information is poorly-defined. This can happen if the data are significantly inter-dependent — a situation which can arise if, for example, two DOFs are included in the analysis and yet the limited resolution available in the measured data mean that those two DOFs yield apparently identical data. The practical significance of this situation is that only one of the two DOFs is strictly necessary and the inclusion of both can cause a deterioration in the analysis. The solution here is to eliminate one of the two DOFs — not always the obvious conclusion. Once again, we see that great care and some experience is required for the successful application of these methods, but, on the other hand, we have seen that successful applications are perfectly accessible when the necessary steps are followed.

6.5 RESPONSE PREDICTION AND FORCE DETERMINATION

6.5.1 Response Prediction

Another reason for deriving an accurate mathematical model for the dynamics of a structure is to provide the means to predict and/or to understand the response of that structure to more complicated and numerous excitations than can readily be measured directly in laboratory tests. Thus the idea that by performing a set of measurements under relatively simple excitation conditions, and analysing these data appropriately (i.e. a modal test), we can then predict the structure's response to several excitations applied simultaneously.

The basis of this philosophy is itself quite simple and is summarised in the standard equation for a single-frequency component of forced vibration with n_1 excitation forces and n_2 response measurements:

$$\{X\}_{n_1 \times 1} e^{i\omega t} = [H(\omega)]_{n_2 \times n_1} \{F\}_{n_1 \times 1} e^{i\omega t} \tag{6.58}$$

where $\{F\}$ is the vector of excitation forces, $\{X\}$ is the vector of responses (or the Operational Deflection Shape — ODS). The required elements in the FRF matrix can be derived from the modal model by the familiar formula:

$$[H(\omega)]_{n_2 \times 1} = [\Phi]_{n_2 \times m} \left[\lambda_r^2 - \omega^2\right]^{-1}_{m \times m} [\Phi]^T_{m \times n_1}$$

A simple example illustrates the concept. Fig. 6.30 shows a

cantilever beam with six responses of interest and two excitation positions (coinciding with two of the responses). A modal model was formed for the beam referred to the six coordinates shown using standard single-point excitation FRF measurement methods followed by curve-fitting. It was then required to predict the responses at each of the six points to an excitation consisting of two sinusoidal forces applied simultaneously, both having the same frequency but with different magnitudes and phases. Thus the following calculation was performed at each of a number of frequencies in the range of interest (30-2000 Hz):

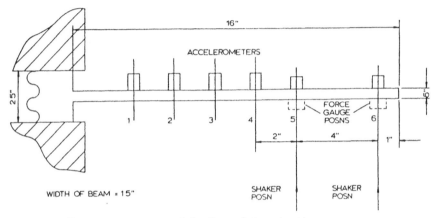

Fig. 6.30 Beam structure used for force determination

$$\{X\}_{6\times1}\, e^{i\omega t} = \left[H(\omega)\right]_{6\times2} \{F\}_{2\times1}\, e^{i\omega t} \tag{6.59}$$

The results are compared with actual measurements (made here for the purpose of assessing the validity of the method) and a typical example of these is shown in Fig. 6.31, where it can be seen that the predictions are very reliable indeed, even though we are calculating six response quantities from only two excitations.

In general, this prediction method is capable of supplying good results provided sufficient modes are included in the modal model from which the FRF data used are derived.

6.5.2 Force Determination
6.5.2.1 Basic method
Although used less frequently, the inverse calculation procedure to that just described is also a potentially powerful application of a modal model: that of deducing unmeasurable excitation forces from knowledge of the responses they generate. The basic idea of force determination is very similar to that of response prediction except that in this case it is

the excitation forces which are predicted or 'determined'. Here, it is proposed to use n_2 measured responses together with knowledge of the dynamic characteristics of the structure in order to determine what must be the n_1 forces which are causing the observed vibration.

The basic equation is simply the inverse of the previous case and is:

$$\{F\}_{n_1 \times 1}\, e^{i\omega t} = \left[H(\omega)\right]^{-1}_{n_1 \times n_2}\, \{X\}_{n_{21} \times 1}\, e^{i\omega t} \tag{6.60}$$

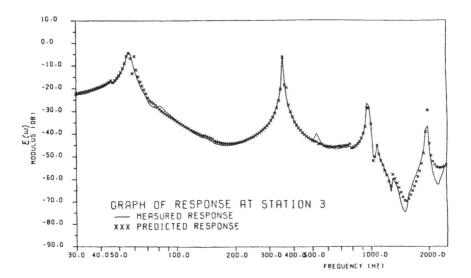

Fig. 6.31 Responses predicted from measured forces

However, once again we encounter the difficulty of inverting the rectangular and possibly rank-deficient FRF matrix in order to apply equation (6.60). To accommodate the former characteristic, we can make use of the pseudo-inverse matrix but in order to proceed it is first essential that there be at least as many responses measured as there are forces to be determined ($n_2 > n_1$). This is a restriction which does not apply to the response prediction case. In fact, it transpires that it is advantageous to use several more responses than there are forces to be determined and to use the consequent redundancy to obtain a least-squares optimisation of the determined forces. Thus we may write our basic equation for finding the excitation forces as:

$$\{F\}_{n_1 \times 1} = \left[H(\omega)\right]^{+}_{n_1 \times n_2}\, \{X\}_{n_{21} \times 1}$$

where

502

$$[H(\omega)]^{+} = \left[[H(\omega)]^{T} [H(\omega)]\right]^{-1} [H(\omega)]^{T} \tag{6.61}$$

(Note: when using complex FRF data for damped systems, it is necessary to use the Hermitian transpose in place of the ordinary transpose shown here.)

6.5.2.2 Practical case study

Once again, we shall use the simple cantilever beam shown in Fig. 6.30 as an example, referring first to a series of measured results and then later to a numerical study made to explore the behaviour of what turns out to be a calculation process prone to ill-conditioning.

In the first example, the beam is excited simultaneously at two points with sinusoidal forces of the same frequency but differing magnitude and relative phase and we measure these forces and the responses at six points. Using the previously-obtained modal model of the beam (see Section 6.5.1) together with the measured responses in equation (6.61), estimates are made for the two forces and the results are shown in Fig. 6.32 alongside the actual measurements of the forces. Close examination of the scales on these plots will show that the prediction of the forces is very poor indeed, especially at lower frequencies; the estimates sometimes being greater than an order of magnitude different from the measured data and of the opposite sign!

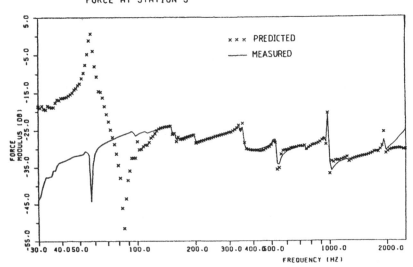

Fig. 6.32 Forces predicted from measured acceleration responses

It was observed in the course of a numerical simulation of this study that if strain gauges were used to measure the response (instead of accelerometers, or displacement transducers), then much better results would be obtained for the force estimates. Turning to the experimental investigation and using strains instead of acceleration, results in the predictions shown in Fig. 6.33 which, while not perfect, are a considerable improvement over those derived from acceleration measurements, Fig. 6.32. Further details may be found in Reference [65].

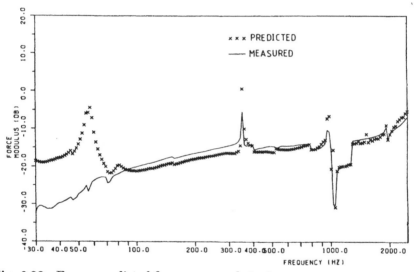

GRAPH OF PREDICTED AND MEASURED FORCES
6 RESPONSES - 2 FORCES
FORCE AT STATION 5

Fig. 6.33 Forces predicted from measured strain responses

Clearly, the determination of forces from response measurements is considerably more difficult an application for a modal model than is the corresponding calculation of responses from given or measured forces. Accordingly, it is appropriate to study the problem in a little more detail in order to understand why this is so, and to see what can be done to ameliorate what appears to be an unacceptable level of error in this important application. This we shall do via a simple numerical example.

6.5.2.3   Simple case study of force determination

Consider the case where we have a system with three inputs ($\{F\}$, excitation forces) and five outputs ($\{X\}$, response 'measurements'), and

504

the FRF matrix which links these excitations and responses is $[H(\omega)]$. Suppose that at a specific frequency of vibration, the three quantities have the numerical values given below:

$$\{F\}_0 = \begin{Bmatrix} -0.825 \\ 0.750 \\ 0.125 \end{Bmatrix} \; ; \; \{X\}_0 = \begin{Bmatrix} 5 \\ 5 \\ 5 \\ 5 \\ 5 \end{Bmatrix} \; ; \; [H] = \begin{bmatrix} 1 & 6 & 11 \\ 2 & 7 & 12 \\ 3 & 8 & 13 \\ 4 & 9 & 14 \\ 5 & 10 & 15 \end{bmatrix}$$

If we use the force vector, $\{F\}_0$, and the FRF matrix, $[H(\omega)]$, to compute the responses, we see that the calculation is correct.

$$\begin{bmatrix} 1 & 6 & 11 \\ 2 & 7 & 12 \\ 3 & 8 & 13 \\ 4 & 9 & 14 \\ 5 & 10 & 15 \end{bmatrix} \begin{Bmatrix} -0.825 \\ 0.750 \\ 0.125 \end{Bmatrix} = \begin{Bmatrix} 5 \\ 5 \\ 5 \\ 5 \\ 5 \end{Bmatrix} \equiv \{X\}_0$$

However, if we seek to perform the reverse calculation, using the given responses, $\{X\}_0$, and the generalised inverse of the FRF matrix, $[H(\omega)]^+$, to determine the forces, $\{F\}_1$, we obtain an unsatisfactory result because when we use $\{F\}_1$ to re-compute the responses, we do not recover the vector used at the outset. We find:

$$\{F\}_1 = \begin{Bmatrix} -0.47 \\ 0.31 \\ 0.31 \end{Bmatrix}$$

which gives:

$$\{X\}_1 = \begin{bmatrix} 1 & 6 & 11 \\ 2 & 7 & 12 \\ 3 & 8 & 13 \\ 4 & 9 & 14 \\ 5 & 10 & 15 \end{bmatrix} \begin{Bmatrix} -0.47 \\ 0.31 \\ 0.31 \end{Bmatrix} = \begin{Bmatrix} 4.84 \\ 5.00 \\ 5.16 \\ 5.31 \\ 5.47 \end{Bmatrix} \neq \{X\}_0$$

Clearly, the complete circle of calculations from forces to responses, and then back to forces, is ill-conditioned, and invalid. A closer examination of the FRF matrix $[H(\omega)]$ reveals that this is singular, and thus the calculation made using its generalised inverse is necessarily ill-conditioned and the resulting matrix $[H]^+$ is unreliable. We can investigate the condition of the matrix via the Singular Value Decomposition (and, indeed, we can use this procedure in order to obtain an inverse of the $[H]$ matrix). If we perform an SVD on $[H]$ we obtain the following singular value matrix:

$$[\Sigma] = \begin{bmatrix} 35.13 & 0 & 0 \\ 0 & 2.465 & 0 \\ 0 & 0 & 2.84 \times 10^{-15} \\ 0 & 0 & 0 \\ 0 & 0 & 0 \end{bmatrix}$$

which clearly shows a matrix having one 'zero' singular value which indicates that the original matrix is of rank 2, even though its order is 3. This means that the $[H]$ matrix is singular and any attempt to perform an inverse calculation with it is doomed to unreliability.

In spite of this unfortunate feature (to which we shall return later for an explanation of what it signifies), the SVD is able to provide an inverse of $[H]$ which we can use for further analysis. This, when used with the observed response vector, $\{X\}_0$, yields a solution for the unknown forces, $\{F\}_2$, of the form:

$$\{F\}_2 = \begin{Bmatrix} -0.5 \\ 0 \\ 0.5 \end{Bmatrix}$$

so that

$$[H]\{F\}_2 = \begin{bmatrix} 1 & 6 & 11 \\ 2 & 7 & 12 \\ 3 & 8 & 13 \\ 4 & 9 & 14 \\ 5 & 10 & 15 \end{bmatrix} \begin{Bmatrix} -0.5 \\ 0 \\ 0.5 \end{Bmatrix} = \{X\}_2 = \begin{Bmatrix} 5 \\ 5 \\ 5 \\ 5 \\ 5 \end{Bmatrix} \equiv \{X\}_0$$

Re-calculation of the responses, using this force vector, shows that the solution obtained, $\{X\}_2$, is, indeed, a valid one, because the computed responses are identical to those used at the start of the analysis, and it must be assumed that this solution is better than the first one obtained, which was not able to reproduce the initial vector of responses.

So, what is it that distinguishes the first, unsuccessful, solution, $\{F\}_1$, from the later, successful one, $\{F\}_2$? And is the second solution really a true solution at all? Also, what is the significance of the rank deficiency of the $[H]$ matrix? and what has been done by using the SVD to obtain an inverse, when strict interpretations indicate that one does not exist? The answers to these questions are surprisingly simple: the essential problem is that the system represented by the FRF matrix, $[H]$, has only two degrees of freedom, even though there are three excitation DOFs and five response DOFs. As a result, the rank of the FRF matrix is only 2, as indicated by the SVD analysis. Also as a result, only two independent forces can or need be defined to specify any possible excitation condition and so a vector which contains three elements is over-determined. In practice, there are an infinity of force vectors which are capable of producing a specified response pattern, such as the one introduced in this example. That means that there is no unique solution to the problem which has been posed, and the result which has been yielded by the SVD analysis is just one from that wide range of possibilities.

So, the answer obtained by direct generalised inverse is unreliable because no account was taken of the ill-condition and singularity of the FRF matrix. The answer obtained by the SVD analysis is a valid solution, but is not the solution. Although a specific force vector was specified at the outset of the problem, it is not possible to derive that solution: it is only possible to derive one which produces the same response. In fact, the solution which is derived by the SVD approach is the force vector which has the lowest norm, and is thus based on a purely numerical property and has nothing whatsoever to do with any physical considerations.

The results from this simple case study can be read across to the more complex cases encountered in practice. Accordingly, considerable attention must be paid to the condition of the matrix of response functions that is to be used in the determination of the unknown forces from measured responses. If study of this matrix reveals that it is heavily rank-deficient, then a reliable solution cannot be obtained for the case being considered and, most probably, measured response data from additional or alternative DOFs would need to be provided in order to improve the condition of the analysis to the point where a reliable solution can be obtained. If this cannot be achieved, then force determination may not be possible, as was the situation in the lower frequency range of the practical case study presented above.

6.6   TEST PLANNING
6.6.1   Test Planning and Virtual Testing

One of the major changes that has taken place in the past two decades
of modal testing practice is a much increased awareness of the cost of
testing and of the need to optimise each test. In the past, modal tests —
like many other types of test — were often planned on the basis that as
many parameters as possible would be measured and as much
information as possible would be extracted from the resulting data.
Nowadays, it is realised that this is not always an efficient strategy and
that there is considerable advantage to be gained from considering
carefully, in advance of any testing, exactly which data should be
measured for maximum benefit; which data are relatively unimportant
to the final application and, generally, how the test should be conducted
for maximum effectiveness. These changes in test philosophy have
given rise to a new application area which we shall call 'Virtual
Testing', in which the general idea is to conduct a rehearsal of each
modal test in a computer simulation before any extensive actual testing
is carried out. It is believed that by this strategy, much more effective
testing can be achieved, with a considerable reduction not only in (i) the
total time and cost of each test, but also in (ii) the instances of poorly-
identified modes, (iii) wasteful redundancy of measurements and (iv)
failures to obtain critical data by poor choices of measurement set up —
all features of more traditional testing which relied so heavily on the
skill of the experimenter.

Thus we can define virtual testing as a set of processes which help
us to decide, first and foremost, which data should be measured and
which data are not required and, secondly, how best to support and
excite the structure so that all the critical data are observed and
accessed with a uniform reliability. These decisions are also referred to
as 'Test Planning' but the subtle difference between straightforward
test planning and virtual testing is that the latter implies that a higher
degree of analysis is involved in the process. 'Planning' can imply little
more than the test engineer making decisions on the basis of his/her
experience: the concept of systematic test planning is more rigorous and
thorough than that and thus warrants a more specific title — hence,
Virtual Testing.

There are three major aspects in which virtual testing can have
significant influence on the progress and prosecution of a modal test
and these are:

(i)   assistance with the basic design of the test configuration: several
        questions should be posed and answered such as whether the
        structure should be tested free-free, or supported in some way?
        whether the boundaries should be loaded or not? which modes of

the tested structure are the most critical for the eventual application? and so on.

(ii) optimisation of the set-up for the test which is finally specified to be conducted, by selection of the best set of excitation, support and response measurement DOFs; and

(iii) selection of the most appropriate post-measurement data analysis algorithms which will be used to extract the required modal (or other) data by assessment of the various alternatives using representative simulated, rather than experimentally-measured, data.

Time spent on these activities will generally be repaid handsomely by more successful practical tests.

6.6.2 Test Planning for Model Validation

Virtual testing, as defined above, is particularly useful for certain applications of modal tests, and especially that of using test data to validate (correlate or update) theoretical models of given structures. This is because in this application (as may also apply in others) there already exists a theoretical model at the time that the test is being planned and so this model can be used as the basis of the 'rehearsal' or computer simulation upon which the test planning is conducted. In this section we shall show some of the methods which can be used in this way although, as usual in these relatively new concepts, there exist a growing number of ideas and techniques which are being developed.

We shall consider essentially three issues which form the basis of most test planning exercises:

(i) how/where to support the structure, so that the suspension has negligible interference with the vibration properties;

(ii) where to excite the structure so that all required modes are equally strongly excited (and any which are not required are not excited); and

(iii) which are the response DOFs that are most important to the application of the modal test results (this is a critical issue, as discussed in Section 6.2 on model correlation).

6.6.2.1 Suspension location

The first of these issues can be readily addressed using a quantity which is variously called the ADPR (average driving point residue) or the ADDOF (average driving point DOF), the definition for each of which is:

$$\text{ADPR}(j) = \text{ADDOF} - D(j) = \sum_r \frac{\phi_{rj}^2}{\omega_r^2} \tag{6.62}$$

This function can be computed, using the initial theoretical model for the test structure, and displayed as a contour plot, such as that illustrated in Fig. 6.34. Clearly, regions where the average response level is low will be better locations to attach whatever suspension is to be used. Care must be taken to compute, display and consider motion in all three translational directions since most suspension devices will have influence in all of these but it is generally a straightforward matter to select good and bad locations for supporting the structure so as to minimise any unwanted influence on the structure's dynamic behaviour from the suspension.

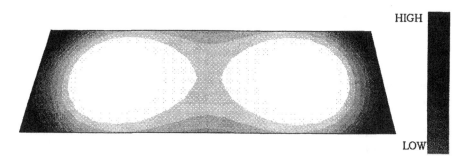

HIGH

LOW

Fig. 6.34 Example of ADDOF contour plot

6.6.2.2 Excitation location

The second consideration will usually be to the location of the best point(s) to excite the structure so as to generate approximately equal levels of response in the various mode of interest. Although, in general, we shall be interested in all modes in a given frequency range, there are several situations where we may wish to limit response to certain modes only: those which are symmetrical (as opposed to those which are skew-symmetrical); horizontal modes (as opposed to those in the vertical plane), for example. In order to ensure that a given mode is excited, it is necessary to ensure that the excitation point is not at or close to a nodal line and so this feature forms the basis of our criterion for good and bad excitation points, the ODP (optimum driving point) which is defined as :

$$ODP(j) = \prod_r \left| \phi_{rj} \right| \qquad (6.63)$$

This function is in effect a measure of the cumulative observability of all the modes in the selection and, once again, a contour plot can be prepared for this parameter, based on the initial theoretical model which is being validated — see Fig. 6.35. Plots such as this can be used

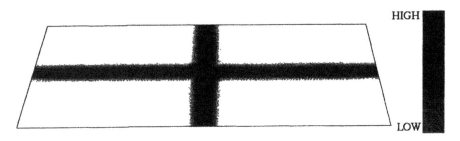

Fig. 6.35 Example of ODP contour plot

to select those points which are best suited to the task of exciting all the modes of interest — essentially, those DOFs which have a large value of ODP. However, this aspect of the test planning is not quite so straightforward because there is a counter consideration which relates to the risk of the exciter interacting with the structure to such an extent that the measurement is compromised. For example, it was shown in Chapter 3 how there is a strong shaker-structure interaction effect when the apparent mass (or stiffness) of the structure is small by comparison with the actual mass of the exciter and this effect makes it very difficult to obtain a reliable force measurement just around resonance. This phenomenon introduces a counter requirement for the exciter to be placed close to a nodal line in order to minimise its severity. At the same time, excitation by an impactor prefers driving points which are not so mobile that they invite a double hit, or bounce of the excitation device. Thus we must consider several, sometimes conflicting, demands on the choice of site(s) for our excitation and to this end a second test planning parameter can be introduced, the Non-optimum Driving Point, NODP(j):

$$NODP(j) = \min_r \left(\left\| \phi_{jr} \right\| \right) \qquad (6.64)$$

Once again, contour plots (Fig. 6.36) can be used to help visualise preferred and undesirable regions of the structure and, together with the previous functions, the test engineer can be guided in his/her task of selecting the best site(s) for applying the excitation.

Fig. 6.36   Example of NODP contour plot

6.6.2.3   Response locations

The third of these primary test planning considerations is to the selection of those DOFs at which the response is to be measured. Once again, there are a number of considerations but two major ones: (i) which points should be measured so as to present a visually informative display of the resulting mode shapes (usually by animation)? and (ii) which DOFs are necessary in order to ensure an unambiguous correlation between test and analysis models, and any ensuing updating analysis which might be undertaken? The former consideration essentially calls for a fairly uniform distribution of points with a sufficiently fine mesh that the essential features of the various mode shapes can be seen, preferably without aliasing effects. However, the second consideration is the more critical in many cases for the reasons which have been discussed earlier in this Chapter, in Section 6.2, dealing with the task of correlating two different models. In that Section, it was shown how it is essential that the $n \times m$ submatrix of measured eigenvector data (m modes, each defined at just n DOFs) is non-singular. If this condition is not achieved, then there may be serious difficulties encountered in seeking the correlated mode pairs that are required for correlation and subsequent updating. In the limit, some of the measured and computed data may have to be omitted from the subsequent analysis because of uncertainties that can arise in matching test modes with predicted modes: if this occurs, then it is clear that those measured data are wasted. The essential question is to determine which are the optimum DOFs from the point of view of independence of the incomplete mode shapes of the selected modes. Given the analytical model, and the mode shapes that it generates, it is possible to take a selection of DOFs and to test the resulting incomplete mode shapes for the modes to be included by computing the AutoMAC of those data. If no significant off-diagonal terms are found, then the selection of DOFs is basically a good one but if one or more off-diagonal terms are of the order of 70% or more, then, at least for the modes in question, considerable difficulty will be experienced in seeking to match pairs of test and analysis modes.

The essential requirement is for more than a check on a proposed set of measurement points: it is necessary to be able to specify which DOFs should be included, and which are not necessary, from the point of view of independence of the resulting necessarily-incomplete mode shape vectors. A number of different algorithms have been proposed to achieve this result, but the one of the more promising ones is based on a technique introduced in 1991 [66] and known as 'Effective Independence'. This is summarised next.

It will be clear that if there are m modes to be included in the modal test-derived model, then each mode shape must be defined at a minimum of m DOFs, otherwise, the submatrix of eigenvectors will necessarily be singular. However, the choice of these m DOFs is quite critical and, for a number of reasons (the test model and the analysis model are not expected to be identical; some of the critical DOFs may be difficult to access for measurement, and so on), it is prudent to define each mode shape at more than this minimum number of locations. Thus we require a procedure that can rank the various DOFs in descending order of importance in respect of this particular criterion of ensuring independence of the restricted-length eigenvectors that will result from the modal test. The basis of how this can be done is as follows: a Predictor matrix, $[A]_{m \times N}$, and its associate, $[E]_{N \times N}$, are computed for the incomplete set of m mode shapes that are to be identified:

$$[A]_{m \times m} = [\Phi]_{m \times N}^T [\Phi]_{N \times m} \quad ; \quad [E]_{N \times N} = [\Phi]_{N \times m} [A]_{m \times m}^{-1} [\Phi]_{m \times N}^T \quad (6.65)$$

The matrix $[E]$ is then used to eliminate successively those DOFs from the full set of N which contribute least to the rank of $[E]$, a process which is continued until the rank of the truncated mode shape matrix ceases to be of full rank (i.e. $r < m$). There are several variations which can be introduced to speed up what is otherwise a rather expensive computation (when m is typically a few tens, and N is likely to be tens of thousands), but the essence of the method is shown as being a search for the subset of DOFs which keep the selected modes mutually independent, even when the number of elements in each vector is severely restricted. Without such a specific algorithm, this independence could only be assured by the inclusion of a great many more DOFs in each mode shape description than is strictly necessary. The results of applying this strategy to a practical structure shown in

Fig. 6.37 can be seen in Fig. 6.5, which illustrates the suitability of two sets of measurement DOFs for a practical aerospace structure: (i) a set of 72 points chosen to give a fairly uniform coverage of the visible surface of the structure and (ii) a set of 30 DOFs selected by the EI algorithm just described. Fig. 6.5(b) shows the AutoMAC for the lowest 20 modes using the first set of 72 DOFs; Fig. 6.5(c) shows the corresponding AutoMAC for the EI-selected set of 30 DOFs and Fig. 6.37 shows the corresponding set of measurement points on the test structure. While it is clear that the latter set are very efficient, it is true that they do not necessarily permit a good visualisation of the resulting mode shapes (if that is a desired outcome). However, the alternative set of more than twice as many measured data may reveal a more accessible image but they fail to discriminate between two pairs of modes — it will be difficult to differentiate between modes 5 and 6, using these DOFs, and between modes 15 and 16. In applications of model validation, such a failure may result in the inadmissibility of the data for those four modes for further analysis — and that constitutes a costly waste of testing time.

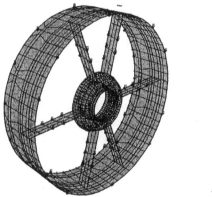

Fig. 6.37 Practical application of EI-selected response DOFs

Further details of these methods can be found in Reference [67].

6.6.3   Other Applications of Virtual Testing
6.6.3.1   Selecting test configurations

While the above three tasks are routinely undertaken in the planning of any modal test (if not always by the advanced methods outlined in the preceding section), there are others which can, and perhaps should, be considered next. These considerations relate mainly to the *configuration* of the structure while it is being tested. It is increasingly the case that modal tests are being performed on components or substructures to validate their respective theoretical models, as well as (or instead of) the full structural assemblies that are the end product of many designs. There are many reasons for this trend; not least is the early availability of individual components which makes validation of some components possible at a conveniently early stage in the design process. However, it is clear that the dynamic properties of a substructure as an isolated component will often be far removed from those that it will experience when installed in an assembly along with several other components. As a result, the relevance of testing an isolated component is frequently questioned — and not without justification. It is being realised that the modes of an individual component that are critical to determining the dynamics of an assembly are not necessarily the first m modes of that component and so the questions which arise here are: which modes of the individual component should be validated to ensure the validity of an assembled structure's dynamics? and how should the component be configured in order to carry out a useful test on it in isolation? In a number of cases, a solution to these questions may be found by adding a non-trivial boundary to the tested component (and to the corresponding theoretical model) so that it experiences loads at the interfaces which are representative of those it will encounter in its final assembled form. Virtual testing can play an important role in determining what these boundary conditions should be and help to design a test which is both convenient and useful.

6.6.3.2 Evaluation of data processing algorithms

As was mentioned earlier, another role that can be played by the virtual test is that of evaluating the performance of the various data analysis methods that are used: from curve-fitting used to extract modal parameters from measured responses, to structural modification and model updating applications. In all of these, it is possible to rehearse the intended practical application using numerically-simulated data which bear a close resemblance to the actual data that are expected to be measured in the real test. Features such as modal density, levels of non-linearity and of noise can be simulated and the proposed analysis routines tested on these simulations so that the analyst can gain valuable experience of which methods to use, and how to use them to greatest effect

6.6.4 A Note on Using Theoretical Models to Simulate Test Data

As a parting comment at the end of this work, it is appropriate to include a cautionary note on the growing (and advocated) use of numerical simulations to help us to understand how best to use our experimental facilities in undertaking modal tests that are both increasingly reliable, efficient and useful. We have proposed the use of numerical simulations derived from theoretical models of the actual testpieces to be studied and this can, indeed, be a powerful way to optimise our testing resources. However, we must at all times remember that the simulations are based on an idealisation of the actual behaviour of our structure(s) and that differences will be found between the expectations and reality. When seeking to simulate these real-life features (such as noise) we should endeavour to introduce such effects into the models in as realistic a way as possible, and not simply to add random noise to the smooth curves that are produced by our theoretical formulae. Thus, if we plan to use numerically-simulated FRF data to evaluate the performance of a given curve-fitting modal analysis routine, we should not simply add x% random noise to the otherwise correct response function: we should seek to introduce noise to the function in a way which emulates the origin of noise in real life — by adding the random perturbations to the original time histories of excitation and response prior to their transformation to the frequency domain. We can approach a more realistic seeding of noise into simulated data by transforming the FRF to the time domain (to create an IRF), then adding x% random noise to the resulting time domain function before transforming back to the frequency domain.

Nothing can quite replace the real-life test!

Notation

This list is extracted from the Notation for Modal Analysis and Testing, [23], proposed and adopted in 1993. Also used by the Dynamic Testing Agency, [12].

1. Basic Terms, Dimensions and Subscripts

x, y, z	translational degrees of freedom/coordinates (time-varying)
$\theta_x, \theta_y, \theta_z$	rotational degrees of freedom/coordinates (time-varying)
$f(t)$	time-varying excitation force
t	time variable
m	number of included/effective modes
N	total number of degrees of freedom/coordinates
n	number of primary/master/measured DOFs
s	number of secondary/slave/unmeasured DOFs
r	current mode number or matrix rank
j, k, l	integers
p	principal/modal coordinate
ω	frequency of vibration (in rad.s^{-1}; Hz)
i	$\sqrt{-1}$

2. Matrices, Vectors and Scalars

$[\ ]$	matrix
$\{\ \}$	column vector
$(\ )$	single element (of matrix or vector)
$\lceil\ \rfloor$	diagonal matrix
$[\ ]^T; \{\ \}^T$	transpose of a matrix; vector (i.e. row vector)
$[\ ]^H$	complex conjugate (Hermitian) transpose of a matrix
$[I]$	identity matrix
$[\ ]^{-1}$	inverse of a matrix

$[\ ]^{+}$	generalised/pseudo inverse of a matrix
$[\ ]^{*}$; $\{ \ \}^{*}$; $(\ )^{*}$	complex conjugate of matrix; vector; single element
$[U]$, $[V]$	matrices of left and right singular vectors
$[\Sigma]$	rectangular matrix of singular values (σ_j is j^{th} singular value)
$[T]$	transformation matrix
$[A^{R}]$, $[A^{E}]$	reduced, expanded matrix
$\| \ \|_{p}$	p-norm of a matrix/vector
ε	value of a norm/error/perturbation

3. Spatial Model Properties

$[M]$, $[K]$, $[C]$, $[D]$	mass, stiffness, viscous damping, structural (hysteretic) damping matrices
$[M_A]$, ...	analytical/theoretical/predicted/FE mass, ... matrix
$[M_X]$, ...	experimentally derived/test mass, ... matrix
$[\Delta M] = [M_X] - [M_A]$, ...	mass, ... error/modification matrix
$\begin{bmatrix} [M]_{11} & : & [M]_{12} \\ ... & : & ... \\ [M]_{21} & : & [M]_{22} \end{bmatrix}$, ...	partitioned mass, ... matrix

4. Modal Properties

ω_r	natural frequency of rth mode (rad.s^{-1})
ζ_r	viscous damping ratio of rth mode
η_r	structural damping loss factor of rth mode
m_r	modal/effective mass of rth mode
k_r	modal/effective stiffness of rth mode
c_r	modal/effective viscous damping of rth mode (proportional damping)
Q	Q factor
δ	logarithmic decrement
$[\lambda_r]$	eigenvalue matrix
$[\Psi]$	mode shape/eigenvector matrix
$[\Phi]$	mass-normalised mode shape/eigenvector matrix
$\{\psi\}_r$, $\{\phi\}_r$	r^{th} mode shape/eigenvector
ψ_{jr}, ϕ_{jr}	j^{th} element of rth mode shape/eigenvector
$[\theta]_{2N \times 2N} = \begin{bmatrix} [\psi] & : & [\psi] \\ ... & : & ... \\ [\psi][\lambda_r] & : & [\psi]^{*}[\lambda_r]^{*} \end{bmatrix}$	eigenvector matrix for viscously damped system

$$\{\theta\}_{r_{(2N \times 1)}} = \left\{ \begin{array}{c} \{\psi\}_r \\ ... \\ \lambda_r \{\psi\}_r \end{array} \right\}$$ r^{th} mode shape/eigenvector for viscously damped system

5. Response Properties

$[H(\omega)]$ Frequency Response Function (FRF) matrix

$H_{jk}(\omega)$ individual FRF element between coordinates j and k (response at coordinate j due to excitation at coordinate k)

$\{H(\omega)\}_k$ kth column of FRF matrix

Response Parameter	Response/Force FRF $H(\omega)$	Force/Response FRF $Z(\omega)$
Displacement	Dynamic Compliance or Receptance $[\alpha(\omega)]$	Dynamic Stiffness
Velocity	Mobility $[Y(\omega)]$	Mechanical Impedance
Acceleration	Accelerance or Inertance $[A(\omega)]$	Apparent/Effective Mass

$_r A_{jk} = \phi_{jr} . \phi_{kr}$ modal constant/residue

$[R]$ residual matrix

$R_{jk} = \displaystyle\sum_{r=m+1}^{N} {_r A_{jk}} / \omega_r^2$ high-frequency residual for FRF between j and k, $(H_{jk}(\omega))$

$[h(t)]$ Impulse Response Function (IRF) matrix

$h_{jk}(t)$ individual IRF element between coordinates j and k (response at coordinate j due to excitation at coordinate k)

APPENDIX 1

Use of Complex Algebra to Describe Harmonic Vibration

If

$$x(t) = x_0 \cos(\omega t + \phi)$$

we can write this as

$$x(t) = \mathrm{Re}\!\left(Xe^{i\omega t}\right) \quad \text{or, usually,} \quad x(t) = Xe^{i\omega t}$$

where X is a COMPLEX AMPLITUDE (independent of time) containing information on MAGNITUDE and PHASE

$$X = x_0 e^{i\phi}$$

(where x_0 is REAL and is the magnitude of the sine wave and ϕ is its phase relative to a chosen datum). Thus:

$$x(t) = \mathrm{Re}\!\left(x_0 e^{i(\omega t + \phi)}\right)$$
$$= \mathrm{Re}\!\left(x_0 \cos(\omega t + \phi) + i\, x_0 \sin(\omega t + \phi)\right)$$
$$= x_0 \cos(\omega t + \phi)$$

Time derivatives are easy:

$$v(t) = \dot{x}(t) = \text{Re}\left(\frac{d}{dt}Xe^{i\omega t}\right)$$

$$= \text{Re}\left(i\omega Xe^{i\omega t}\right) = \text{Re}\left(i\omega x_0 e^{(i\omega t + \phi)}\right)$$

$$= \text{Re}\left(i\omega x_0 \cos(\omega t + \phi) - \omega x_0 \sin(\omega t + \phi)\right)$$

Thus:

$$\dot{x}(t) = -\omega x_0 \sin(\omega t + \phi)$$

and also:

$$\ddot{x}(t) = -\omega^2 x_0 \cos(\omega t + \phi)$$

It is also relevant to note that the ratio of two sinusoids of the frequency can also be expressed conveniently in complex notation. Suppose that we have two sinusoids, $x(t)$ and $y(t)$, both of which are described individually in the form described above, and we wish to describe their ratio, R. We may therefore write:

$$R = \frac{x_0 \sin(\omega t + \phi_x)}{y_0 \sin(\omega t + \phi_y)} = \frac{Xe^{i\omega t}}{Ye^{i\omega t}}$$

so that

$$|R| = \frac{|X|}{|Y|} \quad ; \quad \arg(R) = \left(\phi_x - \phi_y\right)$$

This feature is used throughout the book and is the basis for describing frequency response functions (FRFs) as complex functions of frequency.

APPENDIX 2

Review of Matrix Notation and Properties

(i)   Notation and Definitions

Column vector (N rows $\times$ 1 column)

$$\{x\}_{N \times 1} = \begin{Bmatrix} x_1 \\ x_2 \\ \vdots \\ x_N \end{Bmatrix}$$

Row vector (1 row $\times$ N columns)

$$\{x\}_{1 \times N}^T = \{x_1 \quad x_2 \quad \dots \quad x_N\}$$

Rectangular matrix (N rows $\times$ M columns)

$$[B]_{N \times M} = \begin{bmatrix} b_{11} & b_{12} & \dots & b_{1M} \\ b_{21} & b_{22} & \dots & b_{2M} \\ \vdots & \vdots & \dots & \vdots \\ b_{N1} & b_{N2} & \dots & b_{NM} \end{bmatrix}$$

Square matrix (element $a_{ij} = i^{\text{th}}$ row, j^{th} column)

$$[A]_{N \times N} = \begin{bmatrix} a_{11} & a_{12} & \cdots & a_{1N} \\ a_{21} & a_{22} & \cdots & a_{2N} \\ \vdots & \vdots & \cdots & \vdots \\ a_{N1} & a_{N2} & \cdots & a_{NN} \end{bmatrix}$$

Diagonal matrix

$$[\lambda]_{N \times N} = \begin{bmatrix} 1 & 0 & 0 & \cdots & 0 \\ 0 & 2 & 0 & \cdots & 0 \\ \vdots & \vdots & \vdots & \cdots & \vdots \\ 0 & 0 & \cdots & \cdots & N \end{bmatrix}$$

Transposed matrix

$$[B]_{M \times N}^T$$

$$b_{ij}^T = b_{ji}$$

(ii)   Properties of Square Matrices
Symmetric matrix

$$[A]_{N \times N}^T = [A]_{N \times N}$$

$$a_{ij} = a_{ji}$$

Inverse

$$[A]^{-1}[A] = [A][A]^{-1} = [I]$$

'Unit Matrix'

(iii) Matrix Products

Matrices must conform for multiplication:

$$[A]_{N \times M} [B]_{M \times P} = [C]_{N \times P}$$

Generally,

$$[A][B] \neq [B][A]$$

except for

$$[A]^{-1}[A] = [A][A]^{-1} = [I]$$

Also:

$$([A][B])^{-1} = [B]^{-1}[A]^{-1}$$

and

$$([A][B])^{T} = [B]^{T}[A]^{T}$$

Symmetry

$$[A]_{N \times N} \; [B]_{N \times N} = [C]_{N \times N}$$

Symmetric   NOT necessarily Symmetric

But

$$[B]^{T}[A][B] = [D]$$

Symmetric   Symmetric

-(iv)   Matrix Rank

The product

$$\{\phi\}_{N\times1}\,\{\phi\}^T_{1\times N} = [A]_{N\times N}$$

$$[A]=\begin{bmatrix} \phi_1^2 & \phi_1\phi_2 & \cdots & \phi_1\phi_N \\ \phi_2\phi_1 & \phi_2^2 & \cdots & \phi_2\phi_N \\ \vdots & \vdots & \cdots & \vdots \\ \phi_N\phi_1 & \phi_N\phi_2 & \cdots & \phi_N^2 \end{bmatrix}$$

where $[A]$ has **order** N but **rank** 1.

(v)   Generalised or Pseudo Inverse

Given equations

$$\{x\}_{N\times1} = [A]_{N\times N}\,\{y\}_{N\times1}$$

we can solve for $\{y\}$ by:

$$\{y\} = [A]^{-1}\{x\}$$

However, sometimes equations are **Over-determined** by redundant data. Then,

$$\{x\}_{N\times1} = [B]_{N\times M}\,\{y\}_{M\times1}$$

where $N > M$ and we cannot solve directly (since $[B]^{-1}$ does not exist) so:

$$[B]^T_{M\times N}\,\{x\}_{N\times1} = [B]^T_{M\times N}\,[B]_{N\times M}\,\{y\}_{M\times1}$$

from which

$$\{y\}_{M\times1} = \left([B]^T\,[B]\right)^{-1}_{M\times M}\,[B]^T_{M\times N}\,\{x\}_{N\times1}$$

or

$$\{y\}_{M\times1} = [B]^+_{M\times N}\,\{x\}_{N\times1}$$

$[B]^+$ is the **generalised** or **pseudo inverse** of $[B]$ and gives a **least-squares** solution of the redundant equations. This generalised inverse is defined as:

$$[B]^+ = \left([B]^T [B] \right)^{-1} [B]^T$$

It can be noted that it is a property of the generalised inverse matrix that:

$$[B]^+_{M \times N} [B]_{N \times M} = [I]_{M \times M}$$

but, equally, it must be noted that the alternative combination is not so useful because:

$$[B]_{N \times M} [B]^+_{M \times N} \neq [I]_{N \times N}$$

It can be seen that a similar set of characteristics can be developed for rectangular matrices which have the other aspect: i.e. for which $N < M$. However, matrices of this form are rarely encountered in modal testing and analysis and so they will not be considered further here.

APPENDIX 3

Matrix Decomposition and the SVD

A3.1 INTRODUCTION

At various stages in the book, reference has been made to a matrix procedure known as the Singular Value Decomposition (universally referred to simply as the SVD). This is simply the general version of a basic property of all matrices: that any matrix can be decomposed into a set of standard matrices, such as the eigenvalue and eigenvector matrices, often to considerable advantage. The singular values and singular vectors are a more general version of the common eigenvalues and eigenvectors, as is explained below.

A3.2 EIGENVALUE DECOMPOSITION

By way of introduction to the SVD, we shall first recall various relationships between the spatial model and the modal model of our dynamic system which are defined in terms of the eigenvalues and eigenvectors. It can be recalled that:

$$[\Phi]^T[M][\Phi] = [I] \quad \text{and} \quad [\Phi]^T[K][\Phi] = [\lambda_r]$$

where the symbols meanings are as used throughout the text, with λ_r as the eigenvalue (or square of the natural frequency of mode r: ω_r). It can therefore be seen that the single system matrix, $[A]$, formed by combining the mass and stiffness matrices into a single quantity, can be written as:

$$[A] = [M]^{-1}[K] = \left([\Phi][\Phi]^T\right)\left([\Phi]^{-T}[\lambda_r][\Phi]^{-T}\right) = [\Phi][\lambda_r][\Phi]^{-1}$$

or, using the unitary properties of the eigenvector matrix, we can write:

$$[A] = [\Phi][\lambda_r][\Phi]^T$$

530

which is a general statement of the eigenvalue decomposition of matrix [A] and applies to all symmetric square matrices.

This decomposition can be shown to extend to non-symmetric square matrices for which case two types of eigenvector — left-hand and right-hand, as encountered in the Section concerned with rotating structures (2.8) — are applicable, and in this case the corresponding decomposition can be written as:

$$[A_{non-symm}] = [\Phi_{LH}][\lambda_r][\Phi_{RH}]^T$$

A3.3   SINGULAR VALUE DECOMPOSITION

The most general version of these matrix decompositions is that which applies to a rectangular matrix, [B], which will be taken to be of order $N \times M$ where $N > M$ (although this is not a strict mathematical requirement, it is the usual version of rectangular matrices encountered in this subject). In this most general case, the relevant decomposition is called the Singular Value Decomposition (SVD) and is expressed as:

$$[B]_{N \times M} = [U]_{N \times N}[\sigma]_{N \times M}[V]^T_{M \times M}$$

where [U] and [V] are the left- and right- singular vectors, respectively, and [σ] is the diagonal matrix of singular values. The form of a diagonal rectangular matrix is shown below:

$$[\sigma]_{N \times M} = \begin{bmatrix} \sigma_1 & 0 & \dots & 0 & \dots & 0 \\ 0 & \sigma_2 & \dots & 0 & \dots & 0 \\ \dots & \dots & \dots & \dots & \dots & \dots \\ 0 & 0 & \dots & \sigma_r & \dots & 0 \\ \dots & \dots & \dots & \dots & \dots & \dots \\ 0 & 0 & \dots & 0 & \dots & \sigma_M \\ \dots & \dots & \dots & \dots & \dots & \dots \\ 0 & 0 & \dots & 0 & \dots & 0 \end{bmatrix}$$

It can be shown that the M squared singular values $(\sigma_r)^2$ are identical to the M eigenvalues of the matrix product: $([B]^T[B])_{M \times M}$.

The SVD has special relevance in the field of modal testing because of the rectangular form of many of the matrices which are constructed from measured data. It is standard practice, when using measured data to identify parameters in a model — such as is done in every curve-fit procedure — to use the excess of test data to construct a set of over-determined equations and to use this feature to obtain smoothed or averaged results for the required model parameters. Conventionally, this approach results in more equations (rows) than unknowns (columns); hence the rectangular matrices (in which the number of equations, N, is greater than the number of unknowns, M). We have already seen the use of the generalised or pseudo inverse as a matrix form of least-squares curve-fitting to obtain a 'best' averaged result. However, in problems with many variables, there may be some difficulty in ensuring that all the measured data are independent, and this is an important requirement because if it turns out that there are actually **fewer** independent equations than there are unknowns, then the problem becomes **under**-determined and a unique solution is not possible. In such circumstances, it becomes important to be able to establish the degree of independence of the various items of measured data, and this is something that the SVD is ideally suited to do.

A3.4   MATRIX RANK AND THE SVD
The singular values are usually extracted and presented in descending order of magnitude ($\sigma_1 > \sigma_2 ... > \sigma_r ... > \sigma_M$). The rank of a matrix is a measure of the independence of the rows and columns and, like the condition number (which is the ratio of the highest and the lowest eigenvalues of the matrix), the rank directly reflects its non-singularity. If all the singular values of a matrix $[B]$ are non-zero, then the matrix is of full rank (i.e. M), is non-singular and is essentially well-conditioned from a numerical processing standpoint. If, on the other hand, one of the singular values is 'zero', then in this case the matrix is rank-deficient, has a very high condition number and is singular. In practical applications using real measured data, when singular values become very small — typically of the order of 10^6 or 10^{10} times smaller than their neighbours — they can be considered to be 'zero'. The actual threshold to be used is often the subject of a 'calibration'.

A3.5   MATRIX INVERSION AND THE SVD
It is often found that the SVD is an invaluable aid in the various matrix inversions that we may be required to perform as part of one of the modal analysis procedures. Given that

$$[B]_{N \times M} = [U]_{N \times N} [\sigma]_{N \times M} [V]^T_{M \times M}$$

we can see that the generalised inverse of $[B]$ can be expressed as:

$$[B]^+_{M \times N} = [V]^{-T}_{M \times M} [\sigma]^+_{M \times N} [U]^{-1}_{N \times N} = [V]_{M \times M} [\sigma]^+_{M \times N} [U]^T_{N \times N}$$

once again using the unitary properties of the singular vectors. The diagonal elements of $[\sigma]^+$ are simply obtained form the inverses of the corresponding diagonal elements in the matrix of singular values but a problem is encountered when any of these latter quantities are 'zero'. This, of course, is exactly the situation where the matrix to be inverted is singular and we are simply encountering this fact in one of the steps of the SVD method of obtaining a generalised inverse. Using the SVD does not remove the essential problem presented by a singular matrix but it can provide a way to proceed when such an impasse is reached. What is done in the case of one or more zero singular values in the original matrix, $[B]$, is to replace any element in the $[\sigma]^+$ matrix which would become infinite if defined by $(1/\sigma)$ with a zero. In this way, the numerical condition obstacle to computing an inverse is removed and we may proceed with a solution to the task at hand. Of course, there is a 'cost' associated with such a process. A solution has not been obtained without some concession being made, reflecting the fact that the reason why the original matrix was singular was because the problem that it described was essentially ill-posed (i.e. impossible). The apparent removal of the singularity will often have been achieved by reducing the order of the problem by one or more degrees. While we may certainly benefit from the assistance offered to many difficult computation tasks by this SVD technique, we must equally well understand the compromise that has been made in order to obtain a solution.

A3.6   EXAMPLE OF USE OF SVD IN FORCE DETERMINATION

This type of use of the SVD is best illustrated by a simple example, and the one presented here relates to the application of force determination, discussed in Section 6.5. In this application we seek to determine the excitation forces which are applied to a system from observation of the responses they produce, together with knowledge of the system's dynamic properties in the form of its FRF matrix. The basic relationship involved can be expressed simply as:

$$\{X\}_{5 \times 1}\, e^{i\omega t} = [H(\omega)]_{5 \times 3}\, \{F\}_{3 \times 1}\, e^{i\omega t}$$

For the particular case where:

$$[H] = \begin{bmatrix} 1 & 6 & 11 \\ 2 & 7 & 12 \\ 3 & 8 & 13 \\ 4 & 9 & 14 \\ 5 & 10 & 15 \end{bmatrix} \quad ; \quad \{F\} = \begin{Bmatrix} 2 \\ -5 \\ 3 \end{Bmatrix}$$

it is found that:

$$\{X\} = \begin{Bmatrix} 5 \\ 5 \\ 5 \\ 5 \\ 5 \end{Bmatrix}$$

If we now consider the inverse problem and seek to deduce the three excitation forces from knowledge of the system matrix, $[H]$ and the responses, $\{X\}$, using the standard generalised inverse formula presented in the previous appendix to invert the above equation to solve for $\{F\}$, we find that the sought force vector is:

$$\{F\} = \begin{Bmatrix} -0.47 \\ +0.31 \\ +0.31 \end{Bmatrix}$$

which is not only different to the one presented in the problem definition but is not capable of reproducing the response vector, $\{X\}$, which has been used in its derivation:

$$\{X\} = \begin{bmatrix} 1 & 6 & 11 \\ 2 & 7 & 12 \\ 3 & 8 & 13 \\ 4 & 9 & 14 \\ 5 & 10 & 15 \end{bmatrix} \begin{Bmatrix} -0.47 \\ +0.31 \\ +0.31 \end{Bmatrix} = \begin{Bmatrix} 4.84 \\ 5.00 \\ 5.16 \\ 5.31 \\ 5.47 \end{Bmatrix} \neq \begin{Bmatrix} 5 \\ 5 \\ 5 \\ 5 \\ 5 \end{Bmatrix}$$

If we now use the SVD to compute the generalised inverse of $[H]$, we find that one of the three singular values is extremely small when compared with the other two, and we conclude that this original matrix is singular (indeed, inspection of the matrix reveals that the third column can be derived directly from the second and the third).

$$
\begin{bmatrix} 1 & 6 & 11 \\ 2 & 7 & 12 \\ 3 & 8 & 13 \\ 4 & 9 & 14 \\ 5 & 10 & 15 \end{bmatrix} = \begin{bmatrix} -0.36 & -0.69 & +0.56 & .. & .. \\ -0.40 & -0.38 & -0.64 & .. & .. \\ -0.44 & -0.06 & -0.00 & .. & .. \\ -0.49 & +0.25 & -0.33 & .. & .. \\ -0.53 & +0.56 & \_0.41 & .. & .. \end{bmatrix} \begin{bmatrix} 35.13 & 0 & 0 \\ 0 & 2.465 & 0 \\ 0 & 0 & 2.84 \times 10^{-15} \\ 0 & 0 & 0 \\ 0 & 0 & 0 \end{bmatrix}
$$

$$
\begin{bmatrix} -0.20 & +0.81 & +0.41 \\ -0.52 & +0.26 & -0.82 \\ -0.83 & -0.38 & \_0.41 \end{bmatrix}
$$

Proceeding with the SVD-based inversion as described above, we find that a satisfactory solution for the excitation forces **can** be obtained and that it is:

$$
\{F\} = \begin{Bmatrix} 0.5 \\ 0 \\ -0.5 \end{Bmatrix} \quad \text{resulting in} \quad \{X\} = \begin{Bmatrix} 5 \\ 5 \\ 5 \\ 5 \\ 5 \end{Bmatrix}
$$

Now, it is necessary to understand how this solution has been obtained, and what — if any — conditions are attached to it. Indeed, a solution seemed impossible from an earlier direct analysis.

What has happened in this solution is that the problem has been reduced from one with three unknowns to one with just two. This is the direct implication of the discovery that the original $[H]$ matrix was rank-deficient, and of the actions that were taken using the SVD approach to inverting the matrix. By this method, we have found **a** solution for the three unknown forces but not **the** solution. Indeed, there is no unique solution: that is the sense in which the problem was ill-posed at the outset. The non-uniqueness becomes evident when we realise that an infinite number of solutions can be found: for example, the vector:

$$
\{F_2\} = \begin{Bmatrix} 0 \\ -1 \\ 1 \end{Bmatrix}
$$

is also a possible solution, as is

$$\{F_3\} = \begin{Bmatrix} 1.5 \\ -4 \\ 2.5 \end{Bmatrix}$$

It is easily seen that many combinations of these solutions can also satisfy the requirement of the set problem.

So what is it that results in this slightly unexpected situation? The clue to the answer to this question lies in the zero singular value of $[H]$. That feature indicated that the system described by this FRF matrix has only 2 degrees of freedom, and not 3 as implied by the statement of the problem. As such, a unique solution for three unknown forces is clearly an impossibility, even though a solution is possible, as we have seen. The physics behind this situation, which is typical of many encountered in various aspects of modal analysis, can be explained by a simple practical interpretation of the set problem, illustrated in the accompanying figure. Here we see a system with just

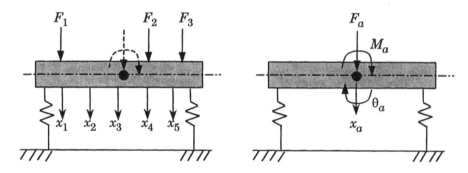

2DOFs, but one whose motion is described using several more DOFs than that. In the system shown here, it is perfectly feasible to calculate the response of the system at 5 degrees of freedom, as shown, when it is subject to three forces, also shown. There is no difficulty in determining these responses in this case. However, if we try to undertake an inverse analysis, we encounter difficulties. The motion of the system — any motion — can be determined by a single force together with a single moment, both applied through the centre of gravity of the mass. These constitute the excitation in the 2 degrees of freedom that the system possesses. The complication arises in the set problem because it is possible to generate these same excitation conditions using any number of combinations of three or more forces. As long as the net effect of these three forces is equal to the single force and the single moment, they will always generate the same response. Now we can see the origin of the non-unique solution yielded by the SVD solution to the set problem.

The power of the SVD approach is that it can determine the true order of the problem, and this is already a valuable contribution. As in this case, it can also go further and reveal a solution where none was available by other means. It must be noted that the solution thus obtained is not **the** solution, but just one of many. In fact, it yields the force vector which has the minimum norm (sum of the squares of the elements). Unfortunately, there is no particular reason why this particular solution should be any more likely to be close to the true solution than any other. But, still, it **is** a solution.

APPENDIX 4

Transformations of Equations of Motion between Stationary and Rotating Axes

In Chapter 2 we have shown how the coordinates in the stationary frame of reference and those in the rotating frame of reference are linked by a 'simple' transformation, $[T]$:

$$\begin{Bmatrix} x_S \\ y_S \end{Bmatrix} \equiv \begin{Bmatrix} x \\ y \end{Bmatrix} = \begin{bmatrix} c & -s \\ s & c \end{bmatrix} \begin{Bmatrix} x_R \\ y_R \end{Bmatrix} \quad \text{and} \quad \begin{Bmatrix} x_R \\ y_R \end{Bmatrix} = \begin{bmatrix} c & s \\ -s & c \end{bmatrix} \begin{Bmatrix} x \\ y \end{Bmatrix}$$

where

$$c = \cos(\Omega t) \quad ; \quad s = \sin(\Omega t)$$

so that:

$$\{x\} = [T]\{x_R\} \quad ; \quad \{x_R\} = [T]^T \{x\}$$

In seeking to transform equations of motion between one of these sets of coordinates and the other, it is useful to note the following properties:

$$\begin{Bmatrix} x_R \\ y_R \end{Bmatrix} = \begin{bmatrix} c & s \\ -s & c \end{bmatrix} \begin{Bmatrix} x \\ y \end{Bmatrix} = [T_1]\{x\}$$

$$\begin{Bmatrix} \dot{x}_R \\ \dot{y}_R \end{Bmatrix} = \begin{bmatrix} c & s \\ -s & c \end{bmatrix} \begin{Bmatrix} \dot{x} \\ \dot{y} \end{Bmatrix} + \Omega \begin{bmatrix} -s & c \\ -c & -s \end{bmatrix} \begin{Bmatrix} x \\ y \end{Bmatrix} = [T_1]\{\dot{x}\} + \Omega[T_2]\{x\}$$

$$\begin{Bmatrix} \ddot{x}_R \\ \ddot{y}_R \end{Bmatrix} = \begin{bmatrix} c & s \\ -s & c \end{bmatrix} \begin{Bmatrix} \ddot{x} \\ \ddot{y} \end{Bmatrix} + 2\Omega \begin{bmatrix} -s & c \\ -c & -s \end{bmatrix} \begin{Bmatrix} \dot{x} \\ \dot{y} \end{Bmatrix} - \Omega \begin{bmatrix} c & s \\ -s & c \end{bmatrix} \begin{Bmatrix} x \\ y \end{Bmatrix}^2$$

$$= [T_1]\{\ddot{x}\} + 2\Omega[T_2]\{\dot{x}\} - \Omega^2[T_1]\{x\}$$

and the following:

$$[T_1]^T \begin{bmatrix} A & 0 \\ 0 & A \end{bmatrix} [T_1] = \begin{bmatrix} A & 0 \\ 0 & A \end{bmatrix} \quad \text{and} \quad [T_1]^T \begin{bmatrix} 0 & B \\ -B & 0 \end{bmatrix} [T_1] = \begin{bmatrix} 0 & B \\ -B & 0 \end{bmatrix}$$

$$[T_1]^T \begin{bmatrix} A & 0 \\ 0 & A \end{bmatrix} [T_2] = \begin{bmatrix} 0 & A \\ -A & 0 \end{bmatrix} \quad \text{and} \quad [T_1]^T \begin{bmatrix} 0 & B \\ -B & 0 \end{bmatrix} [T_1] = \begin{bmatrix} -B & 0 \\ 0 & -B \end{bmatrix}$$

APPENDIX 5

Fourier Analysis

In this appendix, we list some of the more important features of Fourier Analysis, as required for most modal testing applications. These will be summarised in three categories:

(i)   Fourier Series
(ii)   Fourier Transform
(iii) Discrete Fourier Series (or Transform)

All share the common feature of being the means of describing a time-varying quantity in terms of a set of sinusoids and, conversely, of reconstituting a time history from a set of frequency components.

Only a summary will be included here. For a more detailed treatment, refer to a specialist text such as Bendat and Piersol (Reference [34]) or Newland (Reference [33]). However, it must be noted that there exist slight differences in definition and terminology which can make cross referencing from one source to the other somewhat confusing!

(i)   Fourier Series
A function $x(t)$ which is periodic in time T can be represented as an infinite series of sinusoids:

$$x(t) = \tfrac{1}{2}a_0 + \sum_{n=1}^{\infty} (a_n \cos \omega_n t + b_n \sin \omega_n t)$$

where

$$\omega_n = \frac{2\pi n}{T}$$

in which the coefficients are given by

$$a_0 = \frac{2}{T} \int_0^T x(t)dt$$

$$a_n = \frac{2}{T} \int_0^T x(t)\cos\omega_n\, t dt \quad ; \quad b_n = \frac{2}{T} \int_0^T x(t)\sin\omega_n\, t dt$$

Alternative forms

(a)

$$x(t) = c_0 + \sum_{n=1}^{\infty} c_n \cos(\omega_n t + \phi_n)$$

where

$$c_n = \sqrt{a_n^2 + b_n^2} \quad ; \quad \phi_n = \tan^{-1}\left(-\frac{b_n}{a_n}\right)$$

(b)

$$x(t) = \sum_{-\infty}^{\infty} X_n e^{i\omega_n t}$$

where

$$X_n = \frac{1}{T} \int_0^T x(t)e^{-i\omega_n t}\, dt$$

NOTE:

$$X_{-n} = X_n^* \quad ; \quad \text{Re}(X_n) = \frac{a_n}{2}$$

$$\text{Im}(X_n) = -\frac{b_n}{2}$$

$$|X_n| = \frac{c_n}{2}$$

(ii) Fourier Transform

A nonperiodic function $x(t)$ which satisfies the condition

$$\int_{-\infty}^{\infty} |x(t)| dt < \infty$$

can be represented by the integral:

$$x(t) = \int_{-\infty}^{\infty} (A(\omega)\cos\omega t + B(\omega)\sin\omega t) d\omega$$

where

$$A(\omega) = \frac{1}{\pi} \int_{-\infty}^{\infty} x(t)\cos\omega t\, dt \quad ; \quad B(\omega) = \frac{1}{\pi} \int_{-\infty}^{\infty} x(t)\sin\omega t\, dt$$

Alternative form

The alternative complex form is more convenient, and familiar, as:

$$x(t) = \int_{-\infty}^{\infty} X(\omega)e^{i\omega t} d\omega$$

where

$$X(\omega) = \frac{1}{2\pi} \int\limits_{-\infty}^{\infty} x(t) e^{-i\omega t}\, dt$$

NOTES:

$$\mathrm{Re}\big(X(\omega)\big) = \frac{A(\omega)}{2}$$

$$\mathrm{Im}\big(X(\omega)\big) = \frac{B(\omega)}{2}$$

also

$$X(-\omega) = X^{*}(\omega)$$

(iii)   Discrete Fourier Series/Transform (DFT)

A function which is defined **only** at N discrete points (at $t = t_k$, $k = 1$, N) can be represented by a **finite** series:

$$x(t_k) \equiv (x_k) = \frac{1}{2} a_0 + \sum_{n=1}^{\frac{N}{2} \text{ or } \frac{(N-1)}{2}} \left(a_n \cos\frac{2\pi n k}{N} + b_n \sin\frac{2\pi n k}{N} \right)$$

where

$$a_0 = \frac{2}{N}\sum_{k=1}^{N} x_k \quad ; \quad a_n = \frac{1}{N}\sum_{k=1}^{N} x_k \cos\frac{2\pi n k}{N} \quad ; \quad b_n = \frac{1}{N}\sum_{k=1}^{N} x_k \sin\frac{2\pi n k}{N}$$

Alternative form

$$x(t_k) \equiv (x_k) = \sum_{n=0}^{N-1} X_n e^{2\pi i n k/N}$$

where

$$X_n = \frac{1}{N} \sum_{k=1}^{N} x_k e^{-2\pi i n k/N} \quad , \quad n = 1, N$$

NOTE:

$$X_{N-r} = X_r^*$$

Notes on the discrete Fourier Transform

(a) This is the form of Fourier Analysis most commonly used on digital spectrum analysers.

(b) The DFT necessarily assumes that the function $x(t)$ is periodic.

(c) The DFT representation is only valid for the specific values x_k ($x(t)$ at $t = t_k$) used in the discretised description of $x(t)$.

(d) It is important to realise that in the DFT, there are just a discrete number of items of data in either form: there are just N values x_k and, correspondingly, the Fourier Series is described by just N values.

Example
Let $N = 10$.

In the time domain, we have

$$x_1, x_2, ..., x_{10}$$

In the frequency domain, we have

$$a_0, a_1, a_2, a_3, a_4, a_5, b_1, b_2, b_3, b_4$$

or

$$X_0 (= \text{Real}), \text{Re}(X_1), \text{Im}(X_1), \text{Re}(X_2), \text{Im}(X_2),$$
$$\text{Re}(X_3), \text{Im}(X_3), \text{Re}(X_4), \text{Im}(X_4), X_5 (= \text{Real})$$

(NOTE: $X_5 = \text{Real}$ because X_{10-5} $(= X_5) = X_5^*$.)

References

[1]   Kennedy, C.C., & Pancu, C.D.P. 1947. 'Use of vectors in vibration measurement and analysis'. *J. Aero. Sci.*, **14**(11)

[2]   Bishop, R.E.D., & Gladwell, G.M.L. 1963. 'An investigation into the theory of resonance testing'. *Proc. R. Soc. Phil. Trans.*, **255**(A), p. 241

[3]   Salter, J.P. 1969. "Steady State Vibration". Kenneth Mason Press

[4]   Allemang, R.J. 1984. 'Experimental modal analysis bibliography'. Proc. IMAC2

[5]   Mitchell, L.D., & Mitchell, L.D. 1984. 'Modal analysis bibliography — An update — 1980–1983'. Proc. IMAC2

[6]   Maia, N., Silva, J., He, J., Lieven, N., Lin, R-M., Skingle, G., To, W., & Urgueira, A. 1997. "Theoretical and Experimental Modal Analysis". Research Studies Press Ltd.

[7]   Heylen, W., Lammens, S., & Sas, P. 1998. "Modal Analysis Theory and Testing". Katholieke Universiteit Leuven, Belgium

[8]   Natke, H.G. 1992. "Fundamentals and Advances in the Engineering Sciences". Friedr. Vieweg and Sons

[9]   Fu, Z. 1995. "Vibration Modal Analysis and Parameter Identification". Mechanical Industry Publishing Co., People's Republic of China

[10]   Buzdugan, G., Mihailesci, E., and Rade, M. 1986. "Vibration Measurement". Martinus Nijhoff Publishers

[11]   Dynamic Testing Agency. "Primer on Best Practice in Dynamic Testing"

[12]   Dynamic Testing Agency. 1993. "Handbook on Modal Testing"

[13]   McConnell, K.G. 1995. "Vibration Testing Theory and Practice". John Wiley & Sons

[14]   International Modal Analysis Conference Proceedings. 1982-2000. Society for Experimental Mechanics, USA, **1-18**

[15]   International Seminar on Modal Analysis Proceedings. 1975-2000. Katholieke Universiteit Leuven, Belgium, **1-25**

546

[16] *The International Journal of Analytical and Experimental Modal Analysis*. Society of Experimental Mechanics, USA

[17] *Journal of Mechanical Systems and Signal Processing*. Academic Press, UK

[18] *Journal of Vibration and Control*. Sage Science Press

[19] Ewins, D.J., & Griffin, J. 1981. 'A state-of-the-art assessment of mobility measurement techniques — Results for the mid-range structures'. *J. Sound & Vib.*, 78(2), pp. 197–222

[20] Balmes, E. 1996. "Garteur Group on Ground Vibration Testing, Results from the Test of a Single Structure by 12 Laboratories in Europe". ONERA

[21] Smallwood, D.O., & Gregory, D.L. 1985. 'A rectangular plate as a proposed modal test structure', *J. Environmental Sciences*, pp. 41-46

[22] Ahlin, K. 1996. 'Round robin exercise on modal parameter extraction', Proc. 2nd Int. Conf. on Structural Dynamics Modelling, Test, Analysis and Correlation, DTA/NAFEMS

[23] Lieven, N.A.J., & Ewins, D.J. 1992. 'A proposal for standard notation and terminology in modal analysis', Proc. IMAC 10, San Diego, USA, pp. 1414-1419

[24] Gasch, R., & Pfutzner, H. 1975. "Rotordynamics". Springer

[25] To, W.M., & Ewins, D.J. 1991. 'A closed-loop model for single/multi-shaker modal testing', *Mechanical Systems and Signal Processing*, 5(4), pp. 305-316

[26] Skingle, G.W., & Ewins, D.J. 1989. 'Sensitivity analysis using resonance and anti-resonance frequencies — a guide to structural modification', Proc. Aachen Conf., 17-19 April

[27] Virgin, L.V. 2000. "Introduction to Experimental Nonlinear Dynamics — A Case Study in Mechanical Vibration". Cambridge University Press

[28] Förch, P., Gähler, C., & Nordmann, R. 1996. 'AMB systems for rotordynamic experiments: calibration results and control', Proc. 5th Int. Symp. on Magnetic Bearings, Kanazawa, Japan

[29] Mitchell, L.D., West, R.L., & Wicks, A.L. 1998. 'An emerging trend in experimental dynamics: merging of laser-based three dimensional structural imaging and modal analysis', *J. Sound & Vibration*, 211(3), pp. 323-333

[30] Stanbridge, A.B., & Ewins, D.J. 1999. 'Modal testing using a scanning laser doppler vibrometer', *Mechanical Systems and Signal Processing*, 13(2), pp. 255-270

[31] Cooley, J.W., & Tukey, J.W. 1965. 'An algorithm for the machine calculation of complex Fourier series'. *Maths. of Comput.*, 19(90), pp. 297–301

[32] Randall, R.B. 1987. "Frequency Analysis". Bruel & Kjaer Publications

[33] Newland, D.E. 1984. "Random Vibrations and Spectral Analysis". Longman Press

[34] Bendat, J.S., & Piersol, A.G. 1986. "Random Data: Analysis and Measurement Procedures". 2nd Edition, John Wiley & Sons

[35] International Organisation for Standardisation (ISO). "Methods for the Experimental Determination of Mechanical Mobility".
Part 1: Basic definitions
Part 2: Measurements using single-point translation excitation with an attached vibration exciter
Part 3: Measurement of rotational mobility
Part 4: Measurement of the complete mobility matrix using attached exciters
Part 5: Measurements using impact excitation with an exciter which is not attached to the structure

[36] Cawley, P. 1984. 'The reduction of bias error in transfer function estimates using FFT-based analysers'. *Trans. ASME, J. Vib. Ac. Stress & Rel.*, **106**(1)

[37] Tomlinson, G.R. 1998. 'Nonlinearity in modal analysis', Proc. NATO Advanced Study Institute on Modal Analysis and Testing, pp. 569-597

[38] Zaveri, K. 1984. "Modal Analysis of Large Structures — Multiple Exciter Systems". Bruel and Kjaer Publications

[39] Ewins, D.J. 1973. 'Vibration characteristics of bladed disc assemblies', *J. Mech. Eng. Sci.*, **15**(3), pp. 165-186

[40] Pickrel, C. 1996. 'Estimating the rank of measured response data using SVD and principal response functions', Proc. 2nd Int. Conf. on Structural Dynamics Modelling, Test Analysis and Correlation, DTA/NAFEMS, Lake District, UK, pp. 89-100

[41] Rades, M. 1994. 'A comparison of some mode indicator functions', *Mechanical Systems and Signal Processing*, **8**(4), pp. 459-474

[42] Gaukroger, D.R., Skingle, C.W., & Heron, K.H. 1973. 'Numerical analysis of vector response loci'. *J. Sound & Vib.*, **29**(3), pp. 263-273

[43] Gimenez, J.G., & Carrascosa, L.I. 1982. 'Global fitting: an efficient method of experimental modal analysis of mechanical systems', Proc. IMAC1, pp. 528-533

[44] Richardson, M.H., & Formenti, D.L. 1985. 'Global curve-fitting of frequency response measurements using the rational fraction polynomial method', Proc. IMAC3, Orlando, Florida, pp. 390-397

[45] Ewins, D.J., & Gleeson, P.T. 1989. 'A method for modal identification of lightly damped structures', *Mechanical Systems and Signal Processing*, **3**(2), pp. 173-193

548

[46] Goyder, H.G.D. 1980. 'Methods and application of structural modelling from measured structural frequency response data'. *J. Sound & Vib.*, **68**(2), pp. 209–230

[47] Ibrahim, S.R., & Mikulcik, E.C. 1976. 'The experimental determination of vibration parameters from time responses', *The Shock and Vibration Bulletin*, **46**(5), pp. 187-196

[48] Zhang, L., Kanda, H., Brown, D., & Allemang, R. 1985. 'A polyreference frequency domain method for modal parameter identification', ASME Paper No. 85-DET-106, pp. 1-6

[49] Prony, R. 1795. 'Essai expérimental et analitique sur les lois de la dilatabilité des fluides elastiques et sur celles de la force expansive de la vapeur de l'eau et de la vapeur de l'alkool à différentes températures', *J. l'Ecole Polytechnique (Paris)*, **1**(2), Floréal et Prairial, An, III, pp. 24-76

[50] Allemang, R.J., & Brown, D.L. 1998. 'A unified matrix polynomial approach to modal identification', *Journal of Sound and Vibration*, **211**(3), pp. 301-322

[51] Asher G.W. 1958. 'A method of normal mode excitation utilizing admittance measurements', Proc. National Specialists Meeting in Dynamics and Aeroelasticity, Institute of Aeronautical Sciences, pp. 69-76

[52] Niedbal, N. 1994. 'Analytical determination of real normal modes from measured complex responses', Proc. 25th Structures, Structural Dynamics and Materials Conf., California, USA, pp. 292-295

[53] O'Callahan, J., Avitabile, P., & Riemer, R. 1989 'System equivalent reduction expansion process', Proc. 7th Int. Modal Analysis Conf., Las Vegas, pp. 29-37

[54] Friswell, M.I., & Mottershead, J.E. 1995. "Finite Element Model Updating in Structural Dynamics". Kluwer Academic Publishers

[55] Imregun, M., & Visser, W.J. 1991. 'A review of model updating techniques', *Shock & Vibration Digest*, **3**(1), pp 9-20

[56] Mottershead, J.E., & Friswell, M.I. 1993, 'Model updating in structural dynamics: a survey', *Journal Sound and Vibration*, **167**(2), pp. 347-375

[57] Berman, A., & Nagy, E.J. 1983. 'Improvement of a large analytical model using test data', *AIAA Journal*, **21**(8), pp. 1168-1173

[58] He, J. 1987. "Identification of Structural Dynamics Characteristics", PhD Thesis, Imperial College of Science, Technology and Medicine, University of London

[59] Lin, R.M. 1991. "Identification of the Dynamic Characteristics of Non-linear Structures", PhD Thesis, Imperial College of Science, Technology and Medicine, University of London

[60] Visser, W.J. 1992. "Updating Structural Dynamic Models Using Frequency Response Data", PhD Thesis, Imperial College of Science, Technology and Medicine, University of London

[61] Ewins, D.J., Silva, J.M.M.,& Maleci, G. 1980. 'Vibration analysis of a helicopter with an externally-attached carrier structure', *Shock and Vibration Bulletin*, **50**(2), pp. 155-171

[62] Skingle, G.W. 1989. "Structural Dynamic Models Using Experimental Data", PhD Thesis, Imperial College of Science, Technology and Medicine, University of London

[63] Hillary, B., & Ewins, D.J. 1984. 'The use of strain gauges in force determination and frequency response function measurements', Proc. IMAC2

[64] Kammer, D.C. 1992. 'Effect of noise on sensor placement for on-orbit modal identification and correlation of large space structures', *Trans. ASME, Journal of Dynamical Systems, Measurement and Control*, **114**(3), pp. 435-443

[65] Imamovic, N. 1998. "Validation of Large Structural Dynamics Models Using Modal Test Data", PhD Thesis, Imperial College of Science, Technology and Medicine, University of London

PhD Theses — Imperial College, University of London, 1970-2000

Antonio, J. 1984. "Power Flow in Structures During Steady-State Forced Vibration"

Ashory, M.R. 1999. "High quality modal testing methods"

Blashke, P. 1996. "Vibro-acoustic Design Tool for Noise Optimization of Rotating Machinery"

Burgess, A. 1988. "Transient Response of Mechanical Structures Using Modal Analysis Techniques"

Duarte, M.L.M. 1996. "Experimentally-derived Structural Models for use in Further Dynamic Analysis"

Ferreira, J. 1998. "Dynamic Response of Structures with Nonlinear Components"

Gleeson, P.T. 1979. "Identification of Spatial Modes"

Grafe, H. 1998. "Model Updating of Large Structure Dynamics Models Using Measured Response Functions"

Hardenberg, J.F. 1971. "The Transmission of Vibrations Through Structures"

He, J. 1987. "Identification of Structural Dynamic Characteristics"

Hillary, B. 1982. "Identification Measurement of Vibration Excitation Forces"

Imamovic, N. 1998. "Validation of Large Structural Dynamic Models using Modal Test Data"

Jung, H. 1992. "Structural Dynamic Model Updating Using Eigensensitivity Analysis"

Kirshenboim, J. 1981. "Theory and Practice of Model Identification"

Liéven-Lieven, N.A.J. 1990. "Validation of Structural Dynamic Models"

Lin, R. 1991. "Identification of the Dynamic Characteristics of Nonlinear Structures"

Maia, N.M.M. 1988. "Extraction of Valid Modal Properties from Measured Data in Structural Vibrations"

Montalvão, E. Silva J.M. 1978 "Measurement and Applications of Structural Mobility Data"

Sainsbury, M.G. 1977. "Experimental and Theoretical Techniques for the Vibration Analysis of Damped Complex Structures"

Salehzadeh Nobari, A. 1992. "Identification of the Dynamic Characteristics of Structural Joints"

Sidhu, J. 1993. "Reconciliation of Predicted and Measured Modal Properties of Structures"

Skingle, G.W. 1989. "Structural Dynamic Modification Using Experimental Data"

To, W.M. 1990. "Sensitivity Analysis of Mechanical Structures Using Experimental Data"

Urgueira, A.P.V. 1989. "Dynamic Analysis of Coupled Structures Using Experimental Data"

Visser, W.Y. 1992. "Updating Structural Dynamic Models Using Frequency Response Data"

Ziaei Rad, S. 1997. "Methods for Updating Numerical Models in Structural Dynamics"

Index

Printed and bound in the UK by
CPI Antony Rowe, Eastbourne

Printed and bound by CPI Group (UK) Ltd, Croydon, CR0 4YY

27/10/2024

14580190-0005